Mathematical Analysis for Engineers

T0350089

Mathematical Analysis for Engineers

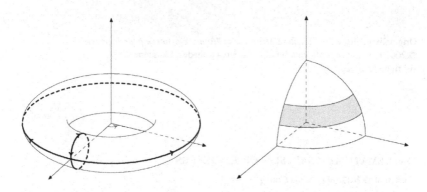

Bernard Dacorogna
Chiara Tanteri

Ecole Polytechnique Fédérale de Lausanne (EPFL), Switzerland

Imperial College Press

ICP

Published by

Imperial College Press
57 Shelton Street
Covent Garden
London WC2H 9HE

Distributed by

World Scientific Publishing Co. Pte. Ltd.
5 Toh Tuck Link, Singapore 596224
USA office: 27 Warren Street, Suite 401-402, Hackensack, NJ 07601
UK office: 57 Shelton Street, Covent Garden, London WC2H 9HE

British Library Cataloguing-in-Publication Data
A catalogue record for this book is available from the British Library.

Originally published in French under the title: "*Analyse avancée pour ingénieurs*".
© 2002 Presses polytechniques et universitaires romades, Lausanne.
All rights reserved.

MATHEMATICAL ANALYSIS FOR ENGINEERS

ISBN-13 978-1-84816-912-8
ISBN-10 1-84816-912-4

Printed in Singapore by World Scientific Printers.

Contents

Foreword

This book is a translation of the third French edition. It is intended for engineering students who followed a basic course in analysis (differential and integral calculus). It corresponds to a second year course at the Ecole Polytechnique Fédérale of Lausanne. It can also be useful, as a complement to a more theoretical course, to mathematics and physics students.

There are excellent books on the matters that are discussed here; some of them that we particularly like are mentioned in the bibliography. Our approach is, however, different. We have emphasized the learning of the field through examples and exercises. The theoretical part is short, definitions and theorems are given without comments.

The book is organized as follows. The first three parts (Vector analysis, Complex analysis and Fourier analysis) represent the theoretical part and they are essentially independent of each other. The fourth part gives detailed solutions to all exercises that are proposed in the first three parts. The theoretical discussion follows the following pattern.

1) Definitions and theorems are stated, with mathematical rigor, but without comments. We also mention the precise pages of certain books from the bibliography where the interested reader can find further developments.

2) Some significant examples are discussed in detail.

3) Finally, several exercises are given and, as already said, solved in the fourth part of the book. The first type of exercise will help students to master the concepts and the techniques. A second type (identified with a *) presents some theoretical developments that allow the more motivated students to deepen their understanding of the subject.

We would now like to make some comments on the bibliography. We have selected two types of book.

1) As mathematical references, we particularly like the following books:

- for vector analysis the Protter–Morrey book and the more advanced Fleming book;

- for complex analysis the very classical Ahlfors book;

- for Fourier series the already mentioned Protter–Morrey book, while for Fourier and Laplace transforms the Widder book;

- the two Stein–Shakarchi books cover a large part of the matters discussed here (complex and Fourier analysis);

- finally, in French, the three volumes of Chatterji cover in detail the entire subject of our book.

2) For engineers we recommend the Kreyszig book. The two small books, in French, by Arbenz–Wohlhauser are also nice as a short introduction.

We have benefited from several comments from students and colleagues; notably S. Bandyopadhyay, M. Cibils, G. Croce, G. Csato, J. Douchet, H. Gebran, O. Kneuss, P. Metzener, G. Pisante, A. Ribeiro, L. Rollaz and K. D. Semmler. The translation and the preparation of the English version have been carried out by R. Guglielmetti.

Part I

Vector analysis

Chapter 1

Differential operators of mathematical physics

1.1 Definitions and theoretical results

Definition 1.1 *Let $\Omega \subset \mathbb{R}^n$ be open, $n \geq 2$ and $x = (x_1, \cdots, x_n)$.*

*(i) If $f \in C^1(\Omega)$, we define, for every $x \in \Omega$, the **gradient** of f as*

$$\operatorname{grad} f(x) = \nabla f(x) = \left(\frac{\partial f}{\partial x_1}(x), \cdots, \frac{\partial f}{\partial x_n}(x) \right) \in \mathbb{R}^n.$$

*(ii) If $f \in C^2(\Omega)$, we define, for every $x \in \Omega$, the **Laplacian** of f by*

$$\Delta f(x) = \sum_{i=1}^{n} \frac{\partial^2 f}{\partial x_i^2}(x) \in \mathbb{R}.$$

*(iii) Let $F = F(x) = (F_1(x), \cdots, F_n(x))$, $F \in C^1(\Omega; \mathbb{R}^n)$. We define, for every $x \in \Omega$, the **divergence** of F as*

$$\operatorname{div} F(x) = \sum_{i=1}^{n} \frac{\partial F_i}{\partial x_i}(x) \in \mathbb{R}.$$

We sometimes write $\nabla \cdot F$ for $\operatorname{div} F$.

*(iv) For $n = 2$ and $F(x) = (F_1(x), F_2(x))$ with $F \in C^1(\Omega; \mathbb{R}^2)$, we define, for every $x \in \Omega$, the **curl** of F by*

$$\operatorname{curl} F(x) = \frac{\partial F_2}{\partial x_1}(x) - \frac{\partial F_1}{\partial x_2}(x) \in \mathbb{R}.$$

*For $n = 3$ and $F(x) = (F_1(x), F_2(x), F_3(x))$ with $F \in C^1(\Omega; \mathbb{R}^3)$, we define the **curl** of F, $\operatorname{curl} F \in \mathbb{R}^3$, as*

$$\operatorname{curl} F(x) = \left(\frac{\partial F_3}{\partial x_2}(x) - \frac{\partial F_2}{\partial x_3}(x), \frac{\partial F_1}{\partial x_3}(x) - \frac{\partial F_3}{\partial x_1}(x), \frac{\partial F_2}{\partial x_1}(x) - \frac{\partial F_1}{\partial x_2}(x) \right).$$

We symbolically write

$$\operatorname{curl} F = \begin{vmatrix} e_1 & e_2 & e_3 \\ \frac{\partial}{\partial x_1} & \frac{\partial}{\partial x_2} & \frac{\partial}{\partial x_3} \\ F_1 & F_2 & F_3 \end{vmatrix} = \begin{pmatrix} \dfrac{\partial F_3}{\partial x_2} - \dfrac{\partial F_2}{\partial x_3} \\ \dfrac{\partial F_1}{\partial x_3} - \dfrac{\partial F_3}{\partial x_1} \\ \dfrac{\partial F_2}{\partial x_1} - \dfrac{\partial F_1}{\partial x_2} \end{pmatrix}.$$

We sometimes denote the curl of F as $\nabla \wedge F$.

Remark We can also define the curl of F in any dimension. When $n \geq 2$ and for every $x \in \Omega$ we let (with an appropriate order that we do not specify here)

$$\operatorname{curl} F(x) = \left((-1)^{i+j} \left(\frac{\partial F_i}{\partial x_j}(x) - \frac{\partial F_j}{\partial x_i}(x) \right) \right)_{1 \leq i < j \leq n} \in \mathbb{R}^{\frac{n(n-1)}{2}}.$$

Theorem 1.2 *Let $\Omega \subset \mathbb{R}^n$ be an open set.*

(i) Let $f \in C^2(\Omega)$, then

$$\operatorname{div} \operatorname{grad} f = \Delta f.$$

(ii) Let $n = 3$, $f \in C^2(\Omega)$ and $F \in C^2(\Omega; \mathbb{R}^3)$, then

$$\operatorname{curl} \operatorname{grad} f = 0 \quad and \quad \operatorname{div} \operatorname{curl} F = 0.$$

(iii) Let $f \in C^1(\Omega)$ and $g \in C^2(\Omega)$, then

$$\operatorname{div}(f \operatorname{grad} g) = f \Delta g + \operatorname{grad} f \cdot \operatorname{grad} g$$

where $\cdot$ denotes the scalar product of two vectors, namely if $x, y \in \mathbb{R}^n$, then

$$x \cdot y = \sum_{i=1}^{n} x_i \, y_i \,.$$

(iv) Let $f, g \in C^1(\Omega)$, then

$$\operatorname{grad}(f \, g) = f \operatorname{grad} g + g \operatorname{grad} f.$$

(v) Let $f \in C^1(\Omega)$ and $F \in C^1(\Omega; \mathbb{R}^n)$, then

$$\operatorname{div}(f \, F) = f \operatorname{div} F + F \cdot \operatorname{grad} f.$$

(vi) If $n = 3$ and $F \in C^2(\Omega; \mathbb{R}^3)$, then

$$\operatorname{curl} \operatorname{curl} F = -\Delta F + \operatorname{grad} \operatorname{div} F$$

(where, for $F = (F_1, F_2, F_3)$, we write $\Delta F = (\Delta F_1, \Delta F_2, \Delta F_3)$).

(vii) If $n = 3$, $f \in C^1(\Omega)$ and $F \in C^1(\Omega; \mathbb{R}^3)$, then

$$\operatorname{curl}(f \, F) = \operatorname{grad} f \wedge F + f \operatorname{curl} F$$

where $\wedge$ denotes the exterior product of two vectors, namely if $x, y \in \mathbb{R}^3$, then

$$x \wedge y = \begin{vmatrix} e_1 & e_2 & e_3 \\ x_1 & x_2 & x_3 \\ y_1 & y_2 & y_3 \end{vmatrix} = \begin{pmatrix} x_2 y_3 - x_3 y_2 \\ x_3 y_1 - x_1 y_3 \\ x_1 y_2 - x_2 y_1 \end{pmatrix}.$$

(More details can be found in [2] 4–7, [4] 113–116, [7] 316–317, [10] 485–497, [11] 417–424.)

1.2 Examples

Example 1.3 *Let $x = (x_1, \cdots, x_n)$, $a = (a_1, \cdots, a_n)$ and r be such that*

$$r = \|x - a\| = \sqrt{\sum_{i=1}^{n} (x_i - a_i)^2} \,.$$

Let $f(x) = 1/r$. Without using the above theorem, compute

$$F = \operatorname{grad} f, \quad \Delta f, \quad \operatorname{div} F.$$

For $n = 3$, compute $\operatorname{curl} F$.

Discussion Note first that the domain of f is $\Omega = \mathbb{R}^n \backslash \{a\}$. We first compute $\partial r / \partial x_i = r_{x_i}$. Using the definition of r^2 and taking the derivative of both sides, we find

$$2rr_{x_i} = 2(x_i - a_i) \Rightarrow r_{x_i} = \frac{x_i - a_i}{r}.$$

(i) We have, for every $i = 1, \cdots, n$,

$$\frac{\partial f}{\partial x_i} = \frac{\partial}{\partial x_i}(r^{-1}) = -r^{-2}r_{x_i} = -\frac{x_i - a_i}{r^3}$$

and hence

$$F(x) = \operatorname{grad} f(x) = -\frac{1}{r^3}(x_1 - a_1, \cdots, x_n - a_n) = -\frac{x - a}{r^3}.$$

(ii) On the other hand, we have

$$\frac{\partial^2 f}{\partial x_i^2} = -\frac{\partial}{\partial x_i}\left(\frac{x_i - a_i}{r^3}\right) = -\frac{r^3 - 3(x_i - a_i)r^2 r_{x_i}}{r^6}$$

$$= -\frac{r - 3(x_i - a_i)\frac{x_i - a_i}{r}}{r^4} = \frac{3(x_i - a_i)^2 - r^2}{r^5}.$$

We then deduce that

$$\Delta f = \sum_{i=1}^{n} \frac{\partial^2 f}{\partial x_i^2} = \frac{1}{r^5}\left[3\sum_{i=1}^{n}(x_i - a_i)^2 - \sum_{i=1}^{n} r^2\right]$$

$$= \frac{1}{r^5}\left[3r^2 - nr^2\right] = \frac{3 - n}{r^3}.$$

Note that, when $n = 3$, we have $\Delta f = 0$. (We could have deduced this result from the theorem, the previous calculation for the gradient and the next calculation for the divergence.)

(iii) Since $F_i = f_{x_i}$, we get

$$\frac{\partial F}{\partial x_i} = \frac{\partial}{\partial x_i}(f_{x_i}) = f_{x_i x_i}$$

and we find

$$\operatorname{div} F = \Delta f = \frac{3 - n}{r^3}.$$

(iv) By the above theorem, we must have $\operatorname{curl} F = \operatorname{curl} \operatorname{grad} f = 0$. The calculations give

$$\operatorname{curl} F = \begin{vmatrix} e_1 & e_2 & e_3 \\ \frac{\partial}{\partial x_1} & \frac{\partial}{\partial x_2} & \frac{\partial}{\partial x_3} \\ f_{x_1} & f_{x_2} & f_{x_3} \end{vmatrix} = \begin{pmatrix} f_{x_3 x_2} - f_{x_2 x_3} \\ f_{x_1 x_3} - f_{x_3 x_1} \\ f_{x_2 x_1} - f_{x_1 x_2} \end{pmatrix} = \begin{pmatrix} 0 \\ 0 \\ 0 \end{pmatrix}.$$

Example 1.4 *Let* $F(x, y, z) = (x^2 - e^y, \sin z, y^2 + z)$. *Compute* div F *and* curl F.

Discussion We get

$$\text{div } F = 2x + 0 + 1 = 2x + 1$$

and

$$\text{curl } F = \begin{vmatrix} e_1 & e_2 & e_3 \\ \dfrac{\partial}{\partial x} & \dfrac{\partial}{\partial y} & \dfrac{\partial}{\partial z} \\ x^2 - e^y & \sin z & y^2 + z \end{vmatrix} = \begin{pmatrix} 2y - \cos z \\ 0 - 0 \\ 0 + e^y \end{pmatrix} = \begin{pmatrix} 2y - \cos z \\ 0 \\ e^y \end{pmatrix}.$$

1.3 Exercises

Exercise 1.1 Consider the vector field given by

$$F(x, y, z) = \left(y^2 \sin(xz), e^y \cos(x^2 + z), \log(2 + \cos(xy))\right) = (f_1, f_2, f_3).$$

Compute the following quantities:

(i) grad f_1, grad f_2, grad f_3;

(ii) div F;

(iii) curl F.

Exercise 1.2 Let $f \in C^1(\mathbb{R}^3)$ and $F \in C^1(\mathbb{R}^3; \mathbb{R}^3)$. Which one of the following expressions makes sense?

(i) grad f (ii) f grad f (iii) $F \cdot$ grad f (iv) div f

(v) div(fF) (vi) curl(fF) (vii) curl f (viii) f curl F

(ix) curl div F.

Exercise 1.3 Let $x \in \mathbb{R}^n$ and $r = r(x) = \sqrt{x_1^2 + \cdots + x_n^2}$.

(i) Suppose that $f(x) = \psi(r)$. Show that if $x \neq 0$, then

$$\Delta f = \psi''(r) + \frac{n-1}{r}\psi'(r).$$

(ii) Find a solution to the equation $\Delta f = 0$ in $\Omega = \mathbb{R}^n \backslash \{0\}$.
Hint: use (cf. Example 1.3) the fact that

$$r_{x_i} = \frac{x_i}{r}.$$

Exercise 1.4 Let $x = r\cos\theta$, $y = r\sin\theta$,

$$\Omega = \{(x,y) \in \mathbb{R}^2 : x, y > 0\}$$

and $f \in C^2(\Omega)$, $f = f(x,y)$. Let

$$g = g(r,\theta) = f(r\cos\theta, r\sin\theta),$$

that is

$$f(x,y) = g(r(x,y), \theta(x,y)).$$

Show that

$$\Delta f = \frac{\partial^2 f}{\partial x^2} + \frac{\partial^2 f}{\partial y^2} = \frac{\partial^2 g}{\partial r^2} + \frac{1}{r}\frac{\partial g}{\partial r} + \frac{1}{r^2}\frac{\partial^2 g}{\partial \theta^2}.$$

Compute Δf for

$$f(x,y) = \sqrt{x^2 + y^2} + \left(\arctan\frac{y}{x}\right)^2.$$

Hint: begin by showing that

$$\frac{\partial r}{\partial x} = \frac{x}{r}, \quad \frac{\partial r}{\partial y} = \frac{y}{r}, \quad \frac{\partial \theta}{\partial x} = \frac{-y}{r^2} \quad \text{and} \quad \frac{\partial \theta}{\partial y} = \frac{x}{r^2}.$$

Exercise 1.5 Show (i) and (ii) of Theorem 1.2.

Exercise 1.6 Prove (iii) and (iv) of Theorem 1.2.

Exercise 1.7 Show (v), (vi) and (vii) of Theorem 1.2.

Chapter 2

Line integrals

2.1 Definitions and theoretical results

Definition 2.1 *Let $\Gamma \subset \mathbb{R}^n$ be a regular curve (for a more precise definition, cf. Section 8.3) with parametrization*

$$\gamma : [a,b] \to \Gamma \quad and \quad \gamma(t) = (\gamma_1(t), \cdots, \gamma_n(t)).$$

We write

$$\gamma'(t) = (\gamma_1'(t), \cdots, \gamma_n'(t)) \quad and \quad \|\gamma'(t)\| = \sqrt{\sum_{\nu=1}^{n} (\gamma_\nu'(t))^2}.$$

*(i) Let $f : \Gamma \to \mathbb{R}$ be a continuous function. The **integral of f along** Γ is defined by*

$$\int_\Gamma f \, dl = \int_a^b f(\gamma(t)) \, \|\gamma'(t)\| \, dt.$$

*(ii) Let $F = (F_1, \cdots, F_n) : \Gamma \to \mathbb{R}^n$ be a vector field. The **integral of F along** Γ is defined by*

$$\int_\Gamma F \cdot dl = \int_a^b F(\gamma(t)) \cdot \gamma'(t) \, dt = \int_a^b \sum_{\nu=1}^{n} F_\nu(\gamma(t)) \, \gamma_\nu'(t) \, dt.$$

(iii) If $\Gamma \subset \mathbb{R}^n$ is a piecewise regular curve (cf. Section 8.3; in particular, there exist a continuous function $\gamma : [a,b] \to \Gamma$ and $a = a_1 < a_2 <$

9

$\cdots < a_{N+1} = b$ *with* $\gamma' \in C\left([a_i, a_{i+1}]\right)$, $i = 1, \cdots, N$*), if* $f : \Gamma \to \mathbb{R}$ *is a continuous function and if* $F : \Gamma \to \mathbb{R}^n$ *is a continuous vector field, then*

$$\int_{\Gamma} f \, dl = \sum_{i=1}^{N} \int_{a_i}^{a_{i+1}} f\left(\gamma(t)\right) \|\gamma'(t)\| \, dt$$

$$\int_{\Gamma} F \cdot dl = \sum_{i=1}^{N} \int_{a_i}^{a_{i+1}} F\left(\gamma(t)\right) \cdot \gamma'(t) \, dt.$$

Remark (i) All curves considered in the present book will be piecewise regular simple curves.

(ii) By definition the **length** of a curve Γ is (letting $f \equiv 1$ in the above definition) given by

$$\text{length}\,(\Gamma) = \int_{\Gamma} dl.$$

(iii) The above definitions are independent of the chosen parametrization (for the second one this is understood up to a change of sign).

(For more details, cf. [2] 23, [4] 333–334, [7] 258, [10] 513, [11] 426.)

2.2   Examples

Example 2.2 *Compute the length of the unit circle.*

Discussion If we write

$$\Gamma = \{\gamma : [0, 2\pi] \to \mathbb{R}^2 : \gamma(t) = (\cos t, \sin t)\},$$

we then find

$$\gamma'(t) = (-\sin t, \cos t) \quad \text{and} \quad \|\gamma'(t)\| = 1.$$

We thus have

$$\text{length}\,(\Gamma) = \int_{\Gamma} dl = \int_{0}^{2\pi} dt = 2\pi.$$

Example 2.3 *(i) Compute* $\int_{\Gamma} f \, dl$ *for* $f(x,y) = \sqrt{x^2 + 4y^2}$ *and*

$$\Gamma = \{(x,y) \in \mathbb{R}^2 : 2y = x^2, \, x \in [0,1]\}.$$

(ii) Compute $\int_{\Gamma} F \cdot dl$ *for* $F(x,y) = \left(x^2, 0\right)$ *and*

$$\Gamma = \{(x,y) \in \mathbb{R}^2 : y = \cosh x, \, x \in [0,1]\}.$$

Discussion (i) A parametrization of the curve is given by

$$\gamma(t) = \left(t, \frac{t^2}{2}\right), \quad t \in [0,1].$$

Since $\gamma'(t) = (1,t)$, we get

$$\int_\Gamma f \, dl = \int_0^1 \sqrt{t^2 + 4\left(\frac{t^2}{2}\right)^2} \sqrt{1+t^2} \, dt = \int_0^1 t(1+t^2) \, dt = \frac{1}{2} + \frac{1}{4} = \frac{3}{4}.$$

(ii) In this case, we can consider

$$\gamma(t) = (t, \cosh t) \;\Rightarrow\; \gamma'(t) = (1, \sinh t).$$

We then find

$$\int_\Gamma F \cdot dl = \int_0^1 (t^2, 0) \cdot (1, \sinh t) \, dt = \frac{1}{3}.$$

2.3   Exercises

Exercise 2.1 (i) Let $\Gamma = \left\{(x,y) \in \mathbb{R}^2 : y = f(x), \; x \in [a,b]\right\}$. Show that

$$\text{length}(\Gamma) = \int_a^b \sqrt{1 + (f'(t))^2} \, dt.$$

(ii) Deduce from the previous question the length of the curve given by

$$\Gamma = \left\{(x,y) \in \mathbb{R}^2 : y = \cosh x, \; x \in [0,1]\right\}.$$

(iii) Let

$$\Gamma = \left\{(x,y) \in \mathbb{R}^2 : x(t) = r(t)\cos t, \; y(t) = r(t)\sin t, \; t \in [a,b]\right\}.$$

Find the length of Γ in terms of r.

Exercise 2.2 Compute $\int_{\Gamma_i} F \cdot dl$ when $F(x,y) = (xy, \; y^2 - x)$ and

$$\Gamma_1 = \left\{(t,t) : t \in [0,1]\right\},$$

$$\Gamma_2 = \left\{(t, e^t) : t \in [0,1]\right\},$$

$$\Gamma_3 = \left\{\left(\sqrt{t}, t^2\right) : t \in [1,2]\right\}.$$

Exercise 2.3 Compute $\int_\Gamma F \cdot dl$ in the following two cases

$$\Gamma = \{(x, y, z) \in \mathbb{R}^3 : x^2 + y^2 = 1,\ z = 0\} \quad \text{and} \quad F(x, y, z) = (x, z, y)$$

$$\Gamma = \{(x, y, z) \in \mathbb{R}^3 : y = e^x, z = x, x \in [0, 1]\} \quad \text{and} \quad F(x, y, z) = (x, y, z).$$

Exercise 2.4 Compute $\int_\Gamma f\, dl$ for $f(x, y, z) = x^2 + y^2 + \sqrt{2}z$ and

$$\Gamma = \left\{ \gamma(t) = \left(\cos t,\ \sin t,\ \frac{1}{2} t^2 \right) : t \in [0, 1] \right\}.$$

We recall that

$$\int \sqrt{x^2 + a^2}\, dx = \frac{x}{2}\sqrt{x^2 + a^2} + \frac{a^2}{2} \log\left(x + \sqrt{x^2 + a^2}\right).$$

Exercise 2.5 Let Γ be a regular curve in $\mathbb{R}^3$ joining A to B. Using Newton law (force = mass $\times$ acceleration), compute the work done by the force field to move a particle of constant mass from A to B along Γ.

Exercise 2.6 Let $F(x, y) = (x + y, -x)$ and

$$\Gamma = \{(x, y) \in \mathbb{R}^2 : y^2 + 4x^4 - 4x^2 = 0,\ x \geq 0\}.$$

(i) Show that $\gamma(t) = (\sin t, \sin 2t)$, $t \in [0, \pi]$, is a parametrization of Γ.
(ii) Compute $\int_\Gamma F \cdot dl$.

Chapter 3

Gradient vector fields

3.1 Definitions and theoretical results

Definition 3.1 *Let $\Omega \subset \mathbb{R}^n$ be open and $F : \Omega \to \mathbb{R}^n$ be written as*

$$F = F(x) = (F_1(x), \cdots, F_n(x)).$$

*We say that F is a **gradient vector field** or a **conservative field** in Ω if there exists $f \in C^1(\Omega)$ (f is called the **potential** of F) such that*

$$F(x) = \operatorname{grad} f(x) = \left(\frac{\partial f}{\partial x_1}(x), \cdots, \frac{\partial f}{\partial x_n}(x) \right), \quad \forall \, x \in \Omega.$$

Theorem 3.2 *Let $\Omega \subset \mathbb{R}^n$ be open and $F \in C^1(\Omega; \mathbb{R}^n)$. If F is a gradient vector field in Ω, then*

$$\frac{\partial F_i}{\partial x_j}(x) - \frac{\partial F_j}{\partial x_i}(x) = 0, \quad \forall \, i, j = 1, \cdots, n \text{ and } \forall \, x \in \Omega.$$

$$(3.1)$$

Remark (i) The above condition can be written as

$$\operatorname{curl} F(x) = 0, \quad \forall \, x \in \Omega.$$

(ii) The condition (3.1) is not sufficient (cf. Example 3.6 below). In order to guarantee the existence of such a potential, we have to give some conditions on the domain Ω. If the domain is convex or, more generally, if it is simply connected, then the condition is sufficient. We recall that, in

13

$\mathbb{R}^2$, a simply connected domain is a domain without hole. For more details about connected, convex and simply connected sets, cf. Chapter 8.

(iii) In a domain (that is an open and connected set) the potential is unique up to an additive constant.

Theorem 3.3 *Let $\Omega \subset \mathbb{R}^n$ be a domain and $F \in C(\Omega; \mathbb{R}^n)$. The following three statements are equivalent.*

(i) F is a gradient vector field in Ω.

(ii) For every closed and piecewise regular simple curve $\Gamma \subset \Omega$

$$\int_\Gamma F \cdot dl = 0.$$

(iii) Let A and B be any points in Ω. Let $\Gamma_1, \Gamma_2 \subset \Omega$ be two piecewise regular simple curves joining A to B. Then

$$\int_{\Gamma_1} F \cdot dl = \int_{\Gamma_2} F \cdot dl.$$

(For more details, cf. [4] 341–351, [7] 261–264, [10] 568–576, [11] 429–433.)

3.2   Examples

Example 3.4 *Let $F(x, y) = (4x^3y^2, 2x^4y + y)$. Show that F is a gradient vector field in $\Omega = \mathbb{R}^2$ and find a potential.*

Discussion The domain of F is $\Omega = \mathbb{R}^2$ which is a convex set. We get

$$\frac{\partial F_2}{\partial x} - \frac{\partial F_1}{\partial y} = 8x^3y - 8x^3y = 0.$$

Theorem 3.2 and the remark which follows tell us that there exists a potential $f : \mathbb{R}^2 \to \mathbb{R}$. In order to find it, we write

$$\frac{\partial f}{\partial x} = 4x^3y^2 \quad \text{and} \quad \frac{\partial f}{\partial y} = 2x^4y + y.$$

Integrating the first equation with respect to x, we get

$$f(x, y) = x^4y^2 + \alpha(y).$$

Taking the derivative of this expression with respect to y and inserting the result in the second equation, we are led to

$$\frac{\partial f}{\partial y} = 2x^4y + \alpha'(y) = 2x^4y + y.$$

This implies

$$\alpha'(y) = y \implies \alpha(y) = \frac{y^2}{2} + c,$$

with $c \in \mathbb{R}$. We therefore have found the potential, namely

$$f(x, y) = x^4 y^2 + \frac{y^2}{2} + c.$$

Example 3.5 *Let* $F(x, y, z) = \left(2x \sin z, ze^y, x^2 \cos z + e^y\right)$. *Show that* F *is a gradient vector field in* $\Omega = \mathbb{R}^3$ *and find a potential.*

Discussion We first check that the necessary condition is satisfied, i.e. $\operatorname{curl} F = 0$. We get

$$\operatorname{curl} F = \begin{vmatrix} e_1 & e_2 & e_3 \\ \dfrac{\partial}{\partial x} & \dfrac{\partial}{\partial y} & \dfrac{\partial}{\partial z} \\ 2x \sin z & ze^y & x^2 \cos z + e^y \end{vmatrix}$$

and thus

$$\operatorname{curl} F = \begin{pmatrix} e^y - e^y \\ 2x \cos z - 2x \cos z \\ 0 - 0 \end{pmatrix} = \begin{pmatrix} 0 \\ 0 \\ 0 \end{pmatrix}.$$

Since $\Omega = \mathbb{R}^3$ is convex, F is the gradient of some potential f which is given by

$$\frac{\partial f}{\partial x} = 2x \sin z, \quad \frac{\partial f}{\partial y} = ze^y \quad \text{and} \quad \frac{\partial f}{\partial z} = x^2 \cos z + e^y.$$

We integrate the first equation with respect to x and get

$$f(x, y, z) = x^2 \sin z + \alpha(y, z).$$

Taking the derivative of this expression with respect to y and z and inserting the result in the second and the third equations, we find

$$\begin{cases} \dfrac{\partial f}{\partial y} = ze^y = \dfrac{\partial \alpha}{\partial y} \\ \dfrac{\partial f}{\partial z} = x^2 \cos z + e^y = x^2 \cos z + \dfrac{\partial \alpha}{\partial z}. \end{cases}$$

Integrating the first equation with respect to y we find

$$\alpha(y, z) = ze^y + \beta(z).$$

Using the above result in the second equation we get

$$\frac{\partial \alpha}{\partial z} = e^y + \beta'(z) = e^y \;\Rightarrow\; \beta'(z) = 0 \;\Rightarrow\; \beta(z) = \beta = \text{constant.}$$

We finally find

$$f(x, y, z) = x^2 \sin z + z e^y + \beta.$$

Example 3.6 *Let*

$$F(x, y) = \left(-\frac{y}{x^2 + y^2}, \frac{x}{x^2 + y^2}\right).$$

(i) Find the domain of F.

(ii) Let

$$\Omega_1 = \left\{(x, y) \in \mathbb{R}^2 : y > 0\right\}$$
$$\Omega_2 = \left\{(x, y) \in \mathbb{R}^2 : y < 0\right\}$$
$$\Omega_3 = \mathbb{R}^2 \backslash \left\{(x, y) \in \mathbb{R}^2 : x \le 0 \text{ and } y = 0\right\}$$
$$\Omega_4 = \mathbb{R}^2 \setminus \{(0, 0)\}.$$

Is F a gradient vector field in Ω_i, $i = 1, 2, 3, 4$? If so, find a potential. If not, find a simple closed curve $\Gamma \subset \Omega_i$ such that $\int_\Gamma F \cdot dl \ne 0$.

Discussion (i) The domain of F is $\Omega_4 = \mathbb{R}^2 \backslash \{(0, 0)\}$ and $F \in C^\infty(\Omega_4; \mathbb{R}^2)$. We moreover have

$$\text{curl}\, F = \frac{\partial}{\partial x}\left(\frac{x}{x^2 + y^2}\right) - \frac{\partial}{\partial y}\left(-\frac{y}{x^2 + y^2}\right) = 0, \quad \forall\, (x, y) \in \Omega_4.$$

(ii) Note that $\Omega_1, \Omega_2 \subset \Omega_3 \subset \Omega_4$. Moreover, Ω_1 and Ω_2 are convex sets, Ω_3 is simply connected (but not convex) and Ω_4 is not simply connected. We first find a potential when $y > 0$ (on Ω_1), then when $y < 0$ (on Ω_2) and finally when $y = 0$.

Case 1: $(x, y) \in \Omega_1$. We want to find a potential $f \in C^1(\Omega_1)$. If such a f exists, then we must have

$$\frac{\partial f}{\partial x} = -\frac{y}{x^2 + y^2} \quad \text{and} \quad \frac{\partial f}{\partial y} = \frac{x}{x^2 + y^2}.$$

Integrating the first equation with respect to x gives us

$$f(x, y) = -\arctan\frac{x}{y} + \alpha_+(y).$$

Substituting this in the second equation gives $\alpha'_+ (y) = 0$. When $y > 0$, this implies that the potential is

$$f(x,y) = -\arctan \frac{x}{y} + \alpha_+, \quad \forall (x,y) \in \Omega_1$$

where $\alpha_+ \in \mathbb{R}$ is an arbitrary constant.

Case 2: $(x,y) \in \Omega_2$. Reasoning as above gives

$$f(x,y) = -\arctan \frac{x}{y} + \alpha_-, \quad \forall (x,y) \in \Omega_2$$

where $\alpha_- \in \mathbb{R}$ is an arbitrary constant.

Case 3: $(x,y) \in \Omega_3$. The first two cases imply that if such a potential $f \in C^1(\Omega_3)$ exists, then it must be of the form

$$f(x,y) = \begin{cases} -\arctan \dfrac{x}{y} + \alpha_+ & \text{if } y > 0 \text{ and } x \in \mathbb{R} \\ -\arctan \dfrac{x}{y} + \alpha_- & \text{if } y < 0 \text{ and } x \in \mathbb{R}. \end{cases}$$

We now have to choose α_+ and α_- so that f is continuous on the half-line $(x,0)$, with $x > 0$. Since x is positive, we have

$$\lim_{y \to 0+} f(x,y) = -\frac{\pi}{2} + \alpha_+ \quad \text{and} \quad \lim_{y \to 0-} f(x,y) = \frac{\pi}{2} + \alpha_-.$$

In order to ensure the continuity of f we choose $\alpha_+ = \alpha_- + \pi$. We therefore find that

$$f(x,y) = \begin{cases} -\arctan \dfrac{x}{y} + (\alpha_- + \pi) & \text{if } y > 0 \text{ and } x \in \mathbb{R} \\ \alpha_- + \dfrac{\pi}{2} & \text{if } y = 0 \text{ and } x > 0 \\ -\arctan \dfrac{x}{y} + \alpha_- & \text{if } y < 0 \text{ and } x \in \mathbb{R} \end{cases}$$

is a $C^1(\Omega_3)$ potential of F, where $\alpha_- \in \mathbb{R}$ is an arbitrary constant.

Case 4: $(x,y) \in \Omega_4$. We cannot reason as above to extend f when $y = 0$ and $x < 0$. Indeed, we would have in this case

$$\lim_{y \to 0+} f(x,y) = \frac{\pi}{2} + \alpha_- + \pi = \alpha_- + \frac{3\pi}{2}$$

$$\lim_{y \to 0-} f(x,y) = -\frac{\pi}{2} + \alpha_-,$$

which is impossible. Therefore, F is not the gradient of a potential on Ω_4. This can be shown in a different way. Let

$$\Gamma = \left\{ (x, y) \in \mathbb{R}^2 : x^2 + y^2 = 1 \right\} \subset \Omega_4.$$

We then have

$$\int_\Gamma F \cdot dl = \int_0^{2\pi} (-\sin\theta, \cos\theta) \cdot (-\sin\theta, \cos\theta) \ d\theta = 2\pi \neq 0.$$

This implies that F is not the gradient of a potential on Ω_4 (see Theorem 3.3).

3.3   Exercises

Exercise 3.1 Consider the vector fields $F_i : \mathbb{R}^2 \to \mathbb{R}^2$ with

$$F_1(x, y) = (y, xy - x), \quad F_2(x, y) = (3x^2y + 2x, x^3), \quad F_3(x, y) = (3x^2y, x^2).$$

Is the field F_i a gradient vector field in $\mathbb{R}^2$? If so, find a potential for F_i. If not, find a simple closed curve Γ such that $\int_\Gamma F_i \cdot dl \neq 0$.

Exercise 3.2 (i) Let $F \in C^1\left(\mathbb{R}^2; \mathbb{R}^2\right)$ be such that

$$F = F(u, v) = (f(u, v), g(u, v)).$$

Let

$$\varphi(x, y) = \int_0^1 [x \, f(tx, ty) + y \, g(tx, ty)] \, dt.$$

Show that if

$$\frac{\partial g}{\partial u} = \frac{\partial f}{\partial v},$$

then $F(x, y) = \operatorname{grad} \varphi(x, y)$. Apply the above result to find a potential for $F(x, y) = (2xy, x^2 + y)$.

(ii) Generalize this result to $\mathbb{R}^n$, with

$$F(u) = (F_1(u), \ldots, F_n(u)), \quad \varphi(x) = \int_0^1 F(tx) \cdot x \, dt$$

and

$$\frac{\partial F_i}{\partial u_j} = \frac{\partial F_j}{\partial u_i}, \quad \forall \, i, j = 1, \cdots, n.$$

Exercise 3.3 Let

$$F(x,y,z) = \left(2xy + \frac{z}{1+x^2}, x^2 + 2yz, y^2 + \arctan x \right).$$

Show that F is a gradient vector field in $\mathbb{R}^3$.

Exercise 3.4 Let

$$\Omega = \mathbb{R}^2 \setminus \{(0,y) \in \mathbb{R}^2 : y \geq 0\} \quad \text{and} \quad F(x,y) = \left(\frac{-y}{x^2+y^2}, \frac{x}{x^2+y^2} \right).$$

Show, by finding a potential f, that F is the gradient of a potential in Ω. Hint: proceed as in Example 3.6.

Exercise 3.5 Let $r = \sqrt{x^2+y^2}$ and

$$F(x,y) = (\alpha(r)x, \ \beta(r)y)$$

with $\alpha, \beta \in C^1(0,\infty)$ such that

$$\lim_{r \to 0} (r\alpha(r)) = \lim_{r \to 0} (r\beta(r)) = 0.$$

(i) Find a necessary condition in order that F is the gradient of a potential in $\Omega = \mathbb{R}^2 \setminus \{(0,0)\}$.

(ii) Show (as in Exercise 3.2) that

$$f(x,y) = \int_0^1 [F(tx,ty) \cdot (x,y)] \ dt$$

is a potential for F in Ω.

Exercise 3.6 Consider the differential equation

$$f_2(t,u(t)) u'(t) + f_1(t,u(t)) = 0, \ t \in \mathbb{R}.$$

(i) Let

$$F(x,y) = (f_1(x,y), \ f_2(x,y))$$

be a vector field which is a gradient of a potential f in $\mathbb{R}^2$. Show that a solution $u = u(t)$ to the differential equation is given in an implicit form by

$$f(t,u(t)) = \text{constant}, \quad \forall t \in \mathbb{R}.$$

Hint: compute $\dfrac{d}{dt} f(t,u(t))$.

(ii) Find a solution to

$$\begin{cases} u^2(t)u'(t) + \sin t = 0 \\ \qquad u(0) = 3. \end{cases}$$

Exercise 3.7 (Integrating factor) Consider the differential equation

$$f_2\left(t, u(t)\right) u'(t) + f_1\left(t, u(t)\right) = 0, \quad t \in \mathbb{R}.$$

(i) Let

$$F(x, y) = \left(f_1(x, y), \ f_2(x, y)\right).$$

Show that if there exists $W \in C^1(\mathbb{R}^2)$ with

$$W(x, y) \neq 0, \quad \forall\, (x, y) \in \mathbb{R}^2$$

such that WF is the gradient of a potential Φ in $\mathbb{R}^2$, then a solution $u = u(t)$ of the differential equation is given in an implicit form by

$$\Phi\left(t, u(t)\right) = \text{constant}, \quad \forall\, t \in \mathbb{R}.$$

(ii) Find a solution to

$$4t \sin(tu(t)) + u(t)(t^2 + 1) \cos(tu(t)) + u'(t) \left[(t^2 + 1)t \cos\left(tu(t)\right)\right] = 0.$$

Hint: choose $W(x, y) = 1 + x^2$.

Exercise 3.8 Let $\Omega = \mathbb{R}^2 \backslash \{(0, 0)\}$ and consider the vector fields

$$F(x, y) = \left(\frac{-x}{(x^2 + y^2)^2}, \frac{-y}{(x^2 + y^2)^2}\right)$$

$$G(x, y) = \left(\frac{y^3}{(x^2 + y^2)^2}, \frac{-xy^2}{(x^2 + y^2)^2}\right).$$

Are these fields gradient vector fields in Ω? If so, find a potential. If not, justify your answer.

Chapter 4

Green theorem

4.1 Definitions and theoretical results

Definition 4.1 *(i) We say that $\Omega \subset \mathbb{R}^2$ is a **regular domain** (cf. Figure 4.1) if there exist bounded open sets $\Omega_0, \Omega_1, \cdots, \Omega_m \subset \mathbb{R}^2$ such that*

$$\Omega = \Omega_0 \setminus \bigcup_{j=1}^{m} \overline{\Omega}_j \qquad\qquad \overline{\Omega}_j \subset \Omega_0, \ \forall j = 1, 2, \cdots, m$$

$$\overline{\Omega}_i \cap \overline{\Omega}_j = \emptyset \ \text{if } i \neq j, \ i,j = 1, \cdots, m \qquad \partial\Omega_j = \Gamma_j, \ j = 0, 1, 2, \cdots, m$$

where the Γ_j are piecewise regular simple closed curves.

*(ii) We say that $\partial\Omega = \Gamma_0 \cup \Gamma_1 \cup \cdots \cup \Gamma_m$ is **positively oriented** if, while moving on $\Gamma_0, \Gamma_1, \cdots, \Gamma_m$, the domain Ω is seen on the left.*

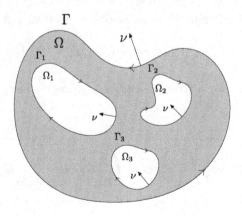

Figure 4.1: Regular domain

Remark The above definition of the orientation of $\partial\Omega$ is only intuitive. A precise definition can be found in Chapter 8. Note that since $\overline{\Omega}_j \subset \Omega_0$, whenever $j = 1, 2, \cdots, m$, the orientation on Γ_0 is positive (i.e. counterclockwise) while the one on $\Gamma_1, \cdots, \Gamma_m$ is negative (i.e. clockwise).

Theorem 4.2 (Green theorem) *Let* $\Omega \subset \mathbb{R}^2$ *be a regular domain with positively oriented boundary* $\partial\Omega$. *Let*

$$F = F(x_1, x_2) = (F_1(x_1, x_2), F_2(x_1, x_2))$$

with $F \in C^1\left(\overline{\Omega}; \mathbb{R}^2\right)$. *Then*

$$\iint_\Omega \operatorname{curl} F(x_1, x_2)\, dx_1 dx_2 = \iint_\Omega \left(\frac{\partial F_2}{\partial x_1} - \frac{\partial F_1}{\partial x_2}\right) dx_1 dx_2 = \int_{\partial\Omega} F \cdot dl.$$

Corollary 4.3 (Divergence theorem in the plane) *Let* Ω, $\partial\Omega$ *and* F *be as in Green theorem. Let* ν *be the vector field of exterior unit normals to* $\partial\Omega$, *then*

$$\iint_\Omega \operatorname{div} F(x_1, x_2)\, dx_1\, dx_2 = \iint_\Omega \left(\frac{\partial F_1}{\partial x_1} + \frac{\partial F_2}{\partial x_2}\right) dx_1\, dx_2 = \int_{\partial\Omega} (F \cdot \nu)\, dl.$$

Corollary 4.4 *Let* Ω *and* $\partial\Omega$ *be as in Green theorem. Let*

$$F(x_1, x_2) = (-x_2, x_1)$$

$$G_1(x_1, x_2) = (0, x_1) \quad and \quad G_2(x_1, x_2) = (-x_2, 0).$$

Then

$$\operatorname{area}(\Omega) = \frac{1}{2} \int_{\partial\Omega} F \cdot dl = \int_{\partial\Omega} G_1 \cdot dl = \int_{\partial\Omega} G_2 \cdot dl.$$

(For more details, cf. [2] 43–44, [4] 363–364, [7] 360, [10] 528–532, [11] 436–445.)

4.2   Examples

Example 4.5 *Let $\Omega = \{(x, y) \in \mathbb{R}^2 : x^2 + y^2 < 1\}$ and $F(x, y) = (y^2, x)$. Verify Green theorem.*

Discussion (i) We first compute $\iint_\Omega \operatorname{curl} F \, dx \, dy$. We easily find that

$$\operatorname{curl} F = 1 - 2y.$$

Setting $x = r \cos \theta$ and $y = r \sin \theta$, we get

$$\iint_\Omega \operatorname{curl} F \, dx \, dy = \int_0^1 \int_0^{2\pi} (1 - 2r \sin \theta) \, r \, dr \, d\theta = \pi.$$

(ii) We next compute $\int_{\partial\Omega} F \cdot dl$. If we set $\gamma(\theta) = (\cos \theta, \sin \theta)$, we find

$$\int_{\partial\Omega} F \cdot dl = \int_0^{2\pi} (\sin^2 \theta, \cos \theta) \cdot (-\sin \theta, \cos \theta) \, d\theta = \int_0^{2\pi} \cos^2 \theta \, d\theta = \pi.$$

Example 4.6 *Let $\Omega = \{(x, y) \in \mathbb{R}^2 : 1 < x^2 + y^2 < 4\}$ and*

$$F(x, y) = (x^2 y, 2xy).$$

Check Green theorem.

Discussion (i) Computation of $\iint_\Omega \operatorname{curl} F \, dx \, dy$. We find

$$\operatorname{curl} F = \frac{\partial F_2}{\partial x} - \frac{\partial F_1}{\partial y} = 2y - x^2.$$

We therefore get, using polar coordinates,

$$\iint_\Omega \operatorname{curl} F \, dx \, dy = \int_1^2 \int_0^{2\pi} (2r \sin \theta - r^2 \cos^2 \theta) \, r \, dr \, d\theta$$

$$= \int_1^2 \int_0^{2\pi} -r^3 \cos^2 \theta \, dr d\theta = -\pi \int_1^2 r^3 dr = -\frac{15}{4}\pi.$$

(ii) Computation of $\int_{\partial\Omega} F \cdot dl$. We set

$$\Gamma_0 = \{\gamma_0(\theta) = (2\cos \theta, 2\sin \theta) : \theta \in [0, 2\pi]\}$$

$$\Gamma_1 = \{\gamma_1(\theta) = (\cos \theta, \sin \theta) : \theta \in [0, 2\pi]\}$$

and we get

$$\int_{\partial\Omega} F \cdot dl = \int_{\Gamma_0} F \cdot dl - \int_{\Gamma_1} F \cdot dl.$$

We then obtain

$$\int_{\Gamma_0} F \cdot dl = \int_0^{2\pi} (8\cos^2\theta\sin\theta, 8\cos\theta\sin\theta) \cdot (-2\sin\theta, 2\cos\theta)\, d\theta$$

$$= -16 \int_0^{2\pi} (\cos^2\theta\sin^2\theta - \cos^2\theta\sin\theta)\, d\theta$$

$$= -16 \int_0^{2\pi} \cos^2\theta\sin^2\theta\, d\theta = -4\pi$$

$$\int_{\Gamma_1} F \cdot dl = \int_0^{2\pi} (\cos^2\theta\sin\theta, \cos\theta\sin\theta) \cdot (-\sin\theta, \cos\theta)\, d\theta = -\frac{\pi}{4}.$$

We finally have

$$\int_{\partial\Omega} F \cdot dl = -\frac{15}{4}\pi.$$

4.3 Exercises

Exercise 4.1 Verify Green theorem for

$$\Omega = \left\{ (x,y) \in \mathbb{R}^2 : x^2 + y^2 < 1 \right\}$$

and $F(x,y) = (xy, y^2)$.

Exercise 4.2 Verify Green theorem for

$$\Omega = \left\{ (x,y) \in \mathbb{R}^2 : 1 < x^2 + y^2 < 4 \right\}$$

and $F(x,y) = (x+y, y^2)$.

Exercise 4.3 Let $\Omega \subset \mathbb{R}^2$ be the triangle with vertices $(0,0)$, $(0,1)$ and $(1,0)$. Let $u(x,y) = y + e^x$.

(i) Compute $\iint_\Omega \Delta u(x,y)\, dx\, dy$.

(ii) Compute

$$\int_{\partial\Omega} \left(\frac{\partial u}{\partial x}\, \nu_1 + \frac{\partial u}{\partial y}\, \nu_2 \right) dl,$$

where $\nu = (\nu_1, \nu_2)$ is the exterior unit normal to $\partial\Omega$ and $\partial\Omega$ is positively oriented.

Exercise 4.4 Verify Green theorem in the following cases:

(i) $\Omega = \left\{ (x,y) \in \mathbb{R}^2 : x^2 + (y-1)^2 < 1 \right\}$ and $F(x,y) = (-x^2 y, xy^2)$;

(ii) $\Omega = \{(x,y) \in \mathbb{R}^2 : x > 0 \text{ and } x^2 + y^2 < 1\}$ and

$$F(x,y) = \left(\frac{x}{2(1+x^2+y^2)}, \varphi(y) \right)$$

where $\varphi \in C^1(\mathbb{R})$.

Exercise 4.5 Verify Green theorem for $F(x,y) = (xy, y)$ and

$$\Omega = \left\{ (x,y) \in \mathbb{R}^2 : x^2 + y^2 > 1 \text{ and } x^2 - 4 < y < 2 \right\}.$$

Exercise 4.6 Verify Green theorem for

$$\Omega = \left\{ (x,y) \in \mathbb{R}^2 : x^2 + y^2 < 1 \right\}.$$

Hint: it is better to use Cartesian rather than polar coordinates and write

$$\Omega = \left\{ (x,y) \in \mathbb{R}^2 : x \in (-1,1) \text{ and } -\sqrt{1-x^2} < y < \sqrt{1-x^2} \right\}$$
$$= \left\{ (x,y) \in \mathbb{R}^2 : y \in (-1,1) \text{ and } -\sqrt{1-y^2} < x < \sqrt{1-y^2} \right\}.$$

Exercise 4.7 Let $\Omega \subset \mathbb{R}^2$ be a regular domain, $\nu = (\nu_1, \nu_2)$ the exterior unit normal to $\partial\Omega$ ($\partial\Omega$ being positively oriented). Let $u, v \in C^2(\overline{\Omega})$. Show that

$$\iint_\Omega \Delta u \, dx \, dy = \int_{\partial\Omega} (\operatorname{grad} u \cdot \nu) \, dl$$

and prove the **Green identities**

$$\iint_\Omega [v\Delta u + \operatorname{grad} u \cdot \operatorname{grad} v] \, dxdy = \int_{\partial\Omega} v \, (\operatorname{grad} u \cdot \nu) \, dl$$

$$\iint_\Omega (u\Delta v - v\Delta u) \, dx \, dy = \int_{\partial\Omega} [u \, (\operatorname{grad} v \cdot \nu) - v \, (\operatorname{grad} u \cdot \nu)] \, dl.$$

Hint: first use Theorem 1.2, namely

$$\operatorname{div}(u \operatorname{grad} v) = u \, \Delta v + \operatorname{grad} u \cdot \operatorname{grad} v.$$

Then, appeal to the divergence theorem (cf. Corollary 4.3).

Exercise 4.8 Using Green theorem, prove the divergence theorem.

Exercise 4.9 (i) Prove Corollary 4.4.

(ii) If $\partial\Omega = \Gamma$ is a closed regular simple curve with parametrization $\gamma(t) = (\gamma_1(t), \gamma_2(t))$, $t \in [a, b]$, prove that

$$\text{area}(\Omega) = \int_a^b \gamma_1(t)\,\gamma_2'(t)\,dt = -\int_a^b \gamma_1'(t)\,\gamma_2(t)\,dt$$

$$= \frac{1}{2}\int_a^b [\gamma_1(t)\,\gamma_2'(t) - \gamma_1'(t)\,\gamma_2(t)]\,dt.$$

Exercise* 4.10 Let $\Omega \subset \mathbb{R}^2$ be a regular domain and $f, \varphi \in C^0(\overline{\Omega})$. Show that the problem

$$(D) \quad \begin{cases} \Delta u(x, y) = f(x, y) & (x, y) \in \Omega \\ u(x, y) = \varphi(x, y) & (x, y) \in \partial\Omega \end{cases}$$

admits at most one solution $u \in C^2(\overline{\Omega})$.

Hint: let u and v be two solutions of (D). Using the first Green identity (cf. Exercise 4.7), show that $w = u - v \equiv 0$.

Chapter 5

Surface integrals

5.1 Definitions and theoretical results

We state here only the basic definitions about surfaces. For more details, see Section 8.4.

Definition 5.1 *(i) We say that $\Sigma \subset \mathbb{R}^3$ is a **regular surface** if (among others) there exist*

– a bounded open set $A \subset \mathbb{R}^2$ whose boundary ∂A is a piecewise regular simple closed curve;

– a one-to-one parametrization $\sigma \in C^1\left(\overline{A}; \mathbb{R}^3\right)$,

$$\sigma : \overline{A} \to \Sigma = \sigma\left(\overline{A}\right) \subset \mathbb{R}^3,$$

$$\sigma\left(u, v\right) = \left(\sigma^1\left(u, v\right), \sigma^2\left(u, v\right), \sigma^3\left(u, v\right)\right),$$

such that the normal vector satisfy

$$\sigma_u \wedge \sigma_v = \begin{vmatrix} e_1 & e_2 & e_3 \\ \sigma_u^1 & \sigma_u^2 & \sigma_u^3 \\ \sigma_v^1 & \sigma_v^2 & \sigma_v^3 \end{vmatrix} = \begin{pmatrix} \sigma_u^2 \sigma_v^3 - \sigma_u^3 \sigma_v^2 \\ \sigma_u^3 \sigma_v^1 - \sigma_u^1 \sigma_v^3 \\ \sigma_u^1 \sigma_v^2 - \sigma_u^2 \sigma_v^1 \end{pmatrix} \neq 0, \quad \forall\, (u, v) \in \overline{A}.$$

*(ii) A surface is said to be **piecewise regular** if (intuitively) it is a finite union of regular surfaces (if necessary, see Definition 8.13).*

Remark All the surfaces considered in this book are piecewise regular surfaces and, moreover, orientable (we do not need to define this concept here; if necessary, see Chapter 8). Furthermore, in all examples, we will be able to deal with piecewise regular surfaces as if they were regular surfaces without going into too much detail.

Example 5.2 *(i) (Sphere) Let*

$$\Sigma = \left\{ (x, y, z) \in \mathbb{R}^3 : x^2 + y^2 + z^2 = 1 \right\}.$$

A piecewise regular parametrization of Σ is given by

$$\sigma(\theta, \varphi) = (\cos\theta \sin\varphi, \sin\theta \sin\varphi, \cos\varphi), \quad (\theta, \varphi) \in \overline{A}$$

where $A = (0, 2\pi) \times (0, \pi)$. The associated normal is

$$\sigma_\theta \wedge \sigma_\varphi = -\sin\varphi \left(\cos\theta \sin\varphi, \sin\theta \sin\varphi, \cos\varphi \right).$$

(ii) (Cylinder) Let

$$\Sigma = \left\{ (x, y, z) \in \mathbb{R}^3 : x^2 + y^2 = 1 \text{ and } 0 < z < 1 \right\}.$$

We choose a piecewise regular parametrization

$$\sigma(\theta, z) = (\cos\theta, \sin\theta, z), \quad (\theta, z) \in \overline{A}$$

with $A = (0, 2\pi) \times (0, 1)$. The associated normal is

$$\sigma_\theta \wedge \sigma_z = (\cos\theta, \sin\theta, 0).$$

Definition 5.3 *(i) Let $\Sigma \subset \mathbb{R}^3$ be a regular surface (with a parametrization $\sigma : \overline{A} \to \mathbb{R}^3$, $\sigma = \sigma(u, v)$) and a continuous function $f : \Sigma \to \mathbb{R}$. We define the **integral of f on Σ** as*

$$\iint_\Sigma f\, ds = \iint_A f\left(\sigma(u, v)\right) \|\sigma_u \wedge \sigma_v\|\, du\, dv.$$

(ii) If Σ is a piecewise regular surface such that $\Sigma = \bigcup_{i=1}^{m} \Sigma_i$ with each Σ_i regular and two by two disjoint, then

$$\iint_\Sigma f\, ds = \sum_{i=1}^{m} \iint_{\Sigma_i} f\, ds.$$

Definition 5.4 *(i) Let $\Sigma \subset \mathbb{R}^3$ be an orientable regular surface (with parametrization $\sigma : \overline{A} \to \mathbb{R}^3$, $\sigma = \sigma(u, v)$). Let $F : \Sigma \to \mathbb{R}^3$ be a continuous vector field. We define the **integral of F on Σ** in the direction $\nu = \sigma_u \wedge \sigma_v$ as*

$$\iint_\Sigma F \cdot ds = \iint_A \left[F\left(\sigma(u, v)\right) \cdot \sigma_u \wedge \sigma_v \right] du\, dv.$$

(ii) Let $\Sigma = \bigcup_{i=1}^{m} \Sigma_i$ *with each* Σ_i *regular and two by two disjoint. Then*

$$\iint_{\Sigma} F \cdot ds = \sum_{i=1}^{m} \iint_{\Sigma_i} F \cdot ds.$$

Remark (i) If $f \equiv 1$, then, by definition, we let

$$\boxed{\text{area}\,(\Sigma) = \iint_{\Sigma} ds.}$$

(ii) The integral $\iint_{\Sigma} F \cdot ds$ is also called the flux across Σ in the direction of the normal ν.

(iii) The above definitions are independent of the chosen parametrization (the second definition being independent only up to a sign).

(For more details, cf. [2] 29–31, [4] 376–389, [7] 334, [10] 540–550, [11] 452–459.)

5.2 Examples

Example 5.5 *Find the area of* Σ, *where*

$$\Sigma = \left\{ (x, y, z) \in \mathbb{R}^3 : x^2 + y^2 + z^2 = R^2 \right\}.$$

Discussion We define $A = (0, 2\pi) \times (0, \pi)$ and

$$\sigma(\theta, \varphi) = (R \cos \theta \sin \varphi, R \sin \theta \sin \varphi, R \cos \varphi).$$

We then find

$$\sigma_\theta \wedge \sigma_\varphi = -R^2 \sin \varphi (\cos \theta \sin \varphi, \sin \theta \sin \varphi, \cos \varphi)$$

and so

$$\|\sigma_\theta \wedge \sigma_\varphi\| = R^2 \sin \varphi.$$

Therefore, the area of Σ is given by

$$\text{area}(\Sigma) = \iint_{\Sigma} ds = \int_0^{2\pi} \int_0^{\pi} R^2 \sin \varphi \, d\theta \, d\varphi = 2\pi R^2 \int_0^{\pi} \sin \varphi \, d\varphi = 4\pi R^2.$$

Example 5.6 *Compute $\iint_\Sigma f\,ds$, where $f(x,y,z) = x^2 + y^2 + 2z$ and*

$$\Sigma = \left\{ (x,y,z) \in \mathbb{R}^3 : x^2 + y^2 = 1 \text{ and } 0 \leq z \leq 1 \right\}.$$

Discussion We define $A = (0, 2\pi) \times (0, 1)$ and

$$\sigma(\theta, z) = (\cos\theta, \sin\theta, z).$$

We find

$$\sigma_\theta \wedge \sigma_z = (\cos\theta, \sin\theta, 0),$$

which implies that

$$\|\sigma_\theta \wedge \sigma_z\| = 1.$$

We thus obtain

$$\iint_\Sigma f\,ds = \int_0^1 \int_0^{2\pi} (\cos^2\theta + \sin^2\theta + 2z)\,d\theta\,dz = 2\pi \int_0^1 (1 + 2z)\,dz = 4\pi.$$

Example 5.7 *Let $F(x,y,z) = (y, -x, z^2)$ and*

$$\Sigma = \left\{ (x,y,z) \in \mathbb{R}^3 : z^2 = x^2 + y^2 \text{ and } 0 \leq z \leq 1 \right\}.$$

Compute the flux across Σ in the upward direction (i.e. in the direction $z > 0$).

Discussion In this case, we set $A = (0, 2\pi) \times (0, 1)$ and

$$\sigma(\theta, z) = (z\cos\theta, z\sin\theta, z).$$

We get

$$\sigma_\theta \wedge \sigma_z = (z\cos\theta, z\sin\theta, -z).$$

Since $-z < 0$, we choose $\nu = -(\sigma_\theta \wedge \sigma_z)$. We therefore obtain

$$\iint_\Sigma F \cdot ds = -\int_0^1 \int_0^{2\pi} (z\sin\theta, -z\cos\theta, z^2) \cdot (z\cos\theta, z\sin\theta, -z)\,dz\,d\theta$$

$$= -\int_0^1 \int_0^{2\pi} (z^2\cos\theta\sin\theta - z^2\cos\theta\sin\theta - z^3)\,dz\,d\theta$$

$$= 2\pi \int_0^1 z^3\,dz = \frac{\pi}{2}.$$

5.3 Exercises

Exercise 5.1 Let $f(x, y, z) = xy + z^2$ and

$$\Sigma = \left\{(x, y, z) \in \mathbb{R}^3 : x^2 + y^2 = z^2 \text{ and } 0 \le z \le 1\right\}.$$

Compute $\iint_\Sigma f \, ds$.

Exercise 5.2 Let $F(x, y, z) = (x^2, y^2, z^2)$ and

$$\Sigma = \left\{(x, y, z) \in \mathbb{R}^3 : z^2 = x^2 + y^2 \text{ and } 0 \le z \le 1\right\}.$$

Compute the flux across Σ in the upward direction (i.e. in the direction $z > 0$).

Exercise 5.3 Let

$$\Sigma = \left\{(x, y, z) \in \mathbb{R}^3 : x^2 + y^2 - z^2 = 0, \ 0 \le z \le 1\right\}.$$

and the density $\rho(x, y, z) = \sqrt{x^2 + y^2}$. Find the mass of the surface.

Exercise 5.4 Let $F(x, y, z) = (0, z, z)$ and

$$\Sigma = \left\{(x, y, z) \in \mathbb{R}^3 : z = 6 - 3x - 2y; \ x, y, z \ge 0\right\}.$$

Compute the flux across Σ going out from the origin.

Exercise 5.5 Let

$$\Omega = \left\{(x, y, z) \in \mathbb{R}^3 : x^2 + y^2 \le z \le 1\right\}.$$

Find the area of $\partial \Omega$.

Exercise 5.6 Let $0 < a < R$. Consider the torus generated by the rotation of the circle $(x - R)^2 + z^2 \le a^2$ around the $z-$axis (cf. Example 8.19).

(i) Show that

$$x = (R + r \cos \varphi) \cos \theta, \quad y = (R + r \cos \varphi) \sin \theta \quad \text{and} \quad z = r \sin \varphi$$

with $0 < r < a$ and $0 \le \theta, \varphi < 2\pi$ is a parametrization of Ω. What is the meaning (cf. Example 8.19) of r, θ and φ? Find the Jacobian of the transformation and compute the volume of Ω.

(ii) Find a parametrization of the boundary $\partial \Omega$ of Ω and a normal vector to $\partial \Omega$.

(iii) Find the area of $\partial \Omega$.

(iv) Compute $\iiint_\Omega z^2 dx \, dy \, dz$.

Chapter 6

Divergence theorem

6.1 Definitions and theoretical results

Definition 6.1 *We say that $\Omega \subset \mathbb{R}^3$ is a **regular domain** (cf. Figure 6.1) if there exist bounded open sets $\Omega_0, \Omega_1, \cdots, \Omega_m \subset \mathbb{R}^3$ such that*

$$\Omega = \Omega_0 \setminus \bigcup_{j=1}^{m} \overline{\Omega}_j \qquad\qquad \overline{\Omega}_j \subset \Omega_0 \,, \ \forall j = 1, 2, \cdots, m$$

$$\overline{\Omega}_i \cap \overline{\Omega}_j = \emptyset \ \text{if} \ i \neq j, \ i, j = 1, \cdots, m \qquad \partial\Omega_j = \Sigma_j \,, \ j = 0, 1, 2, \cdots, m$$

*where Σ_j, $j = 0, 1, 2, \cdots, m$ are orientable piecewise regular surfaces such that $\partial\Sigma_j = \emptyset$. Furthermore, there exists a (piecewise continuous) vector field of **exterior normals** ν to Ω.*

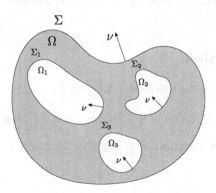

Figure 6.1: Regular domain

Remark The last condition is understood as follows. For every $x \in \Sigma_j$, $j = 0, 1, 2, \cdots, m$, where the field of normals is continuous, and for every $\epsilon > 0$ sufficiently small

$$x + \epsilon \nu(x) \in \Omega_j, \quad x - \epsilon \nu(x) \in \overline{\Omega}_j^c, \quad j = 1, 2, \cdots, m$$

$$x - \epsilon \nu(x) \in \Omega_0, \quad x + \epsilon \nu(x) \in \overline{\Omega}_0^c.$$

Theorem 6.2 (Divergence theorem) *Let $\Omega \subset \mathbb{R}^3$ be a regular domain and ν the exterior unit normal to Ω. Let $F \in C^1\left(\overline{\Omega}; \mathbb{R}^3\right)$, then*

$$\iiint_\Omega \operatorname{div} F(x_1, x_2, x_3)\, dx_1\, dx_2\, dx_3 = \iint_{\partial\Omega} (F \cdot \nu)\, ds.$$

Corollary 6.3 *If Ω and ν are as in the theorem and if*

$$F(x_1, x_2, x_3) = (x_1, x_2, x_3), \quad G_1(x_1, x_2, x_3) = (x_1, 0, 0),$$
$$G_2(x_1, x_2, x_3) = (0, x_2, 0), \quad G_3(x_1, x_2, x_3) = (0, 0, x_3),$$

then

$$\operatorname{vol}(\Omega) = \frac{1}{3}\iint_{\partial\Omega} (F \cdot \nu)\, ds = \iint_{\partial\Omega} (G_i \cdot \nu)\, ds, \quad i = 1, 2, 3.$$

(For more details, cf. [2] 33–36, [4] 397–407, [7] 340–349, [10] 550–555, [11] 475–479.)

6.2 Examples

Example 6.4 *Check the divergence theorem for*

$$\Omega = \left\{(x, y, z) \in \mathbb{R}^3 : x^2 + y^2 + z^2 < 1\right\} \quad and \quad F(x, y, z) = (xy, y, z).$$

Discussion (i) Computation of $\iiint_\Omega \operatorname{div} F$. We easily find that

$$\operatorname{div} F = y + 2.$$

Then, using spherical coordinates,

$$x = r\cos\theta\sin\varphi, \quad y = r\sin\theta\sin\varphi, \quad z = r\cos\varphi,$$

we get

$$\iiint_\Omega (y+2)\, dx\, dy\, dz = \int_0^1 \int_0^{2\pi} \int_0^\pi (2 + r\sin\theta\sin\varphi) r^2 \sin\varphi\, dr\, d\theta\, d\varphi$$

$$= 2\pi \int_0^1 \int_0^\pi 2r^2 \sin\varphi\, dr\, d\varphi = \frac{4}{3}\pi \int_0^\pi \sin\varphi\, d\varphi = \frac{8}{3}\pi.$$

(ii) Computation of $\iint_{\partial\Omega} (F \cdot \nu)\, ds$. A parametrization of the surface is given by

$$\sigma(\theta, \varphi) = (\cos\theta\sin\varphi, \sin\theta\sin\varphi, \cos\varphi)$$

and thus

$$\partial\Omega = \Sigma = \{\sigma(\theta, \varphi) : (\theta, \varphi) \in [0, 2\pi] \times [0, \pi]\}.$$

A direct calculation gives

$$\sigma_\theta \wedge \sigma_\varphi = -\sin\varphi(\cos\theta\sin\varphi, \sin\theta\sin\varphi, \cos\varphi),$$

which is an interior normal. Observing that we have on $\partial\Omega$

$$F \cdot (-\sigma_\theta \wedge \sigma_\varphi)$$
$$= (\cos\theta\sin\theta\sin^2\varphi, \sin\theta\sin\varphi, \cos\varphi) \cdot (\cos\theta\sin\varphi, \sin\theta\sin\varphi, \cos\varphi)\sin\varphi$$

we get

$$\iint_{\partial\Omega} (F \cdot \nu)\, ds$$

$$= \int_0^{2\pi} \int_0^\pi (\cos^2\theta\sin\theta\sin^4\varphi + \sin^2\theta\sin^3\varphi + \cos^2\varphi\sin\varphi)\, d\varphi\, d\theta$$

$$= \pi \int_0^\pi \sin^3\varphi\, d\varphi + 2\pi \int_0^\pi \cos^2\varphi\sin\varphi\, d\varphi$$

$$= \pi \left[-\cos\varphi + \frac{\cos^3\varphi}{3} \right]_0^\pi - 2\pi \left[\frac{\cos^3\varphi}{3} \right]_0^\pi = \frac{8}{3}\pi.$$

Example 6.5 *Verify the divergence theorem for*

$$\Omega = \{(x, y, z) \in \mathbb{R}^3 : x^2 + y^2 < 1 \text{ and } 0 < z < 1\}$$

and $F(x, y, z) = (x^2, 0, 0).$

Discussion (i) Computation of $\iiint_\Omega \operatorname{div} F$. We find that

$$\operatorname{div} F = 2x.$$

Using cylindrical coordinates $x = r\cos\theta$, $y = r\sin\theta$ and $z = z$, we get

$$\iiint_\Omega (2x)\,dx\,dy\,dz = \int_0^{2\pi} \int_0^1 \int_0^1 2r\cos\theta\, r\,dr\,d\theta\,dz = 0.$$

(ii) Computation of $\iint_{\partial\Omega} (F \cdot \nu)\,ds$. We have

$$\partial\Omega = \Sigma_0 \cup \Sigma_1 \cup \Sigma_2\,,$$

where

$$\Sigma_0 = \{(x,y,z) \in \mathbb{R}^3 : x^2 + y^2 < 1 \text{ and } z = 0\}$$
$$\Sigma_1 = \{(x,y,z) \in \mathbb{R}^3 : x^2 + y^2 < 1 \text{ and } z = 1\}$$
$$\Sigma_2 = \{(x,y,z) \in \mathbb{R}^3 : x^2 + y^2 = 1 \text{ and } 0 < z < 1\}$$

whose parametrizations and normals are given by

$$\sigma^0(r,\theta) = (r\cos\theta, r\sin\theta, 0) \ \Rightarrow\ \sigma_r^0 \wedge \sigma_\theta^0 = (0,0,r)$$
$$\sigma^1(r,\theta) = (r\cos\theta, r\sin\theta, 1) \ \Rightarrow\ \sigma_r^1 \wedge \sigma_\theta^1 = (0,0,r)$$
$$\sigma^2(r,\theta) = (\cos\theta, \sin\theta, z) \ \Rightarrow\ \sigma_\theta^2 \wedge \sigma_z^2 = (\cos\theta, \sin\theta, 0)$$

they are, respectively, interior, exterior and exterior normals. We therefore get

$$\iint_{\Sigma_0} (F \cdot \nu)\,ds \ = \ -\int_0^{2\pi}\int_0^1 (r^2\cos^2\theta, 0, 0) \cdot (0,0,r)\,dr\,d\theta = 0,$$

$$\iint_{\Sigma_1} (F \cdot \nu)\,ds \ = \ \int_0^{2\pi}\int_0^1 (r^2\cos^2\theta, 0, 0) \cdot (0,0,r)\,dr\,d\theta = 0,$$

$$\iint_{\Sigma_2} (F \cdot \nu)\,ds \ = \ \int_0^{2\pi}\int_0^1 (\cos^2\theta, 0, 0) \cdot (\cos\theta, \sin\theta, 0)\,d\theta\,dz = 0.$$

We finally obtain

$$\iint_{\partial\Omega} (F \cdot \nu)\,ds = 0.$$

6.3 Exercises

In Exercises 1 to 10 verify the divergence theorem.

Exercise 6.1 Let $F(x, y, z) = (xz, y, y)$ and

$$\Omega = \left\{ (x, y, z) \in \mathbb{R}^3 : x^2 + y^2 + z^2 < 1 \right\}.$$

Exercise 6.2 Let $F(x, y, z) = (0, 0, xyz)$ and

$$\Omega = \left\{ (x, y, z) \in \mathbb{R}^3 : z^2 > x^2 + y^2 \text{ and } 0 < z < 1 \right\}.$$

Exercise 6.3 Let $F(x, y, z) = (xy, yz, xz)$ and the unit tetrahedron given by

$$\Omega = \left\{ (x, y, z) \in \mathbb{R}^3 : 0 < z < 1 - x - y, \ 0 < y < 1 - x, \ 0 < x < 1 \right\}.$$

Exercise 6.4 Let $F(x, y, z) = (x^2, y^2, z^2)$ and

$$\Omega = \left\{ (x, y, z) \in \mathbb{R}^3 : b^2 \left(x^2 + y^2 \right) < a^2 z^2 \text{ and } 0 < z < b \right\}.$$

Exercise 6.5 Let $F(x, y, z) = (x^2, -y^2, z^2)$ and

$$\Omega = \left\{ (x, y, z) \in \mathbb{R}^3 : x^2 + y^2 < 4 \text{ and } 0 < z < 2 \right\}.$$

Exercise 6.6 Let $F(x, y, z) = (x, y, z)$ and

$$\Omega = \left\{ (x, y, z) \in \mathbb{R}^3 : x^2 + y^2 + z^2 < 4 \text{ and } x^2 + y^2 < 3z \right\}.$$

Exercise 6.7 Let $F(x, y, z) = \left(0, \dfrac{3y}{1 + z^2}, 5 \right)$ and

$$\Omega = \left\{ (x, y, z) \in \mathbb{R}^3 : \tan \sqrt{x^2 + y^2} < z < 1, \ x^2 + y^2 < \frac{\pi^2}{16} \text{ and } x < 0 < y \right\}.$$

Exercise 6.8 Let $F(x, y, z) = (x^2, y^2, z^2)$ and

$$\Omega = \left\{ (x, y, z) \in \mathbb{R}^3 : z^2 < x^2 + y^2 < 1 \text{ and } z > 0 \right\}.$$

Exercise 6.9 Let $F(x, y, z) = (2, 0, xy^2 + z^2)$ and

$$\Omega = \left\{ (x, y, z) \in \mathbb{R}^3 : x > 0, \ 0 < z < 2 \text{ and } 4(x^2 + y^2) < (z - 4)^2 \right\}.$$

Exercise 6.10 Let $F(x, y, z) = (x^2, 0, 0)$ and

$$\Omega = \left\{ (x, y, z) \in \mathbb{R}^3 : x^2 + y^2 + z^2 < 4 \text{ and } x^2 + y^2 + (z - 2)^2 < 1 \right\}.$$

Exercise 6.11 Show Corollary 6.3.

Exercise 6.12 (Green identities) Let $\Omega \subset \mathbb{R}^3$ be a regular domain and $\nu : \partial\Omega \to \mathbb{R}^3$ the exterior unit normal. Let $u, v \in C^2(\overline{\Omega})$. We write $\dfrac{\partial u}{\partial \nu}$ for $\operatorname{grad} u \cdot \nu$. Show that

$$\iiint_\Omega [v\Delta u + \operatorname{grad} u \cdot \operatorname{grad} v]\, dx\, dy\, dz = \iint_{\partial\Omega} v\frac{\partial u}{\partial \nu}\, ds$$

and

$$\iiint_\Omega (v \triangle u - u \triangle v)\, dx\, dy\, dz = \iint_{\partial\Omega} \left(v\frac{\partial u}{\partial \nu} - u\frac{\partial v}{\partial \nu} \right) ds.$$

Hint: first, use Theorem 1.2, namely

$$\operatorname{div}(u \operatorname{grad} v) = u \triangle v + \operatorname{grad} u \cdot \operatorname{grad} v$$

and then apply the divergence theorem.

Exercise* 6.13 Let $\Omega \subset \mathbb{R}^3$ be a regular domain and ν the exterior unit normal to $\partial\Omega$. Let $f, \varphi \in C^0(\overline{\Omega})$ and the problem

$$(N) \quad \begin{cases} \Delta u = f & \text{in } \Omega \\ \frac{\partial u}{\partial \nu} = \varphi & \text{on } \partial\Omega \end{cases}$$

where $\dfrac{\partial u}{\partial \nu}$ stands for $\operatorname{grad} u \cdot \nu$. Using the previous exercise, find necessary conditions on f and φ for the problem (N) to admit a solution $u \in C^2(\overline{\Omega})$. What can we say about the uniqueness of the solutions of (N)?

Chapter 7

Stokes theorem

7.1 Definitions and theoretical results

Theorem 7.1 (Stokes theorem) *Let* $\Sigma \subset \mathbb{R}^3$ *be an orientable piecewise regular surface. Let* $F : \Sigma \to \mathbb{R}^3$, $F = (F_1, F_2, F_3)$, *where the* F_i *are* C^1 *on an open set containing* $\Sigma \cup \partial\Sigma$. *Then*

$$\iint_{\Sigma} \operatorname{curl} F \cdot ds = \int_{\partial\Sigma} F \cdot dl.$$

Remark We recall here only the basic facts about surfaces. For more details, see Chapter 5 and Section 8.4.

(i) The boundary of all piecewise regular surfaces $\Sigma \subset \mathbb{R}^3$ with parametrization $\sigma = \sigma(u, v) : \overline{A} \to \Sigma$ that we will consider will be denoted by $\partial\Sigma$. This boundary is obtained from $\sigma(\partial A)$ by removing points and curves which are traversed twice in opposite directions (cf. the examples below).

(ii) For a given parametrization $\sigma : \overline{A} \to \Sigma$, the normal is given by $\sigma_u \wedge \sigma_v$ and the surface integral is understood in the direction of $\sigma_u \wedge \sigma_v$, that is

$$\iint_{\Sigma} \operatorname{curl} F \cdot ds = \iint_{A} (\operatorname{curl} F(\sigma(u, v))) \cdot (\sigma_u \wedge \sigma_v) \, du \, dv.$$

Moreover, the orientation of $\partial\Sigma$ is induced by the parametrization $\sigma : \overline{A} \to \Sigma$, i.e. ∂A is positively oriented.

Example 7.2 *For more details, see Section 8.4.*

(i) (Sphere) Let

$$\Sigma = \left\{ (x, y, z) \in \mathbb{R}^3 : x^2 + y^2 + z^2 = 1 \right\},$$

then $\partial \Sigma = \emptyset$ (cf. Example 8.16). A piecewise regular parametrization of Σ is given by

$$\sigma(\theta, \varphi) = (\cos \theta \sin \varphi, \sin \theta \sin \varphi, \cos \varphi), \quad (\theta, \varphi) \in \overline{A}$$

where $A = (0, 2\pi) \times (0, \pi)$.

(ii) (Hemisphere) If

$$\Sigma = \left\{ (x, y, z) \in \mathbb{R}^3 : x^2 + y^2 + z^2 = 1 \text{ and } z \geq 0 \right\},$$

then

$$\partial \Sigma = \left\{ (x, y, z) \in \mathbb{R}^3 : x^2 + y^2 = 1 \text{ and } z = 0 \right\}$$

(cf. Example 8.17) and the orientation of ∂A induced by the parametrization

$$\sigma(\theta, \varphi) = (\cos \theta \sin \varphi, \sin \theta \sin \varphi, \cos \varphi), \quad (\theta, \varphi) \in \overline{A}$$

(where $A = (0, 2\pi) \times (0, \pi/2)$) is the negative orientation (that is, $\theta : 2\pi \to 0$). We thus find

$$\sigma_\theta \wedge \sigma_\varphi = -\sin \varphi \left(\cos \theta \sin \varphi, \sin \theta \sin \varphi, \cos \varphi \right).$$

We could have chosen another parametrization for Σ, namely

$$\widetilde{\sigma}(x, y) = \left(x, y, \sqrt{1 - x^2 - y^2} \right), \quad (x, y) \in B = \left\{ (x, y) \in \mathbb{R}^2 : x^2 + y^2 < 1 \right\}.$$

In this case, we would also have found

$$\partial \Sigma = \widetilde{\sigma}(\partial B) = \left\{ (x, y, z) \in \mathbb{R}^3 : x^2 + y^2 = 1 \text{ and } z = 0 \right\}.$$

(iii) (Cylinder) If

$$\Sigma = \left\{ (x, y, z) \in \mathbb{R}^3 : x^2 + y^2 = 1 \text{ and } 0 \leq z \leq 1 \right\},$$

then (cf. Example 8.18)

$$\partial \Sigma = \Gamma_0 \cup \Gamma_1,$$

where

$$\begin{aligned} \Gamma_0 &= \left\{ (x, y, z) \in \mathbb{R}^3 : x^2 + y^2 = 1 \text{ and } z = 0 \right\}, \\ \Gamma_1 &= \left\{ (x, y, z) \in \mathbb{R}^3 : x^2 + y^2 = 1 \text{ and } z = 1 \right\}. \end{aligned}$$

If we choose the parametrization

$$\sigma\left(\theta, z\right) = \left(\cos\theta, \sin\theta, z\right), \quad \left(\theta, z\right) \in \overline{A}$$

with $A = (0, 2\pi) \times (0, 1)$, then the orientation of $\partial\Sigma$ induced by the parametrization σ is the positive orientation on Γ_0 ($\theta : 0 \rightarrow 2\pi$) and the negative one on Γ_1 ($\theta : 2\pi \rightarrow 0$). Moreover, the field of unit normals induced by the parametrization is

$$\nu = \sigma_\theta \wedge \sigma_z = \left(\cos\theta, \sin\theta, 0\right).$$

(For more details, cf. [2] 41–49, [4] 410–417, [7] 362–366, [10] 562–567, [11] 470–474.)

7.2 Examples

Example 7.3 *Verify Stokes theorem for $F\left(x, y, z\right) = \left(z, x, y\right)$ and*

$$\Sigma = \left\{\left(x, y, z\right) \in \mathbb{R}^3 : z^2 = x^2 + y^2 \text{ and } 0 < z < 1\right\}.$$

Discussion (i) Computation of $\iint_\Sigma \operatorname{curl} F \cdot ds$. We have

$$\operatorname{curl} F = \begin{vmatrix} e_1 & e_2 & e_3 \\ \frac{\partial}{\partial x} & \frac{\partial}{\partial y} & \frac{\partial}{\partial z} \\ z & x & y \end{vmatrix} = (1, 1, 1).$$

We choose the following parametrization of Σ

$$\sigma(\theta, z) = (z\cos\theta, z\sin\theta, z), \text{ with } (\theta, z) \in A = (0, 2\pi) \times (0, 1).$$

We then find that the normal to the surface is given by

$$\sigma_\theta \wedge \sigma_z = \begin{vmatrix} e_1 & e_2 & e_3 \\ -z\sin\theta & z\cos\theta & 0 \\ \cos\theta & \sin\theta & 1 \end{vmatrix} = (z\cos\theta, z\sin\theta, -z).$$

We therefore obtain

$$\iint_\Sigma \operatorname{curl} F \cdot ds = \int_0^{2\pi} \int_0^1 (1, 1, 1) \cdot (z\cos\theta, z\sin\theta, -z)\, dz\, d\theta$$

$$= \int_0^{2\pi} \int_0^1 (z\cos\theta + z\sin\theta - z)\, dz\, d\theta = -2\pi \int_0^1 z\, dz = -\pi.$$

(ii) Computation of $\int_{\partial \Sigma} F \cdot dl$. We first note that

$$\sigma(\partial A) = \Gamma_1 \cup \Gamma_2 \cup \Gamma_3 \cup \Gamma_4$$

where

$$\Gamma_1 = \{\gamma_1(\theta) = \sigma(\theta, 0) = (0, 0, 0)\} = \{(0, 0, 0)\}$$
$$\Gamma_2 = \{\gamma_2(z) = \sigma(2\pi, z) = (z, 0, z), \ z : 0 \to 1\}$$
$$\Gamma_3 = \{\gamma_3(\theta) = \sigma(\theta, 1) = (\cos\theta, \sin\theta, 1), \ \theta : 2\pi \to 0\}$$
$$\Gamma_4 = \{\gamma_4(z) = \sigma(0, z) = (z, 0, z), \ z : 1 \to 0\} = -\Gamma_2.$$

We then observe that

$$\partial\Sigma = \Gamma_3 = \{\gamma_3(\theta) = \sigma(\theta, 1) = (\cos\theta, \sin\theta, 1), \ \theta : 2\pi \to 0\},$$

and Γ_3 is negatively oriented (since $\theta : 2\pi \to 0$). Since

$$\gamma_3'(\theta) = (-\sin\theta, \cos\theta, 0),$$

we get

$$\int_{\partial\Sigma} F \cdot dl = -\int_0^{2\pi} (1, \cos\theta, \sin\theta) \cdot (-\sin\theta, \cos\theta, 0)\, d\theta$$

$$= -\int_0^{2\pi} \cos^2\theta = -\pi.$$

Example 7.4 *Verify Stokes theorem for* $F(x, y, z) = (0, 0, y^2)$ *and*

$$\Sigma = \{(x, y, z) \in \mathbb{R}^3 : x^2 + y^2 + z^2 = 1 \text{ and } z \le 0\}.$$

Discussion (i) Computation of $\iint_\Sigma \operatorname{curl} F \cdot ds$. We see that

$$\operatorname{curl} F = \begin{vmatrix} e_1 & e_2 & e_3 \\ \frac{\partial}{\partial x} & \frac{\partial}{\partial y} & \frac{\partial}{\partial z} \\ 0 & 0 & y^2 \end{vmatrix} = (2y, 0, 0).$$

We choose the following parametrization of Σ

$$\sigma(\theta, \varphi) = (\cos\theta \sin\varphi, \sin\theta \sin\varphi, \cos\varphi),$$

with $(\theta, \varphi) \in A = (0, 2\pi) \times (\pi/2, \pi)$. A normal to the surface is then given by

$$\sigma_\theta \wedge \sigma_\varphi = -\sin\varphi(\cos\theta \sin\varphi, \sin\theta \sin\varphi, \cos\varphi).$$

We next compute the surface integral

$$\iint_\Sigma \operatorname{curl} F \cdot ds$$

$$= \int_{\frac{\pi}{2}}^{\pi} \int_0^{2\pi} -\sin\varphi(2\sin\theta\sin\varphi, 0, 0) \cdot (\cos\theta\sin\varphi, \sin\theta\sin\varphi, \cos\varphi)\, d\varphi\, d\theta$$

$$= -2 \int_{\frac{\pi}{2}}^{\pi} \int_0^{2\pi} \sin\theta\cos\theta\sin^3\varphi\, d\varphi\, d\theta = 0.$$

(ii) Computation of $\int_{\partial\Sigma} F \cdot dl$. We remark that

$$\sigma(\partial A) = \Gamma_1 \cup \Gamma_2 \cup \Gamma_3 \cup \Gamma_4$$

where

$$\Gamma_1 = \{\gamma_1(\theta) = \sigma(\theta, \pi/2) = (\cos\theta, \sin\theta, 0),\ \theta : 0 \to 2\pi\}$$

$$\Gamma_2 = \{\gamma_2(\varphi) = \sigma(2\pi, \varphi) = (\sin\varphi, 0, \cos\varphi),\ \varphi : \frac{\pi}{2} \to \pi\}$$

$$\Gamma_3 = \{\gamma_3(\theta) = \sigma(\theta, \pi) = (0, 0, -1),\ \theta : 2\pi \to 0\} = \{(0, 0, -1)\}$$

$$\Gamma_4 = \{\gamma_4(\varphi) = \sigma(0, \varphi) = (\sin\varphi, 0, \cos\varphi),\ \varphi : \pi \to \frac{\pi}{2}\} = -\Gamma_2.$$

We thus have

$$\partial\Sigma = \Gamma_1$$

where Γ_1 is positively oriented (since $\theta : 0 \to 2\pi$). We therefore get

$$\gamma_1'(\theta) = (-\sin\theta, \cos\theta, 0).$$

We finally obtain

$$\int_{\partial\Sigma} F \cdot dl = \int_0^{2\pi} (0, 0, \sin^2\theta) \cdot (-\sin\theta, \cos\theta, 0)\, d\theta = 0.$$

7.3 Exercises

In the following exercises, verify Stokes theorem.

Exercise 7.1 Let $F(x, y, z) = (x^2 y, z^2, 0)$ and

$$\Sigma = \{(x, y, z) \in \mathbb{R}^3 : x^2 + y^2 = z^2,\ 0 \le z \le 1\}.$$

Exercise 7.2 Let $F(x, y, z) = (x^2 y, z, x)$ and

$$\Sigma = \{(x, y, z) \in \mathbb{R}^3 : x^2 + y^2 = z^4,\ 0 \le z \le 1\}.$$

Exercise 7.3 Let $F(x, y, z) = (x^2 y^3, 1, z)$ and

$$\Sigma = \left\{ (x, y, z) \in \mathbb{R}^3 : x^2 + y^2 + z^2 = R^2, \ z \geq 0 \right\}.$$

Exercise 7.4 Let $F(x, y, z) = (-2y, xz, y)$ and

$$\Sigma = \left\{ (x, y, z) \in \mathbb{R}^3 : z = 3 - \frac{3}{2}\sqrt{x^2 + y^2}, \ x \geq 0 \text{ and } \frac{4}{9} \leq x^2 + y^2 \leq 4 \right\}.$$

Exercise 7.5 Let $F(x, y, z) = (0, z^2, 0)$ and

$$\Sigma = \left\{ (x, y, z) \in \mathbb{R}^3 : x^2 + y^2 + z^2 = 4, \ x, y \geq 0, \ 1 \leq z \leq \sqrt{3} \right\}.$$

Exercise 7.6 Let $F(x, y, z) = (0, 0, y + z^2)$ and

$$\Sigma = \left\{ (x, y, z) \in \mathbb{R}^3 : \begin{array}{l} x^2 + y^2 + z^2 = 4, \ x, y, z \geq 0, \\ 0 \leq \arccos \dfrac{z}{2} \leq \arctan \dfrac{y}{x} \leq \dfrac{\pi}{2} \end{array} \right\}.$$

Exercise 7.7 Let $F(x, y, z) = (0, x^2, 0)$ and Σ be the triangle with vertices $A = (1, 0, 0)$, $B = (2, 2, 0)$ and $C = (1, 1, 0)$.

Exercise 7.8 Let $F(x, y, z) = (xy, xz, x^2)$ and Σ be the surface obtained from the intersection of the cylinder $x^2 + y^2 \leq 1$ with the plane $x + z = 1$.

Exercise 7.9 Let $F(x, y, z) = (z, y, 0)$ and

$$\Sigma = \left\{ (x, y, z) \in \mathbb{R}^3 : z = \sin(3x + 2y), \ x \geq 0, \ y \geq 0 \text{ and } x + y \leq \frac{\pi}{2} \right\}.$$

Chapter 8

Appendix

8.1 Some notations and notions of topology

Notation (i) We denote the **scalar product** of two vectors by $\cdot$, namely if $x, y \in \mathbb{R}^n$, then

$$x \cdot y = \sum_{i=1}^{n} x_i y_i \, .$$

(ii) The **Euclidean norm** of $x \in \mathbb{R}^n$ is defined as

$$\|x\| = \left(\sum_{i=1}^{n} x_i^2 \right)^{1/2} .$$

(iii) The **exterior product** of two vectors in $\mathbb{R}^3$ is denoted by $\wedge$, namely if $x, y \in \mathbb{R}^3$, then

$$x \wedge y = \begin{vmatrix} e_1 & e_2 & e_3 \\ x_1 & x_2 & x_3 \\ y_1 & y_2 & y_3 \end{vmatrix} = \begin{pmatrix} x_2 y_3 - x_3 y_2 \\ x_3 y_1 - x_1 y_3 \\ x_1 y_2 - x_2 y_1 \end{pmatrix} .$$

(iv) For $x \in \mathbb{R}^n$ and $R > 0$, we denote the **open ball** centered at x and of radius R by

$$B_R(x) = \{ y \in \mathbb{R}^n : \|y - x\| < R \} \, .$$

The **closed ball** is denoted by

$$\overline{B}_R(x) = \{ y \in \mathbb{R}^n : \|x - y\| \leq R \} \, .$$

45

Definition 8.1 *Let $A \subset \mathbb{R}^n$.*

*(i) The set A is **open** if $\forall x \in A$, $\exists \epsilon > 0$ such that $B_\epsilon(x) \subset A$.*

*(ii) We define the **complement** of A, denoted A^c, as*

$$A^c = \mathbb{R}^n \backslash A = \{x \in \mathbb{R}^n : x \notin A\}.$$

*(iii) The set A is **closed** if A^c is open.*

*(iv) We define the **boundary** of A, denoted ∂A, as*

$$\partial A = \{x \in \mathbb{R}^n : B_\epsilon(x) \cap A \neq \emptyset, \ B_\epsilon(x) \cap A^c \neq \emptyset, \ \forall \epsilon > 0\}.$$

*(v) We write $\overline{A}$ for the **closure** of A, which is by definition the smallest closed set containing A.*

*(vi) We define the **interior** of A, denoted $\text{int}\, A$ (or $\mathring{A}$), as the largest open set contained in A.*

Proposition 8.2 *For every subset $A \subset \mathbb{R}^n$ we have the following inclusions*

$$\boxed{\text{int}\, A \subset A \subset \overline{A} \quad and \quad \overline{A} = A \cup \partial A.}$$

Example 8.3 Case $n = 1$.

(i) $\mathbb{R}$ is both open and closed ($\mathbb{R}$ and the empty set are the only sets to have this property).

(ii) The interval $[a, b]$ is closed.

(iii) The interval (a, b), sometimes denoted $]a, b[$, is open.

(iv) The intervals $[a, b)$ and $(a, b]$, sometimes denoted $[a, b[$ and $]a, b]$, are neither closed nor open.

(v) If $-\infty < a < b < \infty$, the boundary of the following intervals $[a, b]$, (a, b), $[a, b)$, $(a, b]$ is $\{a, b\}$.

Case $n \geq 2$.

(i) $\mathbb{R}^n$ is both open and closed.

(ii) A point $x \in \mathbb{R}^n$, viewed as the set $\{x\}$, is closed.

(iii) $B_R(x) \subset \mathbb{R}^n$ and $\mathbb{R}^n \backslash \{0\}$ are open sets.

(iv) $\overline{B}_R(x) = \{y \in \mathbb{R}^n : ||x - y|| \leq R\}$ is closed.

(v) $\partial B_R(x) = \{y \in \mathbb{R}^n : ||x - y|| = R\}$ is the sphere of radius R and it is closed.

Definition 8.4 *(i) We say that a set $A \subset \mathbb{R}^n$ is **convex** (cf. Figure 8.1) if, $\forall t \in [0,1]$, $\forall x, y \in A$, we have*

$$tx + (1-t)y \in A$$

(in more geometrical terms, this means that $\forall x, y \in A$ the line segment $[x, y]$ with endpoints x and y is contained in A).

*(ii) $A \subset \mathbb{R}^n$ is (path-)**connected** if, $\forall x, y \in A$, there exists some curve Γ, with continuous parametrization $\gamma : [a, b] \to \Gamma$, from $x = \gamma(a)$ to $y = \gamma(b)$ which is entirely contained in A.*

*(iii) We say that $A \subset \mathbb{R}^n$ is a **domain** if it is both open and connected.*

*(iv) We say that $A \subset \mathbb{R}^n$ is **simply connected** if it is (path-)connected and if any pair of curves Γ_0 and Γ_1 having the same endpoints and entirely contained in A can be continuously deformed one into the other without going out of A. More precisely, if*

$$\gamma_0 : [a, b] \to \Gamma_0 \subset A \quad \text{and} \quad \gamma_1 : [a, b] \to \Gamma_1 \subset A$$

are two continuous parametrizations of Γ_0 and Γ_1 ($\gamma_0(a) = \gamma_1(a)$, $\gamma_0(b) = \gamma_1(b)$), then there exists a continuous

$$\gamma : [a, b] \times [0, 1] \to A$$

such that

$$\gamma(t, 0) = \gamma_0(t), \quad \gamma(t, 1) = \gamma_1(t), \quad \forall t \in [a, b]$$
$$\gamma(a, s) = \gamma_0(a) = \gamma_1(a), \quad \forall s \in [0, 1]$$
$$\gamma(b, s) = \gamma_0(b) = \gamma_1(b), \quad \forall s \in [0, 1]$$
$$\gamma(t, s) \in A, \quad \forall t \in [a, b], \ \forall s \in [0, 1].$$

Remark (i) The case $n = 1$ is slightly different from the case $n \geq 2$. Indeed, a connected set is necessarily convex, while the notion of a simply connected set is irrelevant.

(ii) We always have

$$A \text{ convex } \underset{\not\Leftarrow}{\Rightarrow} A \text{ simply connected } \underset{\not\Leftarrow}{\Rightarrow} A \text{ connected.}$$

(We should warn the reader that some authors do not require, in the definition of "simply connected", that the set is connected. For them, the

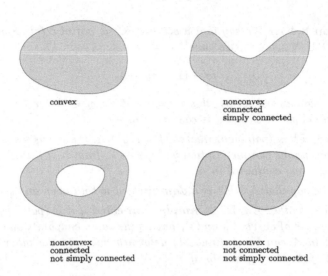

convex

nonconvex
connected
simply connected

nonconvex
connected
not simply connected

nonconvex
not connected
not simply connected

Figure 8.1: Topological notions

implication A simply connected implies A connected is, therefore, in general, false.)

(iii) Intuitively, a set of $\mathbb{R}^2$ is simply connected if it has no holes (however in $\mathbb{R}^3$ this is no longer the case, cf. Example 8.5 (v)).

Example 8.5 *The examples below are constantly used.*

(i) $[a,b]$, (a,b), $[a,b)$, $(a,b]$ *and* $\mathbb{R}$ *are convex sets, contrary to* $[0,1] \cup [2,3]$.

(ii) $B_R(x) \subset \mathbb{R}^n$ *and* $\mathbb{R}^n$ *are convex (and thus simply connected and connected).*

(iii) $A = \mathbb{R}^2 \backslash \{(0,0)\}$ *is neither convex nor simply connected, it is, however, connected.*

(iv) $A = \mathbb{R}^2 \backslash \{(x,y) \in \mathbb{R}^2 : y = 0 \text{ and } x \leq 0\}$ *is not convex, but it is simply connected (and thus connected).*

(v) $A = \mathbb{R}^3 \backslash \{(0,0,0)\}$ *is not convex, however, it is simply connected (and hence connected).*

(vi) $A = \mathbb{R}^3 \backslash \{(x,y,z) \in \mathbb{R}^3 : x = y = 0\}$ *is neither convex, nor simply connected, but it is connected.*

(For more details, cf. [4] 53, 369, [7] 12, 56–58, 372, [10] 572, [11] 160, 373, 431.)

8.2   Some notations for functional spaces

Definition 8.6 *Let $\Omega \subset \mathbb{R}^n$ be open and $k \geq 0$ be an integer (including $k = \infty$).*

(i) $C^k(\Omega)$ is the set of functions $f : \Omega \to \mathbb{R}$ that are k times continuously differentiable functions in Ω. (When $k = 0$ we, sometimes, only write $C(\Omega)$ instead of $C^0(\Omega)$; hence it is the set of continuous functions in Ω.)

(ii) $C^k(\Omega; \mathbb{R}^m)$ is the set of vector fields F such that

$$F = F(x) = (F_1(x), \cdots, F_m(x)) \quad \text{with } F_i \in C^k(\Omega), \forall i = 1, \cdots, m.$$

(iii) We write $f \in C^0(\overline{\Omega})$ if f is continuous in $\overline{\Omega}$.

(iv) If $k \geq 1$, we say that $f \in C^k(\overline{\Omega})$ if there exists an open set $A \supset \overline{\Omega}$ and a function $g \in C^k(A)$ so that the restriction of g to $\overline{\Omega}$ is f (we write $g|_{\overline{\Omega}} = f$).

(v) A function $f : [a, b] \to \mathbb{R}$ is said to be piecewise continuous if there exist $a = a_1 < a_2 < \cdots < a_{N+1} = b$ such that

$$f|_{(a_i, a_{i+1})} \text{ is continuous for every } i = 1, \cdots, N$$

and

$$\lim_{\substack{x \to a_i \\ x > a_i}} f(x) = f(a_i + 0), \quad \lim_{\substack{x \to a_{i+1} \\ x < a_{i+1}}} f(x) = f(a_{i+1} - 0)$$

exist and are finite for every $i = 1, \cdots, N$.

Remark When $k \geq 1$, the definition of $C^k(\overline{\Omega})$ that we adopt is not the definition unanimously accepted by other authors (it is, however, the one found in [4] 172 and [7] 96). Several authors prefer a more intrinsic definition (we write this other definition only in the case $k = 1$), namely, a function $f \in C^1(\overline{\Omega})$ if $f \in C^0(\overline{\Omega}) \cap C^1(\Omega)$ and if there exists $F \in C^0(\overline{\Omega}; \mathbb{R}^n)$ such that $F(x) = \mathrm{grad} f(x)$. It is obvious that our definition is more stringent, but it can be shown that if the domain Ω is sufficiently regular, for example convex, then these two definitions coincide (cf. [7] 97).

We now give two definitions of convergence of sequences.

Definition 8.7 *Let $\Omega \subset \mathbb{R}^n$ and $f_k, f : \Omega \to \mathbb{R}$.*

*(i) We say that f_k (pointwise) **converges** to f in Ω (and we write $f_k \to f$) if $\forall \epsilon > 0$ and $\forall x \in \Omega$, $\exists k_{\epsilon, x} \in \mathbb{N}$ such that*

$$|f_k(x) - f(x)| \leq \epsilon, \ \forall k \geq k_{\epsilon, x}.$$

*(ii) We say that f_k **converges uniformly** to f in Ω (and we write $f_k \to f$ uniformly) if $\forall \epsilon > 0$, $\exists k_\epsilon \in \mathbb{N}$ such that*

$$|f_k(x) - f(x)| \leq \epsilon, \ \forall k \geq k_\epsilon, \ \forall x \in \Omega.$$

8.3   Curves

Definition 8.8 *Let $n \geq 2$ (cf. Figure 8.2).*

*(i) We say that $\Gamma \subset \mathbb{R}^n$ is a **simple** curve, if there exist an interval $I \subset \mathbb{R}$ and a continuous function $\gamma : I \to \mathbb{R}^n$ (γ is called a **parametrization** of Γ) such that*

$$\gamma(I) = \Gamma = \{x \in \mathbb{R}^n : \exists t \in I \text{ such that } x = \gamma(t)\}$$

and if $t_1, t_2 \in I$, $t_1 \neq t_2$ and at least one of the $t_i \in \operatorname{int} I$, then $\gamma(t_1) \neq \gamma(t_2)$ (in particular γ is one-to-one in $\operatorname{int} I$).

*(ii) A simple curve is said to be **closed**, if, in addition, the interval I is of the form $[a, b]$ (i.e. it is closed and bounded) and $\gamma(a) = \gamma(b)$.*

*(iii) We say that $\Gamma \subset \mathbb{R}^n$ is a **regular curve**, if there exists a parametrization $\gamma : [a, b] \to \Gamma$ with $\gamma \in C^1([a, b]; \mathbb{R}^n)$, $\gamma(t) = (\gamma_1(t), \cdots, \gamma_n(t))$ and*

$$\|\gamma'(t)\| = \sqrt{(\gamma_1')^2 + \cdots + (\gamma_n')^2} \neq 0, \quad \forall t \in [a, b].$$

*(iv) A set Γ is called a **piecewise regular curve**, if there exist a partition of the interval $[a, b]$ of the form*

$$a = a_1 < a_2 < \cdots < a_{N+1} = b$$

and $\gamma : [a, b] \to \Gamma$ such that $\gamma \in C([a, b])$, $\gamma' \in C([a_i, a_{i+i}])$ and $\gamma'(t) \neq 0$, $\forall t \in [a_i, a_{i+1}]$, $i = 1, \cdots, N$.

(v) If Γ is a simple curve with parametrization $\gamma : [a, b] \to \mathbb{R}^n$, we define the curve $-\Gamma$ through

$$-\Gamma = \{\widetilde{\gamma}(t) = \gamma(a + b - t) : t \in [a, b]\}.$$

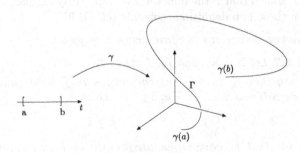

Figure 8.2: Simple curve

Notation (Jordan theorem) If $\Gamma \subset \mathbb{R}^2$ is a simple closed curve, we denote by $\operatorname{int}\Gamma = \Omega$ the open and bounded set $\Omega \subset \mathbb{R}^2$ such that $\partial\Omega = \Gamma$.

Definition 8.9 *We say that a piecewise regular simple closed curve* $\Gamma \subset \mathbb{R}^2$, *with parametrization* $\gamma : [a, b] \to \Gamma$, *is **positively oriented** (cf. Figure 8.3), if at every point* $x \in \Gamma$, $x = \gamma(t) = (\gamma_1(t), \gamma_2(t))$, *where* γ *is* C^1, *the normal vector to the curve* Γ *at the point* x, *namely*

$$\nu(x) = (\gamma_2'(t), -\gamma_1'(t)),$$

*is an **exterior normal** to* $\Omega = \operatorname{int}\Gamma$ *(where* $\Omega \subset \mathbb{R}^2$ *is such that* $\partial\Omega = \Gamma$*). More precisely, for every* $\epsilon > 0$ *sufficiently small*

$$x + \epsilon\nu(x) \in \overline{\Omega}^c \quad \text{and} \quad x - \epsilon\nu(x) \in \Omega.$$

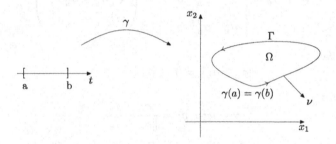

Figure 8.3: Positively oriented curve

Remark A simpler and a more intuitive, but less precise, way of rephrasing the above definition is to say that the curve Γ is positively oriented, if, while moving on Γ, the domain $\Omega = \operatorname{int}\Gamma$ is seen on the left.

Example 8.10 *Let* $I = [a, b]$ *and* $u \in C^1(I; \mathbb{R}^n)$, *then the* graph *of the function* u *defined by*

$$\Gamma = \{(x, u(x)) : x \in I\}$$

is a regular simple curve.

Example 8.11 *The circle*

$$\Gamma = \{(x, y) \in \mathbb{R}^2 : x^2 + y^2 = 1\}$$

is a regular simple closed curve. Indeed it is enough to choose $I = [0, 2\pi]$ *and* $\gamma(\theta) = (\cos\theta, \sin\theta)$.

(For more details, cf. [2] 13–15, [4] 334–335, 360–362 [7] 247, [10] 464–466, [11] 404–409, 436–438.)

8.4   Surfaces

Definition 8.12 *Let $\Sigma \subset \mathbb{R}^3$ (cf. Figure 8.4).*

(i) We say that Σ is a (or an element of a) **regular surface**, *if there exist a domain $A \subset \mathbb{R}^2$ so that ∂A is a piecewise regular simple closed curve and $\sigma : \overline{A} \to \mathbb{R}^3$ (σ is then called* **regular parametrization** *of Σ) such that*

- $\sigma \in C^1\left(\overline{A}; \mathbb{R}^3\right),\ \sigma = \sigma\left(u, v\right) = \left(\sigma^1\left(u, v\right), \sigma^2\left(u, v\right), \sigma^3\left(u, v\right)\right)$
- $\sigma\left(\overline{A}\right) = \Sigma$ *and σ is one-to-one in $\overline{A}$*
- *the vector*

$$
\sigma_u \wedge \sigma_v = \begin{vmatrix} e_1 & e_2 & e_3 \\ \sigma_u^1 & \sigma_u^2 & \sigma_u^3 \\ \sigma_v^1 & \sigma_v^2 & \sigma_v^3 \end{vmatrix} = \begin{pmatrix} \sigma_u^2 \sigma_v^3 - \sigma_u^3 \sigma_v^2 \\ \sigma_u^3 \sigma_v^1 - \sigma_u^1 \sigma_v^3 \\ \sigma_u^1 \sigma_v^2 - \sigma_u^2 \sigma_v^1 \end{pmatrix}
$$

is such that

$$
\|\sigma_u \wedge \sigma_v\| \neq 0, \quad \forall\, (u, v) \in \overline{A}.
$$

The vector

$$
\nu = \nu\left(u, v\right) = \frac{\sigma_u \wedge \sigma_v}{\|\sigma_u \wedge \sigma_v\|}
$$

is called a **unit normal** *to the surface Σ at the point $\sigma\left(u, v\right)$.*

(ii) The **boundary** *of such a surface Σ, denoted $\partial\Sigma$, is the image by σ of ∂A, namely*

$$
\partial\Sigma = \sigma\left(\partial A\right).
$$

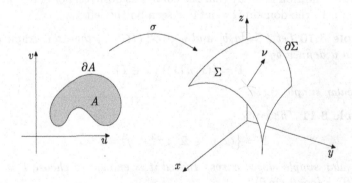

Figure 8.4: Regular surface

Remark (i) The unit normal vector is independent, up to a sign, of the choice of the parametrization. Note that if, for example, the order in which we choose the parameters (u, v) is reversed into (v, u), we obtain the opposite normal (since $\sigma_v \wedge \sigma_u = -\sigma_u \wedge \sigma_v$).

(ii) The boundary $\partial\Sigma$ is also independent of the choice of the parametrization.

Definition 8.13 *We say that $\Sigma \subset \mathbb{R}^3$ is a **piecewise regular surface** (cf. Figure 8.5) if there exist regular surfaces $\Sigma_1, \cdots, \Sigma_m$ such that*

(i) $\Sigma = \bigcup_{i=1}^{m} \Sigma_i$.

(ii) If $i \neq j$, then $\Sigma_i \cap \Sigma_j$ has no point lying in the interior of Σ_i or Σ_j.

(iii) If $i \neq j$, then $\partial\Sigma_i \cap \partial\Sigma_j$ is either empty, or consists of a single point, or is a piecewise regular simple curve.

(iv) If i, j, k are all different, then $\partial\Sigma_i \cap \partial\Sigma_j \cap \partial\Sigma_k$ is either empty or a single point.

(v) Any two points $x, y \in \Sigma$ can be joined by a piecewise regular simple curve lying in Σ.

*(vi) The union of all curves belonging to only one of the $\partial\Sigma_i$ consist of a finite number of disjoint piecewise regular simple closed curves. The **boundary** of the surface Σ, denoted $\partial\Sigma$, is the union of these curves.*

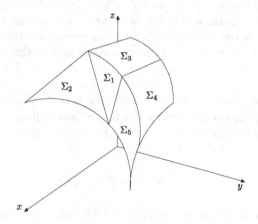

Figure 8.5: Piecewise regular surface

Definition 8.14 *(i) A regular surface Σ is said to be **orientable** if there exists a vector field ν of unit normals which is continuous all over Σ. Such* *a vector field is called an* **orientation** *of* Σ. *The oriented surface is some-times denoted* (Σ, ν).

(ii) If (Σ, ν) *is such a surface, there exist* $A \subset \mathbb{R}^2$ *and* σ, *a regular parametrization of* Σ, *such that*

$$\nu = \frac{\sigma_u \wedge \sigma_v}{\|\sigma_u \wedge \sigma_v\|}.$$

The **orientation** *on* $\partial \Sigma \, (= \sigma\,(\partial A))$ *induced by the parametrization* σ *(we equivalently can say induced by the vector field* ν *of normals) is the one induced by the positive orientation of the simple closed curve* $\partial A \subset \mathbb{R}^2$.

Remark (i) In an analogous way, one defines the notion of an orientable piecewise regular surface (Σ, ν) as well as the orientation induced by the parametrization. One just has to remember that the vector field of normals is not necessarily continuous (it is, however, piecewise continuous, cf. Example 8.21 below). For more details see [11] 464–465.

(ii) An informal way of finding the orientation of $\partial \Sigma \, (= \sigma\,(\partial A))$ induced by the vector field ν of normals is the following. Someone travelling on $\partial \Sigma$ with his head pointing in the direction ν should see the surface Σ on his left.

We would now like to present a very useful "trick" to determine the boundary of a surface and the orientation induced by the vector field of normals. It is not a completely rigorous procedure, but it holds true and can be justified in all the examples considered in this book. It uses the intuitive remark below.

Remark We have seen how to define the boundary of a regular surface Σ and the orientation induced by the vector field ν of normals to Σ. We now proceed in a similar way for a surface Σ which is only piecewise regular but that admits a global parametrization

$$\sigma : \overline{A} \to \Sigma = \sigma\left(\overline{A}\right)$$

where ∂A is a simple closed curve of $\mathbb{R}^2$. We define the boundary of Σ, denoted $\partial \Sigma$, as $\sigma\,(\partial A)$ where we have removed points as well as curves that are traversed twice in opposite directions (once in one direction and once in the opposite direction). The orientation on $\partial \Sigma$ induced by the parametrization σ is the one induced by the positive orientation of the simple closed curve ∂A.

Example 8.15 (Graph of a function) *Let $A \subset \mathbb{R}^2$ be a domain such that ∂A is a regular simple closed curve. Let $f \in C^1\left(\overline{A}\right)$ then the surface*

$$\Sigma = \{(x, y, f(x, y)) : (x, y) \in \overline{A}\}$$

(cf. Figure 8.6) is an orientable regular surface. Its boundary is

$$\partial \Sigma = \{(x, y, f(x, y)) : (x, y) \in \partial A\}.$$

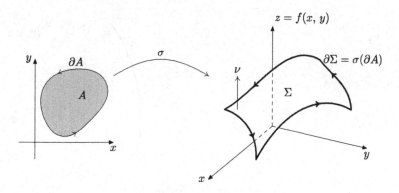

Figure 8.6: Graph of a function

Discussion We choose the parametrization

$$\sigma(x, y) = (x, y, f(x, y)), \quad (x, y) \in \overline{A}$$

which clearly has all the desired properties of being a regular parametrization of the surface. Furthermore,

$$\sigma_x \wedge \sigma_y = (-f_x, -f_y, 1)$$

and a unit normal to Σ is given by

$$\nu = \frac{(-f_x, -f_y, 1)}{\sqrt{1 + f_x^2 + f_y^2}}.$$

The boundary of Σ is by definition

$$\partial \Sigma = \sigma(\partial A) = \{(x, y, f(x, y)) : (x, y) \in \partial A\}$$

and the orientation on $\partial \Sigma$ induced by the parametrization σ is the usual positive orientation.

Example 8.16 (Sphere) *Let*

$$\Sigma = \left\{ (x, y, z) \in \mathbb{R}^3 : x^2 + y^2 + z^2 = 1 \right\}.$$

It is an orientable piecewise regular surface (cf. Figure 8.7) whose boundary is $\partial \Sigma = \emptyset$.

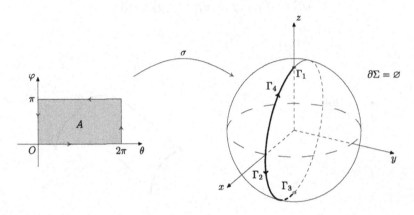

Figure 8.7: Sphere

Discussion We parametrize Σ by $\sigma : \overline{A} \to \Sigma$ where $A = (0, 2\pi) \times (0, \pi)$ and

$$\sigma(\theta, \varphi) = (\cos\theta \sin\varphi, \sin\theta \sin\varphi, \cos\varphi), \quad (\theta, \varphi) \in \overline{A}.$$

Note that this parametrization is not regular in $\overline{A}$ but in view of the above remark this is not a problem. A normal is then given by

$$\sigma_\theta \wedge \sigma_\varphi = -\sin\varphi \, (\cos\theta \sin\varphi, \sin\theta \sin\varphi, \cos\varphi)$$

and, hence,

$$\nu = \frac{\sigma_\theta \wedge \sigma_\varphi}{\|\sigma_\theta \wedge \sigma_\varphi\|}$$

is a unit normal pointing inward to the volume

$$\left\{ (x, y, z) \in \mathbb{R}^3 : x^2 + y^2 + z^2 < 1 \right\}.$$

We have

$$\sigma(\partial A) = \Gamma_1 \cup \Gamma_2 \cup \Gamma_3 \cup \Gamma_4$$

where

$$\Gamma_1 = \{\sigma(\theta, 0) : \theta : 0 \to 2\pi\} = \{(0, 0, 1)\}$$
$$\Gamma_2 = \{\sigma(2\pi, \varphi) : \varphi : 0 \to \pi\} = \{(\sin\varphi, 0, \cos\varphi) : \varphi : 0 \to \pi\}$$
$$\Gamma_3 = \{\sigma(\theta, \pi) : \theta : 2\pi \to 0\} = \{(0, 0, -1)\}$$
$$\Gamma_4 = \{\sigma(0, \varphi) : \varphi : \pi \to 0\} = \{(\sin\varphi, 0, \cos\varphi) : \varphi : \pi \to 0\} = -\Gamma_2.$$

We thus find that $\partial\Sigma = \emptyset$.

Example 8.17 (Hemisphere) *Let*

$$\Sigma = \left\{(x, y, z) \in \mathbb{R}^3 : x^2 + y^2 + z^2 = 1 \text{ and } z \geq 0\right\}.$$

It is an orientable piecewise regular surface (cf. Figure 8.8) whose boundary is given by

$$\partial\Sigma = \left\{(x, y, z) \in \mathbb{R}^3 : x^2 + y^2 = 1 \text{ and } z = 0\right\}.$$

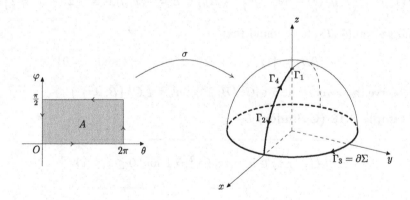

Figure 8.8: Hemisphere

Discussion We let $A = (0, 2\pi) \times \left(0, \dfrac{\pi}{2}\right)$ and

$$\sigma(\theta, \varphi) = (\cos\theta \sin\varphi, \sin\theta \sin\varphi, \cos\varphi).$$

We then have

$$\sigma_\theta \wedge \sigma_\varphi = -\sin\varphi \, (\cos\theta \sin\varphi, \sin\theta \sin\varphi, \cos\varphi)$$

and
$$\sigma\left(\partial A\right) = \Gamma_1 \cup \Gamma_2 \cup \Gamma_3 \cup \Gamma_4$$
where

$$\Gamma_1 = \left\{\sigma\left(\theta, 0\right) : \theta : 0 \to 2\pi\right\} = \left\{(0, 0, 1)\right\}$$
$$\Gamma_2 = \left\{\sigma\left(2\pi, \varphi\right) : \varphi : 0 \to \frac{\pi}{2}\right\} = \left\{(\sin\varphi, 0, \cos\varphi) : \varphi : 0 \to \frac{\pi}{2}\right\}$$
$$\Gamma_3 = \left\{\sigma\left(\theta, \frac{\pi}{2}\right) : \theta : 2\pi \to 0\right\} = \left\{(\cos\theta, \sin\theta, 0) : \theta : 2\pi \to 0\right\}$$
$$\Gamma_4 = \left\{\sigma\left(0, \varphi\right) : \varphi : \frac{\pi}{2} \to 0\right\} = \left\{(\sin\varphi, 0, \cos\varphi) : \varphi : \frac{\pi}{2} \to 0\right\} = -\Gamma_2 .$$

We therefore find
$$\partial\Sigma = \Gamma_3$$
and, hence, the orientation on $\partial\Sigma$ induced by the parametrization σ is the negative orientation (since $\theta : 2\pi \to 0$) on the simple closed curve Γ_3 seen as a curve in the plane $z = 0$.

Note that we could have taken as a parametrization of Σ

$$\widetilde{\sigma}\left(x, y\right) = \left(x, y, \sqrt{1 - x^2 - y^2}\right), \quad (x, y) \in B = \left\{(x, y) \in \mathbb{R}^2 : x^2 + y^2 < 1\right\}$$

and we would also have found that

$$\partial\Sigma = \widetilde{\sigma}\left(\partial B\right) = \left\{(x, y, z) \in \mathbb{R}^3 : x^2 + y^2 = 1 \text{ and } z = 0\right\}.$$

(Observe, however, that $\widetilde{\sigma} \in C^1\left(B; \mathbb{R}^3\right)$ but $\widetilde{\sigma} \notin C^1\left(\overline{B}; \mathbb{R}^3\right)$.)

Example 8.18 (Cylinder) *Let*

$$\boxed{\Sigma = \left\{(x, y, z) \in \mathbb{R}^3 : x^2 + y^2 = 1 \text{ and } 0 \le z \le 1\right\}.}$$

It is an orientable piecewise regular surface (cf. Figure 8.9) whose boundary is
$$\partial\Sigma = \Gamma_1 \cup \Gamma_3$$
where

$$\Gamma_1 = \left\{(x, y, z) : x^2 + y^2 = 1 \text{ and } z = 0\right\}$$
$$\Gamma_3 = \left\{(x, y, z) : x^2 + y^2 = 1 \text{ and } z = 1\right\}.$$

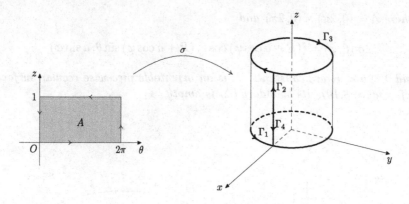

Figure 8.9: Cylinder

Discussion We set $A = (0, 2\pi) \times (0, 1)$ and

$$\sigma(\theta, z) = (\cos\theta, \sin\theta, z).$$

The vector field of unit normals induced by the parametrization is

$$\nu = \sigma_\theta \wedge \sigma_z = (\cos\theta, \sin\theta, 0).$$

We also have

$$\sigma(\partial A) = \Gamma_1 \cup \Gamma_2 \cup \Gamma_3 \cup \Gamma_4$$

where

$$\Gamma_1 = \{\sigma(\theta, 0) = (\cos\theta, \sin\theta, 0) : \theta : 0 \to 2\pi\}$$
$$\Gamma_2 = \{\sigma(2\pi, z) = (1, 0, z) : z : 0 \to 1\}$$
$$\Gamma_3 = \{\sigma(\theta, 1) = (\cos\theta, \sin\theta, 1) : \theta : 2\pi \to 0\}$$
$$\Gamma_4 = \{\sigma(0, z) = (1, 0, z) : z : 1 \to 0\} = -\Gamma_2.$$

We thus deduce that

$$\partial\Sigma = \Gamma_1 \cup \Gamma_3$$

and the orientation on $\partial\Sigma$ induced by the parametrization σ is positive on Γ_1 (seen as a curve in the plane $z = 0$) and negative on Γ_3 (seen as a curve in the plane $z = 1$).

Example 8.19 (Torus) *Let*

$$\Sigma = \left\{ (x, y, z) \in \mathbb{R}^3 : (x, y, z) = \sigma(\theta, \varphi), \ (\theta, \varphi) \in \overline{A} \right\}.$$

where $A = (0, 2\pi) \times (0, 2\pi)$ and

$$\sigma(\theta, \varphi) = ((R + a\cos\varphi)\cos\theta, (R + a\cos\varphi)\sin\theta, a\sin\varphi)$$

and $0 < a < R$ are constants. It is an orientable piecewise regular surface (cf. Figure 8.10). Its boundary $\partial\Sigma$ is empty.

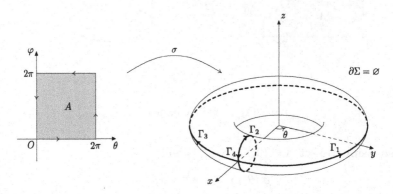

Figure 8.10: Torus

Discussion The normal induced by the parametrization is

$$\sigma_\theta \wedge \sigma_\varphi = a(R + a\cos\varphi)(\cos\theta\cos\varphi, \sin\theta\cos\varphi, \sin\varphi).$$

We have
$$\sigma(\partial A) = \Gamma_1 \cup \Gamma_2 \cup \Gamma_3 \cup \Gamma_4$$

where

$\Gamma_1 = \{\sigma(\theta, 0) = ((R + a)\cos\theta, (R + a)\sin\theta, 0) : \theta : 0 \to 2\pi\}$
$\Gamma_2 = \{\sigma(2\pi, \varphi) = (R + a\cos\varphi, 0, a\sin\varphi) : \varphi : 0 \to 2\pi\}$
$\Gamma_3 = \{\sigma(\theta, 2\pi) = ((R + a)\cos\theta, (R + a)\sin\theta, 0) : \theta : 2\pi \to 0\} = -\Gamma_1$
$\Gamma_4 = \{\sigma(0, \varphi) = (R + a\cos\varphi, 0, a\sin\varphi) : \varphi : 2\pi \to 0\} = -\Gamma_2$

and thus the result $\partial\Sigma = \emptyset$.

Example 8.20 (Cone) *Let*

$$\Sigma = \left\{(x, y, z) \in \mathbb{R}^3 : x^2 + y^2 = z^2 \text{ and } 0 \le z \le 1\right\}.$$

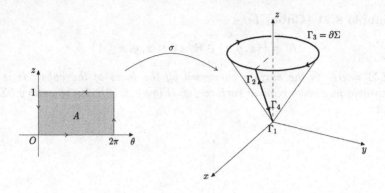

Figure 8.11: Cone

It is an orientable piecewise regular surface (cf. Figure 8.11). Its boundary is

$$\partial\Sigma = \left\{(x,y,z) \in \mathbb{R}^3 : x^2 + y^2 = 1 \text{ and } z = 1\right\}.$$

Discussion We set $A = (0, 2\pi) \times (0, 1)$ and

$$\sigma(\theta, z) = (z\cos\theta, z\sin\theta, z).$$

A normal is given by

$$\sigma_\theta \wedge \sigma_z = (z\cos\theta, z\sin\theta, -z).$$

We have

$$\sigma(\partial A) = \Gamma_1 \cup \Gamma_2 \cup \Gamma_3 \cup \Gamma_4$$

with

$$\Gamma_1 = \{\sigma(\theta, 0) = (0, 0, 0)\}$$
$$\Gamma_2 = \{\sigma(2\pi, z) = (z, 0, z) : z : 0 \to 1\}$$
$$\Gamma_3 = \{\sigma(\theta, 1) = (\cos\theta, \sin\theta, 1) : \theta : 2\pi \to 0\}$$
$$\Gamma_4 = \{\sigma(0, z) = (z, 0, z) : z : 1 \to 0\} = -\Gamma_2.$$

We therefore find $\partial\Sigma = \Gamma_3$. The orientation on Γ_3 (seen as a curve in the plane $z = 1$) is negative.

We can also take the Cartesian coordinates as parametrization, namely

$$\tilde{\sigma}(x,y) = \left(x, y, \sqrt{x^2 + y^2}\right) \quad \text{and} \quad B = \left\{(x,y) \in \mathbb{R}^2 : x^2 + y^2 < 1\right\}.$$

We then find that $\tilde{\sigma}(\partial B) = \partial\Sigma$ (note that $\tilde{\sigma}$ is not differentiable at $(0,0)$ and hence $\tilde{\sigma} \notin C^1(B; \mathbb{R}^3)$).

Example 8.21 (Cube) *Let*

$$K = \left\{ (x,y,z) \in \mathbb{R}^2 : 0 \leq x,y,z \leq 1 \right\}$$

and $\Sigma = \partial K$ *is the surface composed by the faces of the cube. It is an orientable piecewise regular surface (cf. Figure 8.12). Its boundary* $\partial \Sigma$ *is empty.*

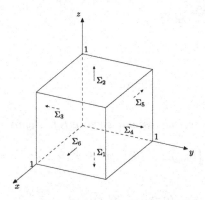

Figure 8.12: Cube

Discussion We have $\Sigma = \bigcup\limits_{i=1}^{6} \Sigma_i$ where

$$\Sigma_1 = \{(x,y,0) : 0 < x,y < 1\}$$
$$\Sigma_2 = \{(x,y,1) : 0 < x,y < 1\}$$
$$\Sigma_3 = \{(x,0,z) : 0 < x,z < 1\}$$
$$\Sigma_4 = \{(x,1,z) : 0 < x,z < 1\}$$
$$\Sigma_5 = \{(0,y,z) : 0 < y,z < 1\}$$
$$\Sigma_6 = \{(1,y,z) : 0 < y,z < 1\}.$$

All of them are regular surfaces. The exterior unit normal is given by

$$\nu = \begin{cases} (0,0-1) & \text{on } \Sigma_1 \\ (0,0,1) & \text{on } \Sigma_2 \\ (0,-1,0) & \text{on } \Sigma_3 \\ (0,1,0) & \text{on } \Sigma_4 \\ (-1,0,0) & \text{on } \Sigma_5 \\ (1,0,0) & \text{on } \Sigma_6. \end{cases}$$

Note that the field of exterior unit normals is piecewise C^1. We easily find that $\partial\Sigma = \emptyset$.

Example 8.22 (Möbius band) *Let*

$$\Sigma = \left\{ (x,y,z) \in \mathbb{R}^3 : (x,y,z) = \sigma\left(\theta,r\right), \ (\theta,r) \in \overline{A} \right\}$$

where $A = (0, 2\pi) \times (-1/2, 1/2)$ and

$$\sigma\left(\theta,r\right) = \left(\left(1 + r\sin\frac{\theta}{2}\right)\cos\theta, \left(1 + r\sin\frac{\theta}{2}\right)\sin\theta, r\cos\frac{\theta}{2} \right).$$

It is a piecewise regular surface (cf. Figure 8.13) which is not orientable.

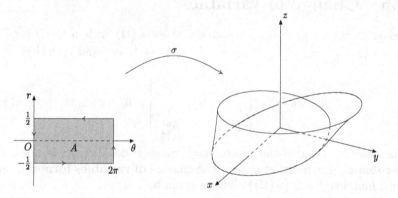

Figure 8.13: Möbius band

Discussion Note that

$$\sigma\left(0,r\right) = (1,0,r) = \sigma\left(2\pi, -r\right).$$

A normal is given by

$$\sigma_\theta \wedge \sigma_r = \begin{pmatrix} \left(1 + r\sin\dfrac{\theta}{2}\right)\cos\dfrac{\theta}{2}\cos\theta + \dfrac{r}{2}\sin\theta \\[2mm] \left(1 + r\sin\dfrac{\theta}{2}\right)\cos\dfrac{\theta}{2}\sin\theta - \dfrac{r}{2}\cos\theta \\[2mm] -\left(1 + r\sin\dfrac{\theta}{2}\right)\sin\dfrac{\theta}{2} \end{pmatrix}$$

and

$$\|\sigma_\theta \wedge \sigma_r\|^2 = \left(1 + r\sin\frac{\theta}{2}\right)^2 + \frac{r^2}{4} > 0, \quad \forall\, (\theta, r) \in \overline{A}.$$

We therefore have

$$\nu\,(\theta, r) = \frac{\sigma_\theta \wedge \sigma_r}{\|\sigma_\theta \wedge \sigma_r\|} \quad \text{is continuous for every } (\theta, r) \in A.$$

It is, however, not continuous for every $(\theta, r) \in \overline{A}$. Indeed, we have

$$\nu\,(0, 0) = (1, 0, 0) \neq \nu\,(2\pi, 0) = (-1, 0, 0)$$

while $\sigma\,(0, 0) = (1, 0, 0) = \sigma\,(2\pi, 0)$. The surface is therefore not orientable. (For more details, cf. [2] 25–29, [4] 373–376, [10] 534–538, [11] 445–466.)

8.5   Change of variables

Let $\Omega \subset \mathbb{R}^n$ be an open set and $u : \overline{\Omega} \to u\,(\overline{\Omega})$ with $u \in C^1\,(\overline{\Omega}; \mathbb{R}^n)$ $(u = u\,(x_1, \cdots, x_n) = (u^1\,(x), \cdots, u^n\,(x)))$ one-to-one and such that

$$\det \nabla u\,(x) = \begin{vmatrix} u^1_{x_1} & \cdots & u^1_{x_n} \\ \vdots & \ddots & \vdots \\ u^n_{x_1} & \cdots & u^n_{x_n} \end{vmatrix} \neq 0, \quad \forall x \in \overline{\Omega} \tag{8.1}$$

The absolute value of the determinant, namely $|\det \nabla u\,(x)|$, is called the **Jacobian** of the transformation. The **change of variables formula**, valid for a function $f \in C\,(u\,(\overline{\Omega}))$, is then given by

$$\int_{u(\Omega)} f(y)\, dy = \int_\Omega f\,(u\,(x))\,|\det \nabla u\,(x)|\ dx$$

(cf. [11] 357). The most common examples given below do not satisfy all the properties so as to apply the change of variables formula (notably, they do not verify (8.1) at every point), but it can easily be shown that the formula is still valid in these examples.

Example 8.23 (Polar coordinates) *If $n = 2$, $r > 0$, $\theta \in [0, 2\pi]$ (cf. Figure 8.14) and*

$$x_1 = u^1\,(r, \theta) = r\cos\theta$$
$$x_2 = u^2\,(r, \theta) = r\sin\theta$$

we have

$$\det \nabla u = \begin{vmatrix} u_r^1 & u_\theta^1 \\ u_r^2 & u_\theta^2 \end{vmatrix} = \begin{vmatrix} \cos\theta & -r\sin\theta \\ \sin\theta & r\cos\theta \end{vmatrix} = r \geq 0.$$

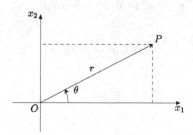

Figure 8.14: Polar coordinates

Example 8.24 (Cylindrical coordinates) *If $n = 3$, $r \geq 0$, $\theta \in [0, 2\pi]$, $z \in \mathbb{R}$ (cf. Figure 8.15) and*

$$x_1 = u^1(r, \theta, z) = r\cos\theta$$
$$x_2 = u^2(r, \theta, z) = r\sin\theta$$
$$x_3 = u^3(r, \theta, z) = z$$

we find

$$\det \nabla u = r \geq 0.$$

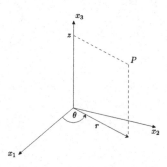

Figure 8.15: Cylindrical coordinates

Example 8.25 (Spherical coordinates) *If $n = 3$, $r \geq 0$, $\theta \in [0, 2\pi]$, $\varphi \in [0, \pi]$ (cf. Figure 8.16) and*

$$x_1 = u^1(r, \theta, \varphi) = r \cos \theta \sin \varphi$$
$$x_2 = u^2(r, \theta, \varphi) = r \sin \theta \sin \varphi$$
$$x_3 = u^3(r, \theta, \varphi) = r \cos \varphi$$

we obtain

$$|\det \nabla u| = r^2 \sin \varphi.$$

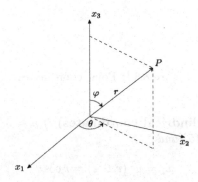

Figure 8.16: Spherical coordinates

Part II

Complex analysis

Chapter 9

Holomorphic functions and Cauchy–Riemann equations

9.1 Definitions and theoretical results

Notation (i) We recall that a complex number z is written as $z = x + iy$ with $x, y \in \mathbb{R}$ (where $i^2 = -1$) and we set

$$x = \operatorname{Re} z \quad \text{and} \quad y = \operatorname{Im} z.$$

(ii) The **complex conjugate** of z is

$$\bar{z} = x - iy.$$

(iii) The **modulus** of z is defined as

$$|z| = |z\bar{z}|^{1/2} = \sqrt{x^2 + y^2}.$$

(iv) Any $z \neq 0$ can be written as

$$z = |z|\, e^{i\theta} = |z| \left(\cos\theta + i\sin\theta \right).$$

The real number θ is called the **argument** of z and is denoted $\arg z$. Note that it is uniquely defined up to 2π, we usually choose $-\pi < \arg z \leq \pi$ (it is called the **principal value** of the argument). When $-\pi < \arg z < \pi$, we can also write

$$\arg z = \begin{cases} \arctan\left(y/x\right) & \text{if } x > 0 \text{ and } y \in \mathbb{R} \\ \pi/2 & \text{if } x = 0 \text{ and } y > 0 \\ -\pi/2 & \text{if } x = 0 \text{ and } y < 0 \\ \pi + \arctan\left(y/x\right) & \text{if } x < 0 \text{ and } y > 0 \\ -\pi + \arctan\left(y/x\right) & \text{if } x < 0 \text{ and } y < 0. \end{cases}$$

(v) An open set O in $\mathbb{C}$ is seen as an open set of $\mathbb{R}^2$. More precisely, the set O is open if for every $z_0 \in O$, there exists $\epsilon > 0$ such that

$$B_\epsilon(z_0) = \{z \in \mathbb{C} : |z - z_0| < \epsilon\} \subset O.$$

(vi) Let $O \subset \mathbb{C}$. For a function $f : O \to \mathbb{C}$, $f = f(z)$, we write

$$u(x, y) = \operatorname{Re} f(z) \quad \text{and} \quad v(x, y) = \operatorname{Im} f(z)$$

where $z = x + iy$.

(vii) For functions $u : \mathbb{R}^2 \to \mathbb{R}$, $u = u(x, y)$, we write, as usual, the partial derivatives of u

$$u_x = \frac{\partial u}{\partial x} \quad \text{and} \quad u_y = \frac{\partial u}{\partial y}.$$

Definition 9.1 *Let $O \subset \mathbb{C}$ be open. We say that $f : O \to \mathbb{C}$ is **holomorphic** (or complex analytic) in O if for every $z_0 \in O$,*

$$\lim_{z \to z_0} \frac{f(z) - f(z_0)}{z - z_0}$$

*exists and is finite. The limit is called the **derivative** of f at z_0 and it is denoted by $f'(z_0)$.*

Remark All the usual properties of derivatives in $\mathbb{R}$ apply in this context. For example,

$$(f + g)'(z_0) = f'(z_0) + g'(z_0)$$
$$(f \cdot g)'(z_0) = f'(z_0)g(z_0) + f(z_0)g'(z_0)$$

and, similarly, for the composition or the quotient.

Theorem 9.2 *Let $O \subset \mathbb{C}$ be open and $f : O \to \mathbb{C}$. The two following statements are then equivalent.*

(i) f is holomorphic in O.

*(ii) The functions $u, v \in C^\infty(O)$ and satisfy, $\forall (x, y) \in O$, the **Cauchy–Riemann equations***

$$u_x = v_y \quad \text{and} \quad u_y = -v_x.$$

In particular, if f is holomorphic in O, then

$$f'(z) = u_x(x, y) + iv_x(x, y) = v_y(x, y) - iu_y(x, y).$$

(For more details, cf. [1] 24–26, [3] 13–14, [5] 36–38, [10] 741–742, [15] 10–13.)

9.2 Examples

Example 9.3 *The function $f(z) = e^z$ is holomorphic in $\mathbb{C}$ and $f'(z) = e^z$.*

Discussion (i) We have

$$e^z = e^{x+iy} = e^x \cos y + ie^x \sin y = u(x,y) + iv(x,y),$$

where u and v are $C^\infty(\mathbb{R}^2)$ and verify Cauchy–Riemann equations, i.e.

$$\begin{cases} u_x = v_y = e^x \cos y \\ u_y = -v_x = -e^x \sin y \end{cases}$$

and, hence, $f'(z) = u_x + iv_x = e^z$.

(ii) An equivalent way of defining the exponential function is through

$$e^z = \sum_{n=0}^{\infty} \frac{z^n}{n!}.$$

Example 9.4 *The function defined by*

$$\sin z = \frac{e^{iz} - e^{-iz}}{2i}$$

is holomorphic in $\mathbb{C}$ and its derivative is

$$\cos z = \frac{e^{iz} + e^{-iz}}{2}.$$

Discussion (i) We first find the real and imaginary parts of the function $\sin z$, namely

$$\begin{aligned} \sin z &= \frac{e^{iz} - e^{-iz}}{2i} = \frac{e^{-y}e^{ix} - e^{y}e^{-ix}}{2i} \\ &= \frac{e^{-y}(\cos x + i\sin x) - e^{y}(\cos x - i\sin x)}{2i} \\ &= -\frac{e^y - e^{-y}}{2}\frac{\cos x}{i} + \frac{e^y + e^{-y}}{2}\sin x \\ &= \cosh y \sin x + i\sinh y \cos x. \end{aligned}$$

(ii) We next find

$$u_x = \cosh y \cos x \qquad u_y = \sinh y \sin x$$
$$v_x = -\sinh y \sin x \qquad v_y = \cosh y \cos x.$$

We therefore have $u_x = v_y$ and $u_y = -v_x$ and thus

$$(\sin z)' = u_x + iv_x = \cosh y \cos x - i\sinh y \sin x = \cos z.$$

Example 9.5 *Let*

$$f(z) = \log z$$

where we define the (principal branch of the) **complex logarithm** *of the non-zero complex number z as follows*

$$\log z = \log |z| + i(\arg z) \quad with \quad -\pi < \arg z \leq \pi$$

where $\log |z|$ *stands for the usual logarithm (in particular* $\log e = 1$*) of the non-zero real number* $|z|$*. The function is holomorphic in*

$$O = \mathbb{C} \setminus \{z \in \mathbb{C} : \operatorname{Im} z = 0 \text{ and } \operatorname{Re} z \leq 0\}.$$

Its derivative is, for every $z \in O$,

$$f'(z) = \frac{1}{z}.$$

Discussion (i) Recall that the argument of a non-zero complex number is only defined up to a multiple of 2π. We have chosen here the principal value of the argument. With a different choice of the argument, we would have had other definitions of the complex logarithm.

(ii) We now prove that $\log z$ is not continuous on $\mathbb{C} \setminus \{0\}$. In order that the function be continuous we have to restrict it to O. Let us check the lack of continuity on $\mathbb{C} \setminus \{0\}$. Let $t \in \mathbb{R} \setminus \{0\}$. We therefore have

$$\log(-1 + it) = \log \sqrt{1 + t^2} + i \arg(-1 + it).$$

From the definition we find

$$\lim_{t \to 0^+} \log(-1 + it) = \log 1 + i\pi = i\pi$$

$$\lim_{t \to 0^-} \log(-1 + it) = \log 1 - i\pi = -i\pi.$$

Thus, f is not continuous on $\mathbb{C} \setminus \{0\}$ (more precisely we have shown that f is not continuous at $z = -1$).

(iii) Let us now show that $f'(z) = 1/z$, $\forall z \in O$. We prove this only in the half plane $\operatorname{Re} z > 0$. Writing $z = x + iy$, we have there

$$-\frac{\pi}{2} < \arg z = \arctan \frac{y}{x} < \frac{\pi}{2}.$$

We proceed in a very similar manner if

$$-\pi < \arg z < -\frac{\pi}{2} \quad \text{or} \quad \frac{\pi}{2} < \arg z < \pi.$$

We therefore have, if $f = u + iv$, that

$$u(x,y) = \log \sqrt{x^2 + y^2} \quad \text{and} \quad v(x,y) = \arctan \frac{y}{x} \,.$$

We therefore have $u, v \in C^\infty \left(O \cap \{(x,y) \in \mathbb{R}^2 : x > 0\} \right)$ and

$$u_x = \frac{x}{x^2 + y^2} \qquad u_y = \frac{y}{x^2 + y^2}$$

$$v_x = \frac{-y}{x^2 + y^2} \qquad v_y = \frac{x}{x^2 + y^2}$$

and, hence,

$$(\log z)' = u_x + iv_x = \frac{x}{x^2 + y^2} + i \frac{-y}{x^2 + y^2} = \frac{x - iy}{(x + iy)(x - iy)} = \frac{1}{z} \,.$$

(iv) One can easily prove that

$$\log e^z = z, \text{ if } \operatorname{Im} z \in (-\pi, \pi] \quad \text{and} \quad e^{\log z} = z, \text{ if } z \in \mathbb{C} - \{0\}$$

$$\log (zw) = \begin{cases} \log z + \log w & \text{if } \arg z + \arg w \in (-\pi, \pi] \\ \log z + \log w - 2\pi i & \text{if } \arg z + \arg w \in (\pi, 2\pi] \\ \log z + \log w + 2\pi i & \text{if } \arg z + \arg w \in (-2\pi, -\pi] \,. \end{cases}$$

Example 9.6 *Let $\gamma \in \mathbb{C}$. We define (the principal branch) of*

$$f(z) = z^\gamma$$

through

$$f(z) = e^{\gamma \log z} \,.$$

Then f is holomorphic in

$$O = \mathbb{C} \setminus \{z \in \mathbb{C} : \operatorname{Im} z = 0 \text{ and } \operatorname{Re} z \leq 0\}$$

and its derivative is

$$f'(z) = \gamma z^{\gamma - 1} \,.$$

Discussion (i) The function f is holomorphic in O since it is a composition of holomorphic functions. We therefore find that

$$f'(z) = e^{\gamma \log z} \cdot \gamma z^{-1} = \gamma z^{\gamma - 1} \,.$$

Note that if $\gamma \in \mathbb{N}$, the function f is holomorphic in $\mathbb{C}$ and not only in O. Similarly, if $(-\gamma) \in \mathbb{N}$, then f is holomorphic in $\mathbb{C} \setminus \{0\}$.

(ii) One can easily prove that if $\beta, \gamma \in \mathbb{C}$ and $z, w \in \mathbb{C} \setminus \{0\}$, then

$$z^{\beta + \gamma} = z^\beta z^\gamma$$

$$z^\gamma w^\gamma = \begin{cases} (zw)^\gamma & \text{if } \arg z + \arg w \in (-\pi, \pi] \\ (zw)^\gamma e^{2\pi\gamma i} & \text{if } \arg z + \arg w \in (\pi, 2\pi] \\ (zw)^\gamma e^{-2\pi\gamma i} & \text{if } \arg z + \arg w \in (-2\pi, -\pi]. \end{cases}$$

9.3   Exercises

Exercise 9.1 Show that the functions

$$\cos z = \frac{e^{iz} + e^{-iz}}{2}, \quad \cosh z = \frac{e^z + e^{-z}}{2}, \quad \sinh z = \frac{e^z - e^{-z}}{2}$$

are holomorphic in $\mathbb{C}$ and compute their derivatives.

Exercise 9.2 Is the function $f(z) = (\operatorname{Re} z)^2$ holomorphic? Justify your answer.

Exercise 9.3 Let $f(z) = \log(1 + z^2)$. Find the largest domain in $\mathbb{C}$ where the function f is holomorphic.

Exercise 9.4 Find a holomorphic function $f : \mathbb{C} \to \mathbb{C}$ whose real part is

$$u(x, y) = e^{(x^2 - y^2)} \cos(2xy).$$

Exercise 9.5 Find a holomorphic function $f : \mathbb{C} \to \mathbb{C}$ whose real part is

$$u(x, y) = x^2 - y^2 + e^{-x} \cos y.$$

Exercise 9.6 Show that if $z = x + iy$ and $f(z) = u(x, y) + iv(x, y)$ is holomorphic in an open set Ω, then u and v are harmonic in Ω meaning that

$$\Delta u = \Delta v = 0.$$

Prove also that

$$u_x v_x + u_y v_y = 0 \quad \text{and} \quad |f'(z)|^2 = u_x v_y - u_y v_x.$$

Exercise 9.7 Prove that if $f : \mathbb{C} \to \mathbb{C}$ is holomorphic, then the three following claims are equivalent.

(i) f is constant.

(ii) $\operatorname{Re}(f)$ is constant.

(iii) $\operatorname{Im}(f)$ is constant.

Deduce that $f(z) = |z|$ is not holomorphic.

Exercise 9.8 Show that the Cauchy–Riemann equations can be written in polar coordinates as

$$\frac{\partial u}{\partial r} = \frac{1}{r}\frac{\partial v}{\partial \theta} \quad \text{and} \quad \frac{\partial v}{\partial r} = -\frac{1}{r}\frac{\partial u}{\partial \theta}.$$

Exercise 9.9 Let f be holomorphic in $\mathbb{C}$ and verifying

$$u_x + v_y = 0.$$

Prove that there exists a real constant c and a complex constant d so that

$$f(z) = -icz + d.$$

Exercise 9.10 Let g be harmonic in $\mathbb{R}^2$ (i.e. $\Delta g = 0$) and $f = u + iv$ be holomorphic in $\mathbb{C}$. Verify that

$$h(x,y) = g(u(x,y), v(x,y))$$

is harmonic in $\mathbb{R}^2$.

Exercise 9.11 Let Ω be open and $u \in C^{\infty}(\Omega)$ be harmonic in Ω (i.e. $\Delta u = 0$ in Ω). Let $z = x + iy$ and

$$f(z) = u_x - iu_y.$$

Show that f is a holomorphic function in Ω.

Exercise* 9.12 Prove that if f is holomorphic in an open set $\Omega \subset \mathbb{R}^2$, then the Cauchy–Riemann equations are satisfied in Ω, namely

$$u_x = v_y \quad \text{and} \quad u_y = -v_x.$$

Hint: choose in Definition 9.1 first $z = z_0 + h$, then $z = z_0 + ih$ with $h \in \mathbb{R}$.

Chapter 10

Complex integration

10.1   Definitions and theoretical results

We refer to Chapters 2 and 8 for precise definitions for curves and line integrals.

Definition 10.1 *(i) Let $\Gamma \subset \mathbb{C}$ be a regular curve with regular parametrization $\gamma : [a, b] \to \Gamma$. Let $f : \Gamma \to \mathbb{C}$ be a continuous function. We write*

$$\int_{\Gamma} f(z) \, dz = \int_{\gamma} f(z) \, dz = \int_{a}^{b} f(\gamma(t)) \, \gamma'(t) \, dt.$$

(ii) If Γ is a piecewise regular curve of the form $\Gamma = \bigcup_{k=1}^{m} \Gamma_k$, with regular curves Γ_k, then

$$\int_{\Gamma} f(z) \, dz = \sum_{k=1}^{m} \int_{\Gamma_k} f(z) \, dz.$$

In the sequel, most of the time, we will not distinguish the curve Γ from its parametrization γ.

Theorem 10.2 *Let $D \subset \mathbb{C}$ be a simply connected domain, $f : D \to \mathbb{C}$ a holomorphic function and γ a piecewise regular simple closed curve contained in D. Then the **Cauchy theorem** holds true, namely*

$$\boxed{\int_{\gamma} f(z) \, dz = 0.}$$

Moreover, f is infinitely many times differentiable and if $n \in \mathbb{N}$, then the **Cauchy integral formula** *is satisfied, namely*

$$f^{(n)}(z) = \frac{n!}{2\pi i} \int_\gamma \frac{f(\xi)}{(\xi - z)^{n+1}} \, d\xi \quad \forall\, z \in \operatorname{int} \gamma.$$

In particular when $n = 0$, this formula reads as

$$f(z) = \frac{1}{2\pi i} \int_\gamma \frac{f(\xi)}{\xi - z} \, d\xi \quad \forall\, z \in \operatorname{int} \gamma.$$

(For more details, cf. [1] 102–123, 141, [3] 27–45, [5] 126–148, [10] 767–793, [15] 45–47 and 97.)

10.2   Examples

Example 10.3 *Let $f(z) = z^2$, γ_1 be the upper half circle of radius 1 centered at the origin and γ_2 be the circle of radius 1 centered at the origin. Compute $\int_{\gamma_i} f(z)\, dz$, $i = 1, 2$.*

Discussion (i) We have $\gamma_1 : [0, \pi] \to \mathbb{C}$, where $\gamma_1(\theta) = e^{i\theta}$ and, hence, $\gamma_1'(\theta) = ie^{i\theta}$. We therefore find

$$\int_{\gamma_1} f(z) \, dz = \int_0^\pi (e^{i\theta})^2 \, ie^{i\theta} \, d\theta = i \int_0^\pi e^{3i\theta} \, d\theta = \left. \frac{e^{3i\theta}}{3} \right|_0^\pi = -\frac{2}{3} \, .$$

(ii) We now have $\gamma_2 : [0, 2\pi] \to \mathbb{C}$, where $\gamma_2(\theta) = e^{i\theta}$ and thus $\gamma_2'(\theta) = ie^{i\theta}$. We therefore obtain

$$\int_{\gamma_2} f(z) \, dz = \int_0^{2\pi} (e^{i\theta})^2 \, ie^{i\theta} \, d\theta = i \int_0^{2\pi} e^{3i\theta} \, d\theta = \left. \frac{e^{3i\theta}}{3} \right|_0^{2\pi} = 0.$$

In this last case we could directly apply Cauchy theorem and immediately conclude.

Example 10.4 *Let γ be a piecewise regular simple closed curve. Find in terms of γ the value of the integral*

$$\int_\gamma \frac{\cos(2z)}{z} \, dz.$$

Discussion Observe that $\xi = 0$ is a singular point for $f(\xi) = \cos(2\xi)/\xi$. We consider three cases.

Case 1: $0 \in \text{int}\,\gamma$. We apply Cauchy integral formula to $g(\xi) = \cos(2\xi)$ and we get

$$1 = g(0) = \frac{1}{2\pi i} \int_\gamma \frac{\cos(2\xi)}{\xi - 0} \, d\xi$$

and thus

$$\int_\gamma \frac{\cos(2z)}{z} \, dz = 2\pi i.$$

Note that a direct computation (without using Cauchy integral formula) would have been difficult, since we would have had to compute

$$\int_\gamma \frac{\cos(2z)}{z} \, dz = \int_0^{2\pi} \frac{\cos(2e^{i\theta})}{e^{i\theta}} ie^{i\theta} \, d\theta = i \int_0^{2\pi} \cos(2e^{i\theta}) \, d\theta.$$

Case 2: $0 \notin \overline{\text{int}\,\gamma}$. Cauchy theorem applied to f leads to

$$\int_\gamma \frac{\cos(2z)}{z} \, dz = 0.$$

Case 3: $0 \in \gamma$. In this case the integral is not well defined.

Example 10.5 *Let* $\gamma = \{z \in \mathbb{C} : |z - 2| = 1\}$ *be a piecewise regular simple closed curve. Evaluate*

$$\int_\gamma \frac{e^{z+2}}{(z-2)^3} \, dz.$$

Discussion Let $f(\xi) = e^{\xi+2}$, $z = 2$ and $n = 2$. Cauchy integral formula immediately gives

$$\int_\gamma \frac{e^{z+2}}{(z-2)^3} \, dz = \frac{2\pi i}{2!} f''(2) = \frac{2\pi i}{2} e^4 = \pi i e^4.$$

10.3   Exercises

Exercise 10.1 Let γ be a piecewise regular simple closed curve. Evaluate, in terms of γ, the integral

$$\int_\gamma \frac{e^{z^2}}{2z} \, dz.$$

Exercise 10.2 Let $\gamma = \{z \in \mathbb{C} : |z - \frac{\pi}{2}| = 1\}$. Compute

$$\int_\gamma \frac{z^2 \sin z}{(z - \frac{\pi}{2})^2} \, dz.$$

Exercise 10.3 Compute the following two integrals.

(i) $\int_\gamma (z^2 + 1) \, dz$ where $\gamma = [1, 1 + i]$ (the segment joining 1 and $1+i$).

(ii) $\int_\gamma \operatorname{Re}(z^2) \, dz$ where γ is the circle of radius 1 centered at 0.

Exercise 10.4 Let γ be a piecewise regular simple closed curve. Find, in terms of γ, the value of the integral

$$\int_\gamma \frac{dz}{z^2}.$$

Exercise 10.5 Discuss, in terms of γ (a piecewise regular simple closed curve), the following quantity

$$\int_\gamma \frac{5z^2 - 3z + 2}{(z - 1)^3} \, dz.$$

Exercise 10.6 Find the value of the following integrals

$$\int_\gamma \frac{e^{2z}}{z} \, dz \quad \text{where} \quad \gamma = \{z \in \mathbb{C} : |z| = 2\}$$

$$\int_\gamma \frac{z^3 + 2z^2 + 2}{z - 2i} \, dz \quad \text{where} \quad \gamma = \left\{z \in \mathbb{C} : |z - 2i| = \frac{1}{4}\right\}$$

$$\int_\gamma \frac{\sin(2z^2 + 3z + 1)}{z - \pi} \, dz \quad \text{where} \quad \gamma = \{z \in \mathbb{C} : |z - \pi| = 1\}.$$

Exercise 10.7 Evaluate

$$\int_\gamma \frac{3z^2 + 2z + \sin(z + 1)}{(z - 2)^2} \, dz \quad \text{where} \quad \gamma = \{z \in \mathbb{C} : |z - 2| = 1\}$$

and

$$\int_\gamma \frac{e^z}{z(z + 2)} \, dz \quad \text{where} \quad \gamma = \{z \in \mathbb{C} : |z| = 1\}.$$

Exercise 10.8 Find the value of

$$\int_\gamma \frac{e^{z^2}}{(z - 1)^2(z^2 + 4)} \, dz$$

in the following cases.

   (i) γ is the circle centered at $(1,0)$ and of radius 1.

   (ii) γ is the boundary of the rectangle $[-1/2, 1/2] \times [0,4]$.

   (iii) γ is the boundary of the rectangle $[-2,0] \times [-1,1]$.

Exercise 10.9 Show that

$$\int_{-\infty}^{\infty} e^{-x^2} \cos(2bx)\, dx = \sqrt{\pi} e^{-b^2} \quad \text{and} \quad \int_{-\infty}^{\infty} e^{-x^2} \sin(2bx)\, dx = 0.$$

Hint: first prove that $f(z) = e^{-z^2}$ is holomorphic. Then consider the curve $\gamma = \gamma_1 \cup \gamma_2 \cup \gamma_3 \cup \gamma_4$ (cf. Figure 10.1).

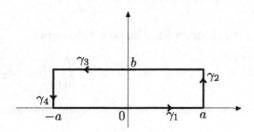

Figure 10.1: Exercise 10.9

Show that

   (i) $\lim\limits_{a \to +\infty} \int_{\gamma_2} f(z)\, dz = \lim\limits_{a \to +\infty} \int_{\gamma_4} f(z)\, dz = 0.$

   (ii) $\int_{\gamma} f(z)\, dz = 0$ (use the fact that f is holomorphic).

   (iii) Recalling that $\int_{-\infty}^{\infty} e^{-x^2} dx = \sqrt{\pi}$, conclude.

Exercise 10.10 Let f be a continuous function in a simply connected domain D. Let $F : D \to \mathbb{C}$ be a holomorphic function and such that $F'(z) = f(z)$. Show that, for every regular curve $\gamma : [a,b] \to D \subset \mathbb{C}$, then

$$\int_{\gamma} f(z)\, dz = F(\gamma(b)) - F(\gamma(a)).$$

In particular, if γ is closed, deduce that

$$\int_{\gamma} f(z)\, dz = 0.$$

(This is a weaker version of Cauchy theorem.)

Exercise 10.11 Let $f(z) = 1/z$, γ_1 be the circle of radius 1 centered at $z = 2$ and γ_2 be the circle of radius 1 centered at the origin. With the help of the previous exercise, compute, for $i = 1, 2$,

$$\int_{\gamma_i} f(z)\, dz.$$

Exercise* 10.12 Prove Cauchy theorem.

Hint: (i) Write $f = u + iv$ and $\gamma = \alpha + i\beta : [a, b] \to \mathbb{C}$, $\gamma = \gamma(t)$. Prove that

$$\int_\gamma f(z)\, dz = \int_a^b \left(u(\alpha, \beta)\alpha' - v(\alpha, \beta)\beta' \right) dt + i \int_a^b \left(v(\alpha, \beta)\alpha' + u(\alpha, \beta)\beta' \right) dt.$$

(ii) Apply Green theorem (cf. Chapter 4) to obtain

$$\int_\gamma f(z)\, dz = \int_{\text{int } \gamma} (-v_x - u_y)\, dx\, dy + i \int_{\text{int } \gamma} (u_x - v_y)\, dx\, dy.$$

(iii) Deduce the result using Cauchy–Riemann equations.

Chapter 11

Laurent series

11.1 Definitions and theoretical results

Theorem 11.1 *Let $D \subset \mathbb{C}$ be a simply connected domain and $z_0 \in D$. Let $f : D \setminus \{z_0\} \to \mathbb{C}$ be a holomorphic function, $N \in \mathbb{N}$ and*

$$L_N f(z) = \sum_{n=-N}^{N} c_n(z - z_0)^n$$

$$= c_N(z - z_0)^N + \cdots + c_1(z - z_0) + c_0 + \frac{c_{-1}}{(z - z_0)} + \cdots + \frac{c_{-N}}{(z - z_0)^N}$$

where

$$\boxed{c_n = \frac{1}{2\pi i} \int_{\gamma} \frac{f(\xi)}{(\xi - z_0)^{n+1}} \, d\xi}$$

(in particular $c_{-1} = \frac{1}{2\pi i} \int_{\gamma} f(\xi) \, d\xi$) and where γ is a piecewise regular simple closed curve contained in D ($z_0 \in \mathrm{int}\,\gamma$). Let $R > 0$ be such that (cf. Figure 11.1)

$$\{z \in \mathbb{C} : |z - z_0| < R\} \subset D.$$

Then, for every $0 < |z - z_0| < R$, $\lim_{N \to \infty} L_N f(z) = L f(z)$ exist, is finite and, moreover,

$$\boxed{L f(z) = f(z) = \sum_{n=-\infty}^{+\infty} c_n(z - z_0)^n = \sum_{n=0}^{+\infty} c_n(z - z_0)^n + \sum_{n=1}^{+\infty} \frac{c_{-n}}{(z - z_0)^n} \, .}$$

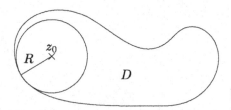

Figure 11.1: Laurent theorem

Definition 11.2 *We adopt the notations of the theorem.*

*(i) The quantity $Lf(z)$ is called the **Laurent series** of f at z_0.*

(ii) The quantity

$$\sum_{n=0}^{+\infty} c_n(z - z_0)^n$$

*is called the **regular part** of the Laurent series. The quantity*

$$\sum_{n=-\infty}^{-1} c_n(z - z_0)^n = \sum_{n=1}^{+\infty} c_{-n}(z - z_0)^{-n}$$

*is the **singular part** of the Laurent series.*

*(iii) We say that z_0 is a **regular point** for f if the singular part of the Laurent series is zero.*

*(iv) We say that z_0 is a **pole of order** m for f if $c_{-m} \neq 0$ and $c_{-k} = 0$, $\forall k \geq m + 1$. We therefore have*

$$Lf(z) = \sum_{n=1}^{m} \frac{c_{-n}}{(z - z_0)^n} + \sum_{n=0}^{+\infty} c_n(z - z_0)^n.$$

*(v) We say that z_0 is an (isolated) **essential singularity** for f if $c_{-k} \neq 0$, for infinitely many k. In this case we have*

$$Lf(z) = \sum_{n=1}^{\infty} \frac{c_{-n}}{(z - z_0)^n} + \sum_{n=0}^{+\infty} c_n(z - z_0)^n.$$

*(vi) The value c_{-1} is called the **residue** of f at z_0 and is denoted $\operatorname{Res}_{z_0}(f)$.*

*(vii) The **radius of convergence** of the series is the largest $R > 0$ such that $\{z \in \mathbb{C} : |z - z_0| < R\} \subset D$.*

Remark If $f : D \to \mathbb{C}$ is holomorphic, then the Laurent series coincide with the **Taylor series**, meaning that $c_n = f^{(n)}(z_0)/n!$ and

$$Lf(z) = Tf(z) = \sum_{n=0}^{+\infty} \frac{f^{(n)}(z_0)}{n!}(z - z_0)^n.$$

Proposition 11.3 *Let*

$$f(z) = \frac{p(z)}{q(z)}$$

where p and q are holomorphic functions in the neighborhood of z_0. Let z_0 be a zero of order k of p, i.e. (if $p(z_0) \neq 0$, we let $k = 0$)

$$p(z_0) = \cdots = p^{(k-1)}(z_0) = 0 \quad but \quad p^{(k)}(z_0) \neq 0$$

and z_0 be a zero of order l of q. Then the following conclusions hold true.

(i) If $l > k$, then z_0 is a pole of order $l-k$ of f.

(ii) If $l \leq k$, then z_0 is a regular point of f (where we have set $f(z_0) = \lim_{z \to z_0} p(z)/q(z)$, cf. Example 11.12).

Proposition 11.4 *If z_0 is a pole of order m, then*

$$\mathrm{Res}_{z_0}(f) = \frac{1}{(m-1)!} \lim_{z \to z_0} \frac{d^{m-1}}{dz^{m-1}} [(z - z_0)^m f(z)].$$

Proposition 11.5 *Let*

$$f(z) = \frac{p(z)}{q(z)}$$

where p and q are holomorphic functions in the neighborhood of $z_0 \in \mathbb{C}$ and such that $p(z_0) \neq 0$ and z_0 is a zero of order 1 of q. Then

$$\mathrm{Res}_{z_0}(f) = \frac{p(z_0)}{q'(z_0)}.$$

(For more details, cf. [1] 184–186, [3] 62, [5] 155–159, [10] 842–844, [15] 109.)

11.2   Examples

In the following examples, we will find the Laurent series of the given function f. We will also specify the nature of the singularity, the radius of convergence of the series and the residue.

Example 11.6 $f(z) = 1/z$ and $z_0 = 0$.

Discussion We immediately have

$$Lf(z) = 0 + \frac{1}{z}$$

i.e. the regular and singular part are respectively 0 and $1/z$. Moreover, $z_0 = 0$ is a pole of order 1 and $\mathrm{Res}_0(f) = 1$. The Laurent series converges $\forall z \in \mathbb{C} \setminus \{0\}$, i.e. $R = \infty$.

Example 11.7 $f(z) = 1/z$ and $z_0 = 1$.

Discussion We first recall the sum of the geometric series. If $q \in \mathbb{C}$ and $|q| < 1$, then

$$\frac{1}{1-q} = \sum_{n=0}^{+\infty} q^n.$$

In our case we have

$$Lf(z) = \frac{1}{z} = \frac{1}{1+(z-1)} = \sum_{n=0}^{+\infty} (-1)^n (z-1)^n.$$

The Laurent series of f therefore coincides with its Taylor series and $z_0 = 1$ is a regular point. The convergence of the series hold in $|z - 1| < 1$, i.e. $R = 1$, and the residue is zero.

Example 11.8 $f(z) = 1/z^2$ and $z_0 = 0$.

Discussion It is easy to see that $z_0 = 0$ is a pole of order 2 and that

$$Lf(z) = \frac{1}{z^2}.$$

The residue is zero and the series converges $\forall z \in \mathbb{C} \setminus \{0\}$, i.e. $R = \infty$.

Example 11.9 $f(z) = \frac{1}{z^3} + \frac{2}{z}$ and $z_0 = 0$.

Discussion We have that

$$Lf(z) = \frac{1}{z^3} + \frac{2}{z},$$

$z_0 = 0$ is a pole of order 3, $\mathrm{Res}_0(f) = 2$ and the series converges $\forall z \in \mathbb{C} \setminus \{0\}$, i.e. $R = \infty$.

Example 11.10 $f(z) = \frac{1}{z^2+z}$ *and* $z_0 = 0$.

Discussion We write (using the formula for the sum of the geometric series)

$$\frac{1}{z^2+z} = \frac{1}{z(z+1)} = \frac{1}{z} - \frac{1}{z+1} = \frac{1}{z} - \sum_{n=0}^{+\infty}(-1)^n z^n = Lf(z)$$

for every $0 < |z| < 1$ (i.e. $R = 1$). Furthermore, $z_0 = 0$ is a pole of order 1 and $\mathrm{Res}_0(f) = 1$.

Example 11.11 $f(z) = e^{\frac{1}{z}}$ *and* $z_0 = 0$.

Discussion Letting $y = 1/z$ we get, since

$$e^y = \sum_{n=0}^{+\infty} \frac{y^n}{n!},$$

that

$$f(z) = e^{\frac{1}{z}} = Lf(z) = \sum_{n=0}^{+\infty} \frac{1}{n!}\frac{1}{z^n} = 1 + \sum_{n=1}^{+\infty} \frac{1}{n!}\frac{1}{z^n}.$$

We therefore have that $z_0 = 0$ is an essential singularity for f, $\mathrm{Res}_0(f) = 1$ and the series converges $\forall z \in \mathbb{C} \setminus \{0\}$, i.e. $R = \infty$.

Example 11.12 *A function f may seem to have a singularity at z_0, but when properly redefined this is not the case. In this case we speak of a **removable singularity**. For example, $z_0 = 0$ is a removable singularity for $f(z) = \sin z/z$, since it is enough to set*

$$f(z) = \begin{cases} \dfrac{\sin z}{z} & \text{if } z \neq 0 \\ 1 = \lim_{z \to 0} \dfrac{\sin z}{z} & \text{if } z = 0. \end{cases}$$

Conclusion: $z_0 = 0$ is a regular point and the function f is holomorphic in $\mathbb{C}$.

Example 11.13 *Give an example showing that not all singularities are poles or essential singularities.*

Discussion Let $z_0 = 0$ and

$$f(z) = \tan\left(\frac{1}{z}\right).$$

We recall that

$$\tan y = \infty \iff y = y_k = (2k+1)\frac{\pi}{2}, \; k \in \mathbb{Z}$$

and, hence,

$$\tan\left(\frac{1}{z}\right) = \infty \iff z = \frac{1}{y_k} = \frac{2}{(2k+1)\pi} \quad (\to z_0 = 0, \text{ whenever } k \to \infty).$$

Therefore, f cannot be written in Laurent series at $z_0 = 0$, since then we should have that the radius of convergence is $R = 0$. In conclusion, $z_0 = 0$ **is not an isolated singularity**.

11.3 Exercises

In Exercises 1 to 13, one should find the Laurent series of f (or at least some of its terms), identify the type of singularities, the residue and the radius of convergence.

Exercise 11.1

(i) $\sin z$ at $z_0 = \dfrac{\pi}{4}$ (ii) $\dfrac{\sin z}{z^3}$ at $z_0 = 0$

(iii) $\dfrac{z}{1+z^2}$ at $z_0 = 1$ (iv) $\sin\left(\dfrac{1}{z}\right)$ at $z_0 = 0$

(v) $\dfrac{z^2 + 2z + 1}{z+1}$ at $z_0 = -1$ (vi) $\dfrac{1}{(1-z)^3}$ at $z_0 = 1$

(vii) $\dfrac{z^2 + z + 1}{z^2 - 1}$ at $z_0 = 1$ (viii) $\dfrac{1}{\sin z}$ at $z_0 = 0$

Exercise 11.2 $f(z) = \dfrac{1}{z(z+2)^3}$ at $z_1 = 0$ and $z_2 = -2$. (Write at least the four first terms.)

Exercise 11.3

(i) $\dfrac{\cos z}{(z - \pi)}$ at $z_0 = \pi$ (ii) $z^2 e^{1/z}$ at $z_0 = 0$

(iii) $\dfrac{z^2}{(z-1)^2(z+3)}$ at $z_0 = 1$.

Exercise 11.4 $f(z) = z^2 e^{(z-1)^{-2}}$ at $z_0 = 1$.

Exercise 11.5 $f(z) = z \cos\left(\frac{1}{z}\right)$ at $z_0 = 0$.

Exercise 11.6 $f(z) = \frac{e^z}{(z-1)^2}$ at $z_0 = 1$.

Exercise 11.7 $f(z) = \frac{\sqrt{z}}{(z-1)^2}$ at $z_0 = 1$.

Exercise 11.8 $f(z) = \dfrac{1}{\cos^2\left(\frac{\pi}{2}z\right)}$ at $z_0 = 1$.

Exercise 11.9 $f(z) = \dfrac{\log(1+z)}{\sin(z^2)}$ at $z_0 = 0$.

Exercise 11.10 $f(z) = e^{(1/z)} \sin\left(\frac{1}{z}\right)$ at $z_0 = 0$.

Exercise 11.11 $f(z) = \dfrac{\sin z}{z(e^z - 1)}$ at $z_0 = 0$.

Exercise 11.12 $f(z) = \dfrac{\sin z}{(z - \pi)^2}$ at $z_0 = \pi$.
Hint: observe that $\sin z = -\sin(z - \pi)$, $\forall z \in \mathbb{C}$.

Exercise 11.13 $f(z) = \dfrac{\sin z}{\sin(z^2)}$ at $z_0 = 0$ (it is sufficient to give the first two terms of the Laurent series).

Exercise 11.14 Let $f(z) = \log(1+z)$.

(i) Find the largest domain in $\mathbb{C}$ where the function f is holomorphic.

(ii) Find its Taylor series at $z_0 = 0$ and $z_0 = i$. Give, in both cases, its radius of convergence.

Exercise 11.15 Let $f(z) = \dfrac{\sin(z^2 + 1)}{(z^2 + 1)^2}$.

(i) Find all singularities of f and determine the nature of these singularities.

(ii) Find the residue at all these points.

(iii) Find the radius of convergence of the Laurent series at these points.

Exercise 11.16 Let $f(z) = \log(1 + z^2)$.

(i) Find the largest domain in $\mathbb{C}$ where the function is holomorphic (justify your answer).

(ii) Compute the Taylor series of f at $z_0 = 0$ (what is the radius of convergence and the terms in z^n when $n = 0, 1, 2, 3$).

Exercise* 11.17 (L'Hôpital rule) Let f and g be holomorphic functions in the neighborhood of z_0 such that

$$f(z_0) = g(z_0) = 0 \quad \text{and} \quad g'(z_0) \neq 0.$$

Prove that

$$\lim_{z \to z_0} \frac{f(z)}{g(z)} = \lim_{z \to z_0} \frac{f'(z)}{g'(z)} = \frac{f'(z_0)}{g'(z_0)}.$$

Exercise* 11.18 (Liouville theorem) (i) Show that if f is holomorphic and bounded in $\mathbb{C}$, then f is constant.

(ii) Give an example of a real analytic function, namely, a function of the form

$$f(x) = \sum_{n=0}^{\infty} a_n x^n, \quad \forall x \in \mathbb{R}$$

which is bounded but not constant.

Hint: (i) Apply the remark following Definition 11.2 to deduce that

$$f(z) = \sum_{n=0}^{\infty} \frac{f^{(n)}(0)}{n!} z^n, \quad \forall z \in \mathbb{C}.$$

(ii) Write Cauchy integral formula for $f^{(n)}(0)$ on a circle of radius R centered at 0.

(iii) With the help of the previous question prove that

$$|f^{(n)}(0)| \leq \frac{n!}{2\pi R^n} \int_0^{2\pi} |f(Re^{it})| \, dt.$$

(iv) Conclude that $f^{(n)}(0) = 0$, for every $n \geq 1$.

Chapter 12

Residue theorem and applications

12.1 Part I

12.1.1 Definitions and theoretical results

Theorem 12.1 *Let $D \subset \mathbb{C}$ be a simply connected domain, γ be a piecewise regular simple closed curve contained in D (cf. Figure 12.1). Let $z_1, \cdots, z_m \in \operatorname{int} \gamma$ ($z_i \neq z_j$, if $i \neq j$) and $f : D \setminus \{z_1, \cdots, z_m\} \to \mathbb{C}$ a holomorphic function, then*

$$
\int_\gamma f(z)\, dz = 2\pi i \sum_{k=1}^{m} \operatorname{Res}_{z_k}(f)
$$

where $\operatorname{Res}_{z_k}(f)$ is the residue of the function f in z_k.

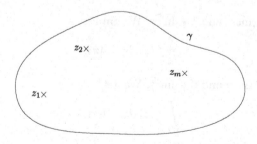

Figure 12.1: Residue theorem

(For more details, cf. [1] 150–151, [3] 64–65, [5] 201–202, [10] 865–866, [15] 77.)

12.1.2   Examples

Example 12.2 *Let D be as in the theorem. If f is a holomorphic function in D, we deduce from Cauchy theorem that*

$$\int_\gamma f(z)\,dz = 0.$$

Example 12.3 *Let D and γ be as in the theorem and $f(z) = 1/z$. Compute $\int_\gamma f(z)\,dz$.*

Discussion *Case 1:* $0 \in \operatorname{int}\gamma$. We have $\operatorname{Res}_0(f) = 1$ and thus

$$\int_\gamma f(z)\,dz = 2\pi i.$$

Case 2: $0 \notin \overline{\operatorname{int}\gamma}$. Cauchy theorem implies

$$\int_\gamma f(z)\,dz = 0.$$

Case 3: $0 \in \gamma$. In this case the integral is not well defined.

Example 12.4 *Let D and γ be as in the theorem and*

$$f(z) = \frac{2}{z} + \frac{3}{z-1} + \frac{1}{z^2}.$$

Evaluate $\int_\gamma f(z)\,dz$.

Discussion We can observe that the singularities of f are in $z = 0$ and $z = 1$. It is clear that $\operatorname{Res}_0(f) = 2$ and $\operatorname{Res}_1(f) = 3$. We consider five cases.

Case 1: $0,1 \in \operatorname{int}\gamma$. We immediately find

$$\int_\gamma f(z)\,dz = 2\pi i(2+3) = 10\pi i.$$

Case 2: $0 \in \operatorname{int}\gamma$ and $1 \notin \overline{\operatorname{int}\gamma}$. We obtain

$$\int_\gamma f(z)\,dz = 4\pi i.$$

Case 3: $1 \in \operatorname{int}\gamma$ and $0 \notin \overline{\operatorname{int}\gamma}$. We get

$$\int_\gamma f(z)\,dz = 6\pi i.$$

Case 4: $0,1 \notin \overline{\operatorname{int}\gamma}$. We find

$$\int_\gamma f(z)\,dz = 0.$$

Case 5: $0 \in \gamma$ or $1 \in \gamma$. In this case the integral is not well defined.

12.2 Part II: Evaluation of real integrals

Application 1 *We first show how to compute integrals of the form*

$$\boxed{\int_0^{2\pi} f\left(\cos\theta, \sin\theta\right) d\theta}$$

where $f : \mathbb{R}^2 \to \mathbb{R}$ *is such that*

$$f(x,y) = \frac{P(x,y)}{Q(x,y)}$$

P and Q being polynomials and

$$Q(\cos\theta, \sin\theta) \neq 0, \quad \forall \theta \in [0, 2\pi].$$

Discussion We set $z = e^{i\theta}$ and we recall that

$$\cos\theta = \frac{e^{i\theta} + e^{-i\theta}}{2} = \frac{1}{2}\left(z + \frac{1}{z}\right)$$

$$\sin\theta = \frac{e^{i\theta} - e^{-i\theta}}{2i} = \frac{1}{2i}\left(z - \frac{1}{z}\right).$$

Note that

$$d\theta = \frac{1}{i}\frac{dz}{z}.$$

Let γ be the circle of radius 1 centered at the origin and

$$\widetilde{f}(z) = \frac{1}{iz} f\left(\frac{1}{2}\left(z + \frac{1}{z}\right), \frac{1}{2i}\left(z - \frac{1}{z}\right)\right).$$

It follows from the residue theorem that

$$\int_\gamma \widetilde{f}(z)\, dz = \int_0^{2\pi} f\left(\cos\theta, \sin\theta\right) d\theta = 2\pi i \sum_{k=1}^{m} \operatorname{Res}_{z_k}(\widetilde{f}),$$

where z_k are the singularities of $\widetilde{f}$ in the interior of γ (cf. Figure 12.2).

(For more details, cf. [1] 155, [3] 65–66, [5] 203–205, [10] 868–869.)

Example 12.5 *Compute* $\int_0^{2\pi} \dfrac{d\theta}{2 + \cos\theta}$.

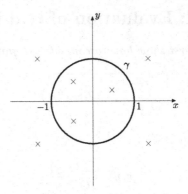

Figure 12.2: Application 1

Discussion We here have

$$f(\cos\theta, \sin\theta) = \frac{1}{2 + \cos\theta}$$

and, recalling that $z = e^{i\theta}$,

$$\widetilde{f}(z) = \frac{2}{i(z^2 + 4z + 1)} = \frac{2}{i(z + 2 + \sqrt{3})(z + 2 - \sqrt{3})}.$$

The singularities of $\widetilde{f}$ are

$$z_1 = -(2 + \sqrt{3}) \quad \text{and} \quad z_2 = \sqrt{3} - 2.$$

Only z_2, which is a pole of order 1, lies in the interior of γ. We therefore get

$$\operatorname{Res}_{\sqrt{3}-2}(\widetilde{f}) = \lim_{z \to \sqrt{3}-2} \frac{2(z + 2 - \sqrt{3})}{i(z + 2 + \sqrt{3})(z + 2 - \sqrt{3})} = \frac{1}{i\sqrt{3}}$$

and thus

$$\int_0^{2\pi} \frac{d\theta}{2 + \cos\theta} = 2\pi i \frac{1}{i\sqrt{3}} = \frac{2\pi}{\sqrt{3}}.$$

Application 2 *Let $a \geq 0$, P and Q be polynomials verifying*

$$Q(x) \neq 0, \ \forall x \in \mathbb{R} \quad \text{and} \quad \deg Q - \deg P \geq 2$$

(deg denoting the degree of the polynomials). We want to evaluate integrals of the form

$$\boxed{\int_{-\infty}^{+\infty} \frac{P(x)}{Q(x)} e^{iax} \, dx.}$$

We set

$$f(z) = \frac{P(z)}{Q(z)} e^{iaz}.$$

Under our hypotheses the integral is well defined and it follows from the residue theorem that

$$\int_{-\infty}^{+\infty} f(x)\, dx = 2\pi i \sum_{k=1}^{m} \operatorname{Res}_{z_k}(f),$$

where z_k are the singularities of f lying in the upper half plane (i.e. $\operatorname{Im} z \geq 0$).

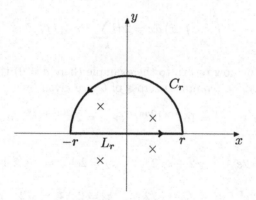

Figure 12.3: Application 2

Example 12.6 *Evaluate* $\int_{-\infty}^{+\infty} \dfrac{x^2}{16 + x^4}\, dx.$

Discussion The discussion is divided into two steps. The first one applies to the general setting, while the second one applies to the example.

Step 1. Let $r > 0$ and

$$\Gamma_r = C_r \cup L_r$$

where (see Figure 12.3)

$$C_r = \{z \in \mathbb{C} : |z| = r \text{ and } \operatorname{Im} z \geq 0\}$$
$$L_r = \{z \in \mathbb{C} : -r < \operatorname{Re} z < r \text{ and } \operatorname{Im} z = 0\}.$$

We choose r sufficiently large so that all singularities of f lying in the upper half plane are in the interior of Γ_r (note that, since $Q(x) \neq 0, \forall x \in \mathbb{R}$, the function f has no singularity on the real axis). The residue theorem leads to

$$\int_{\Gamma_r} f(z)\, dz = 2\pi i \sum_{k=1}^{m} \mathrm{Res}_{z_k}(f),$$

where z_k are the singularities of f lying in the interior of Γ_r. It is fairly clear that our hypotheses imply that

$$\lim_{r\to\infty} \int_{C_r} f(z)\, dz = 0 \quad \text{and} \quad \lim_{r\to\infty} \int_{L_r} f(z)\, dz = \int_{-\infty}^{+\infty} f(x)\, dx.$$

Combining all these results we find

$$\int_{-\infty}^{+\infty} f(x)\, dx = 2\pi i \sum_{k=1}^{m} \mathrm{Res}_{z_k}(f).$$

Step 2. Let us now return to the example (here $a = 0$) that satisfies all the hypotheses. The (complex) zeroes of Q are given by

$$16 + z^4 = 0 \iff z^4 = 16 e^{i(\pi + 2n\pi)} \iff z = 2e^{i(2n+1)\frac{\pi}{4}}, \; n = 0, 1, 2, 3,$$

i.e.

$$z_1 = 2e^{i\frac{\pi}{4}} = \sqrt{2} + i\sqrt{2} \qquad z_2 = 2e^{i\frac{3\pi}{4}} = -\sqrt{2} + i\sqrt{2}$$

$$z_3 = 2e^{i\frac{5\pi}{4}} = -\sqrt{2} - i\sqrt{2} \qquad z_4 = 2e^{i\frac{7\pi}{4}} = \sqrt{2} - i\sqrt{2}.$$

These are poles of order 1 of f where

$$f(z) = \frac{z^2}{16 + z^4}.$$

Only z_1 and z_2 are in the upper half plane and thus

$$\int_{-\infty}^{+\infty} \frac{x^2}{16 + x^4}\, dx = 2\pi i \left[\mathrm{Res}_{z_1}(f) + \mathrm{Res}_{z_2}(f)\right].$$

We next compute the two residues, with the help of Proposition 11.5. We immediately have

$$\mathrm{Res}_{z_1}(f) = \frac{z_1^2}{4z_1^3} = \frac{1}{4z_1} = \frac{1}{4(1+i)\sqrt{2}} = \frac{1+i}{8i\sqrt{2}}$$

$$\mathrm{Res}_{z_2}(f) = \frac{z_2^2}{4z_2^3} = \frac{1}{4z_2} = \frac{1}{4(-1+i)\sqrt{2}} = \frac{1-i}{8i\sqrt{2}}$$

and thus

$$\int_{-\infty}^{+\infty} \frac{x^2}{16 + x^4}\, dx = 2\pi i \left(\operatorname{Res}_{z_1}(f) + \operatorname{Res}_{z_2}(f) \right) = \frac{\pi\sqrt{2}}{4}.$$

(The example could have been treated without the use of the residue theorem.)

(For more details, cf. [1] 156, [3] 69–71, [5] 208–212, [10] 873–877.)

12.3 Exercises

Exercise 12.1 Let $\gamma \subset \mathbb{C}$ be a piecewise regular simple closed curve. Evaluate in terms of γ

$$\int_{\gamma} e^{\frac{1}{z^2}}\, dz.$$

Exercise 12.2 Let γ be a piecewise regular simple closed curve. Compute the following integrals in terms of γ

$$\int_{\gamma} \frac{1}{(z-i)(z+2)^2(z-4)}\, dz$$

$$\int_{\gamma} \frac{z^2 + 2z + 1}{(z-3)^3}\, dz \quad \text{and} \quad \int_{\gamma} \frac{e^{\frac{1}{z}}}{z^2}\, dz.$$

Exercise 12.3 Evaluate

$$\int_{-\infty}^{+\infty} \frac{dx}{\cosh x}.$$

Hint: (i) Let $R > 0$ (cf. Figure 12.4) and γ_R be the boundary of the rectangle $(-R, R) \times (0, \pi)$. Evaluate, with the help of the residue theorem,

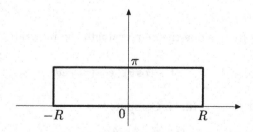

Figure 12.4: Exercise 12.3

$$\int_{\gamma_R} \frac{dz}{\cosh z} \, .$$

(ii) Assuming that

$$\lim_{L \to \pm\infty} \int_0^\pi \frac{dy}{\cosh(L + iy)} = 0,$$

conclude.

Exercise 12.4 Let γ be a piecewise regular simple closed curve. Find the value of

$$\int_\gamma \frac{1 - e^{i\pi z}}{z(z + i)(z - 1)^2} \, dz.$$

Exercise 12.5 Let γ be a regular simple closed curve contained in the disk of radius 2 centered at $z = 0$. Evaluate

$$\int_\gamma \tan z \, dz.$$

Exercise 12.6 Evaluate $\int_0^{2\pi} \dfrac{d\theta}{\sqrt{5} - \sin\theta} \, .$

Exercise 12.7 Compute $\int_0^{2\pi} \dfrac{\cos\theta \sin(2\theta)}{5 + 3\cos(2\theta)} \, d\theta.$

Exercise 12.8 Compute $\int_0^{2\pi} \dfrac{\cos^2\theta}{13 - 5\cos(2\theta)} \, d\theta.$

Exercise 12.9 Evaluate $\int_0^{2\pi} \dfrac{\sin^2(5\theta/2)}{\sin^2(\theta/2)} \, d\theta.$

Exercise 12.10 (i) Let $n \in \mathbb{N}$ and

$$h(z) = \frac{(z - 1)^{2n} \left(z^{2n} + 1\right)}{z^{2n+1}} \, .$$

Show that $\mathrm{Res}_0(h) = 2$.

(ii) Invoking the residue theorem evaluate the integral

$$\int_0^{2\pi} (1 - \cos\theta)^n \cos(n\theta) \, d\theta.$$

Exercise 12.11 Let $p \in (0, 1)$. Compute

$$\int_0^{2\pi} \frac{d\theta}{1 - 2p\cos\theta + p^2} \, .$$

Exercise 12.12 (i) Let $z = re^{i\theta}$. Prove that

$$\left|1 + z^6\right| \geq \left|r^6 - 1\right|.$$

(ii) Let C_r be the upper half circle of radius $r > 1$ and centered at 0. Show, with the help of (i), that

$$\lim_{r \to \infty} \left| \int_{C_r} \frac{z^2}{1 + z^6} \, dz \right| = 0.$$

(iii) Calculate

$$\int_{-\infty}^{+\infty} \frac{x^2}{1 + x^6} \, dx.$$

Exercise 12.13 (i) Let $\theta \in [0, \pi]$, $r > 0$ and $z = re^{i\theta}$. Prove that $\left|e^{iz}\right| \leq 1$.

(ii) Let $z = re^{i\theta} \in \mathbb{C}$. Show that

$$r^4 - 16 \leq \left|16 + z^4\right|.$$

(iii) Establish, invoking (i) and (ii), that if $\theta \in [0, \pi]$, $r > 2$ and $z = re^{i\theta}$, then

$$\left| \frac{e^{iz}}{16 + z^4} \right| \leq \frac{1}{r^4 - 16}.$$

(iv) Let C_r be the upper half circle of radius $r > 2$ and centered at 0. Deduce from (iii) that

$$\lim_{r \to \infty} \left| \int_{C_r} \frac{e^{iz}}{16 + z^4} \, dz \right| = 0.$$

(v) Using (iv), compute

$$\int_{-\infty}^{+\infty} \frac{\cos x}{16 + x^4} \, dx \quad \text{and} \quad \int_{-\infty}^{+\infty} \frac{\sin x}{16 + x^4} \, dx.$$

Chapter 13

Conformal mapping

13.1 Definitions and theoretical results

Definition 13.1 *Let $D \subset \mathbb{C}$ be open. We say that $f : D \to f(D) = D^* \subset \mathbb{C}$ is a conformal mapping from D onto D^* if*

(i) f is one-to-one and onto from D onto D^*,

(ii) f is holomorphic in D,

(iii) $f'(z) \neq 0$, $\forall z \in D$.

Remark The definition of conformal mapping that we adopt here is not the usual definition (cf. [1] 67–76, [3] 20–22, [5] 427–434, [10] 882–886), but it is equivalent. However, in [5] 432, we find the same definition and it is shown (in [5] 429) that (i) + (ii) $\Rightarrow$ (iii).

Definition 13.2 (Möbius transformation) *A Möbius transformation is a map of the form*

$$z \to f(z) = \frac{az + b}{cz + d}$$

where $a, b, c, d \in \mathbb{C}$ with $ad - bc \neq 0$ ($\Rightarrow c \neq 0$ or $d \neq 0$).

Proposition 13.3 *Let $ad - bc \neq 0$ and*

$$f(z) = \frac{az + b}{cz + d}.$$

(i) If $c \neq 0$,

$$\Omega = \mathbb{C} - \{-d/c\} \quad \text{and} \quad \Omega^* = \mathbb{C} - \{a/c\},$$

101

then the Möbius transformation is a conformal mapping from Ω onto Ω^*.

(ii) If $c = 0$ ($\Rightarrow d \neq 0$), then f is conformal from $\mathbb{C}$ onto $\mathbb{C}$.

Proposition 13.4 *Every Möbius transformation maps circles and lines onto circles and lines.*

(For more details, cf. [1] 76–89, [3] 8, [5] 436–459, [10] 887–896, [15] 206–212.)

Theorem 13.5 (Riemann theorem) *Let $\Omega \neq \mathbb{C}$ be a simply connected domain. Then there exists a conformal mapping from Ω onto D, the unit disk (cf. Figure 13.1). Furthermore, if the image of a point $z_0 \in \Omega$ is fixed, for example $f(z_0) = 0$ and $f'(z_0) > 0$ (i.e. $\mathrm{Re}\, f'(z_0) > 0$ and $\mathrm{Im}\, f'(z_0) = 0$), then such a map f is unique.*

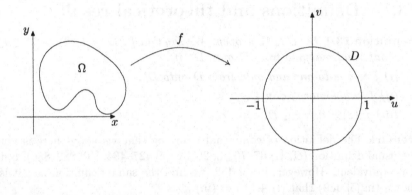

Figure 13.1: Riemann theorem

(For more details, cf. [1] 229–249, [5] 485–504, [15] 224.)

13.2   Examples

Example 13.6 *The function $f(z) = z$ is obviously a conformal mapping from $\mathbb{C}$ onto $\mathbb{C}$.*

Example 13.7 $f(z) = \bar{z}$ *is not a conformal mapping.*

Discussion It is enough to observe that $f(z) = \bar{z}$ is not a holomorphic function.

Example 13.8 *(i) Let $z_i, w_i \in \mathbb{C}$, $i = 1, 2, 3$, with $z_i \neq z_j$, $w_i \neq w_j$, if $i \neq j$. Find a Möbius transformation f such that $f(z_i) = w_i$, $i = 1, 2, 3$*

(i.e. Möbius transformations are fully characterized when one prescribes three different points and their three different images).

(ii) Find a Möbius transformation from $\Omega = \{z \in \mathbb{C} : \operatorname{Re} z > 0\}$ *onto* $D = \{z \in \mathbb{C} : |z| < 1\}$.

Discussion (i) If $f(z_i) = w_i$, $i = 1, 2, 3$, then necessarily

$$\frac{f(z) - w_1}{w_1 - w_2} \cdot \frac{w_3 - w_2}{f(z) - w_3} = \frac{z - z_1}{z_1 - z_2} \cdot \frac{z_3 - z_2}{z - z_3}. \tag{13.1}$$

Set

$$\alpha = \frac{z_3 - z_2}{z_1 - z_2} \quad \text{and} \quad \beta = \frac{w_3 - w_2}{w_1 - w_2}.$$

We deduce from (13.1) that

$$\beta \left[f(z) - w_1 \right] (z - z_3) = \alpha \left[f(z) - w_3 \right] (z - z_1)$$

which implies

$$f(z) \left[(\beta - \alpha)z - \beta z_3 + \alpha z_1 \right] = \beta w_1 (z - z_3) - \alpha w_3 (z - z_1)$$

and thus

$$f(z) = \frac{(\beta w_1 - \alpha w_3)z + (\alpha w_3 z_1 - \beta w_1 z_3)}{(\beta - \alpha)z + \alpha z_1 - \beta z_3}.$$

We hence have the result

$$f(z) = \frac{az + b}{cz + d}$$

setting

$$a = \beta w_1 - \alpha w_3 \quad b = \alpha w_3 z_1 - \beta w_1 z_3$$
$$c = \beta - \alpha \quad\quad d = \alpha z_1 - \beta z_3.$$

(ii) We look for a map of the form

$$f(z) = \frac{az + b}{cz + d}.$$

We choose, for example,

$$z_1 = i, \quad z_2 = -i, \quad z_3 = 0$$

(which all belong to the imaginary axis, i.e. $\operatorname{Re} z = 0$) and

$$w_1 = i, \quad w_2 = -i, \quad w_3 = 1$$

(which all belong to the unit disk, i.e. $|w| = 1$). We then use (i) to get

$$
\begin{cases}
f(i) = i = \dfrac{ai + b}{ci + d} \\[2mm]
f(-i) = -i = \dfrac{-ai + b}{-ci + d} \\[2mm]
f(0) = 1 = \dfrac{b}{d}
\end{cases}
\Rightarrow
\begin{cases}
2b = -2c \\[2mm]
2ai = 2bi \\[2mm]
b = d.
\end{cases}
$$

We therefore find that $a = b = -c = d$ and hence

$$
f(z) = \frac{z+1}{1-z}.
$$

It remains to find out whether f maps Ω onto D or onto the exterior of D. To prove this we take a point belonging to Ω, say $z = 1$, and we find that $f(z) = \infty \notin D$. Thus, the required map, that we call g, is given by

$$
g(z) = \frac{1-z}{z+1}
$$

(since the map $h(\zeta) = \zeta^{-1}$ sends the interior of the unit disk onto the exterior of the unit disk and conversely).

13.3   Exercises

Exercise 13.1 Let $z = x + iy$ and

$$
f(z) = \frac{1}{z}.
$$

(i) Find $u = \operatorname{Re} f$ and $v = \operatorname{Im} f$ as functions of x and y.

(ii) Express x and y as functions of u and v (i.e. find the inverse function of f).

(iii) Let

$$
A_1 = \{z \in \mathbb{C} : \operatorname{Re} z = \operatorname{Im} z\}, \quad A_2 = \{z \in \mathbb{C} : \operatorname{Re} z = 1\}
$$

$$
A_3 = \{z \in \mathbb{C} : |z - 1| = 1\} \quad \text{and} \quad A_4 = \{z \in \mathbb{C} : |z - 1| = 2\}.
$$

Find the image by f of these sets (what is the geometrical meaning of A_i and $f(A_i)$?).

(iv) More generally, show that the map $z \to 1/z$ sends {circles, lines} onto {circles, lines} (this is nothing other than Proposition 13.4 when $f(z) = z^{-1}$).

Exercise 13.2 Find a Möbius transformation that sends

(i) $0 \to -1$, $\quad 1 + i \to 1$, $\quad 1 - i \to -1 + 2i$,

(ii) the unit disk onto the exterior of the disk of radius 2 centered at the origin.

Exercise 13.3 Find a Möbius transformation sending

$$\Omega = \{z \in \mathbb{C} : \operatorname{Im} z > 0\}$$

onto

$$D = \{z \in \mathbb{C} : |z| < 1\}.$$

Exercise 13.4 (i) Find a Möbius transformation sending

$$\Omega = \{z \in \mathbb{C} : \ |z - 2(i + 1)| > 2\}$$

onto

$$D = \{z \in \mathbb{C} : |z| < 1\}.$$

(ii) Find its inverse.

Exercise 13.5 (i) Prove Proposition 13.3. In particular, show that

$$f^{-1}(w) = \frac{-dw + b}{cw - a}.$$

(ii) What happens to a Möbius transformation if $ad = bc$?

Exercise 13.6 The Joukowski transformation is given by

$$f(z) = z + \frac{1}{z}.$$

(i) Find $\operatorname{Re} f$ and $\operatorname{Im} f$.

(ii) Find the image by f of the circle $|z| = a$, $a > 0$.

Exercise 13.7 (i) Show that the function $f(z) = e^z$ is conformal from

$$A = \{z \in \mathbb{C} : 0 < \operatorname{Im} z < \pi\}$$

onto

$$\Omega = \{w \in \mathbb{C} : \operatorname{Im} w > 0\}.$$

(ii) Using Exercise 13.3 and the previous question, find a conformal mapping from A onto the unit disk D.

Exercise 13.8 (i) Prove that if u is harmonic (i.e. $u \in C^\infty\left(\mathbb{R}^2\right)$ with $\Delta u = 0$), then there exists a holomorphic function $g : \mathbb{C} \to \mathbb{C}$ such that $\operatorname{Re} g = u$.

(ii) Appealing to (i), show that if u is harmonic and if $f : \mathbb{R}^2 \to \mathbb{R}^2$ is conformal, then $u \circ f$ is harmonic (where it is understood that $f : \mathbb{R}^2 \to \mathbb{R}^2$, $f(x, y) = (\alpha, \beta)$ is conformal if $\tilde{f}(x + iy) = \alpha + i\beta$ is conformal). Compare with Exercise 9.10.

Exercise 13.9 Let

$$\Omega = \left\{(x, y) \in \mathbb{R}^2 : x, y > 0\right\}$$
$$O = \left\{(x, y) \in \mathbb{R}^2 : y > 0\right\}$$
$$D = \left\{(x, y) \in \mathbb{R}^2 : x^2 + y^2 < 1\right\}.$$

(i) Show that

$$h(z) = \frac{z - i}{z + i}$$

is conformal from O onto D. Find the real and imaginary parts of h^{-1}.

(ii) Prove that the map $g(z) = z^2$ is conformal from Ω onto O. Find the real and imaginary parts of g^{-1}.

(iii) Find a conformal mapping f from Ω onto D.

Exercise* 13.10 Show, invoking Liouville theorem (cf. Exercise 11.18), that there cannot exist a conformal mapping from $\mathbb{C}$ onto the unit disk.

Part III

Fourier analysis

Chapter 14

Fourier series

14.1 Definitions and theoretical results

Definition 14.1 *We say that* $f : \mathbb{R} \to \mathbb{R}$ *is **piecewise regular** (cf. Figure 14.1) if it is piecewise regular on every compact set of the form* $[a, b]$. *This means that*

(i) f *is piecewise continuous on* $[a, b]$, *namely, there exist*

$$a = a_0 < a_1 < \cdots < a_{n+1} = b$$

such that, $\forall i = 0, 1, \cdots, n,$ $f|_{(a_i, a_{i+1})}$ *is continuous,*

$$\lim_{\substack{x \to a_i \\ x > a_i}} f(x) = f(a_i + 0) \quad and \quad \lim_{\substack{x \to a_{i+1} \\ x < a_{i+1}}} f(x) = f(a_{i+1} - 0)$$

exist and are finite;

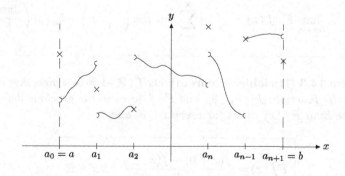

Figure 14.1: Piecewise regular function

109

(ii) f' exists and is piecewise continuous on (a_i, a_{i+1}) and

$$\lim_{\substack{x \to a_i \\ x > a_i}} f'(x) = f'(a_i + 0) \quad \text{and} \quad \lim_{\substack{x \to a_{i+1} \\ x < a_{i+1}}} f'(x) = f'((a_{i+1} - 0)$$

exist and are finite.

Definition 14.2 *Let $T > 0$, $N \geq 1$ be an integer and $f : \mathbb{R} \to \mathbb{R}$ be a bounded function, $T-$periodic (i.e. $f(x + T) = f(x)$, $\forall x$) and integrable on $[0, T]$. Let $n \in \mathbb{N}$ and*

$$a_n = \frac{2}{T} \int_0^T f(x) \cos\left(\frac{2\pi n}{T} x\right) dx, \quad n = 0, 1, 2 \cdots$$

$$b_n = \frac{2}{T} \int_0^T f(x) \sin\left(\frac{2\pi n}{T} x\right) dx, \quad n = 1, 2 \cdots$$

(i) The quantity

$$F_N f(x) = \frac{a_0}{2} + \sum_{n=1}^{N} a_n \cos\left(\frac{2\pi n}{T} x\right) + b_n \sin\left(\frac{2\pi n}{T} x\right)$$

*is called the **partial Fourier series of order N** of f.*

*(ii) The limit of $F_N f(x)$, when it exists, is called the **Fourier series** of f and we denote it by*

$$F f(x) = \lim_{N \to \infty} F_N f(x) = \frac{a_0}{2} + \sum_{n=1}^{\infty} \left\{ a_n \cos\left(\frac{2\pi n}{T} x\right) + b_n \sin\left(\frac{2\pi n}{T} x\right) \right\}.$$

Theorem 14.3 (Dirichlet theorem) *Let $f : \mathbb{R} \to \mathbb{R}$ be a piecewise regular $T-$periodic function. Let a_n, b_n and $F_N f$ be as in the previous definition. Then the limit $F f(x)$ exists for every $x \in \mathbb{R}$ and*

$$F f(x) = \frac{f(x + 0) + f(x - 0)}{2} \quad \forall x \in \mathbb{R}$$

(where $f(x+0) = \lim\limits_{\substack{y \to x \\ y > x}} f(y)$ and $f(x-0) = \lim\limits_{\substack{y \to x \\ y < x}} f(y)$). Hence, in particular, if f is continuous at x, then

$$Ff(x) = f(x).$$

Moreover, if f' is continuous, then the convergence to f of the above series is uniform.

Corollary 14.4 *Let $f : \mathbb{R} \to \mathbb{R}$ be a piecewise regular $T-$periodic function. Then Ff is $T-$periodic and the three following statements hold true.*

(i) If f is **even** *(i.e. $f(x) = f(-x)$, $\forall x$), then $b_n = 0$ and*

$$Ff(x) = \frac{a_0}{2} + \sum_{n=1}^{\infty} a_n \cos\left(\frac{2\pi n}{T}x\right).$$

(ii) If f is **odd** *(i.e. $f(x) = -f(-x)$, $\forall x$), then $a_n = 0$ and*

$$Ff(x) = \sum_{n=1}^{\infty} b_n \sin\left(\frac{2\pi n}{T}x\right).$$

(iii) The series can be written as a **complex series**

$$Ff(x) = \sum_{n=-\infty}^{\infty} c_n e^{i\frac{2\pi n}{T}x} \quad \text{where } c_n = \frac{1}{T} \int_0^T f(x) e^{-i\frac{2\pi n}{T}x} dx.$$

Theorem 14.5 (Term by term differentiation) *Let $f : \mathbb{R} \to \mathbb{R}$ be a continuous $T-$periodic function such that f' is piecewise regular. Let*

$$\frac{a_0}{2} + \sum_{n=1}^{\infty} \left\{ a_n \cos\left(\frac{2\pi n}{T}x\right) + b_n \sin\left(\frac{2\pi n}{T}x\right) \right\}$$

be its Fourier series. Then the series obtained through term by term differentiation of the Fourier series of f converges and $\forall x \in \mathbb{R}$

$$\frac{f'(x+0) + f'(x-0)}{2} = \sum_{n=1}^{\infty} \frac{2\pi n}{T} \left\{ b_n \cos\left(\frac{2\pi n}{T}x\right) - a_n \sin\left(\frac{2\pi n}{T}x\right) \right\}.$$

Theorem 14.6 (Term by term integration) *Let $f : \mathbb{R} \to \mathbb{R}$ be a piecewise regular $T-$periodic function. Let*

$$\frac{a_0}{2} + \sum_{n=1}^{\infty} \left\{ a_n \cos\left(\frac{2\pi n}{T}x\right) + b_n \sin\left(\frac{2\pi n}{T}x\right) \right\}$$

be its Fourier series. Then, for every $x_0, x \in [0, T]$ the following holds

$$\int_{x_0}^{x} f(t)\, dt = \int_{x_0}^{x} \frac{a_0}{2}\, dt + \sum_{n=1}^{\infty} \int_{x_0}^{x} \left\{ a_n \cos\left(\frac{2\pi n}{T}t\right) + b_n \sin\left(\frac{2\pi n}{T}t\right) \right\} dt.$$

Furthermore, for fixed x_0, the convergence is uniform.

Theorem 14.7 (Parseval identity) *Let $f : \mathbb{R} \to \mathbb{R}$ be a piecewise regular $T-$periodic function. Then*

$$\frac{2}{T} \int_0^T (f(x))^2 dx = \frac{a_0^2}{2} + \sum_{n=1}^{\infty} (a_n^2 + b_n^2).$$

Corollary 14.8 (Cosine Fourier series) *Let $f : [0, L] \to \mathbb{R}$ be a piecewise regular function. Then the following series converges*

$$F_c f(x) = \frac{a_0}{2} + \sum_{n=1}^{\infty} a_n \cos\left(\frac{\pi n}{L}x\right)$$

where

$$a_n = \frac{2}{L} \int_0^L f(y) \cos\left(\frac{\pi n}{L}y\right) dy.$$

Moreover, for every $x \in (0, L)$ where f is continuous, the following holds

$$F_c f(x) = f(x).$$

Corollary 14.9 (Sine Fourier series) *Let $f : [0, L] \to \mathbb{R}$ be a piecewise regular function. Then the following series converges*

$$F_s f(x) = \sum_{n=1}^{\infty} b_n \sin\left(\frac{\pi n}{L}x\right)$$

where

$$b_n = \frac{2}{L} \int_0^L f(y) \sin\left(\frac{\pi n}{L} y\right) dy.$$

Furthermore, for every $x \in (0, L)$ where f is continuous the following holds

$$F_s f(x) = f(x).$$

(For more details, cf. [2] 61–73, [6] 239–293, [8] 51–64, [9] 53–69, [10] 582–605, [11] 262–281, [12] Chapters 1 to 4, [14] Chapters 2 and 3.)

14.2 Examples

Example 14.10 *Find the Fourier series of $f(x) = \cos x$ with $T = 2\pi$.*

Discussion We immediately note that the function being even, we have $b_n = 0$, $\forall n$. We also observe that $a_n = 0$, $\forall n \neq 1$ and

$$a_1 = \frac{1}{\pi} \int_0^{2\pi} \cos x \cos x \, dx = 1.$$

We therefore have

$$F_0 f = 0, \quad F_1 f = \cos x, \quad F_N f = F_1 f = F f = \cos x, \ \forall N.$$

Example 14.11 *Let*

$$f(x) = \begin{cases} 1 & \text{if } x \in [0, \pi) \\ 0 & \text{if } x \in [\pi, 2\pi) \end{cases}$$

be extended by 2π−periodicity to $\mathbb{R}$. Find the Fourier series of f, compare Ff and f. Evaluate the series

$$\sum_{n=0}^{\infty} \frac{(-1)^n}{2n+1}.$$

Discussion We first determine the Fourier coefficients, namely

$$a_0 = \frac{1}{\pi} \int_0^{\pi} dx = 1$$

$$a_n = \frac{1}{\pi} \int_0^{\pi} \cos(nx) \, dx = \frac{1}{\pi} \left[\frac{\sin(nx)}{n} \right]_0^{\pi} = 0, \quad \forall n \geq 1,$$

$$b_n = \frac{1}{\pi} \int_0^{\pi} \sin(nx)\, dx = \frac{1}{\pi} \left[\frac{-\cos(nx)}{n} \right]_0^{\pi}$$

$$= \frac{1}{n\pi}[1 - (-1)^n] = \begin{cases} 0 & \text{if } n \text{ is even} \\ \dfrac{2}{n\pi} & \text{if } n \text{ is odd.} \end{cases}$$

We therefore have

$$Ff(x) = \frac{1}{2} + \frac{2}{\pi} \sum_{n=0}^{\infty} \frac{\sin((2n+1)x)}{2n+1} = \begin{cases} 1 & \text{if } x \in (0, \pi) \\ 0 & \text{if } x \in (\pi, 2\pi) \\ 1/2 & \text{if } x = 0, \pi, 2\pi \end{cases}$$

and in particular, if $x = \pi/2$,

$$\sum_{n=0}^{\infty} \frac{(-1)^n}{2n+1} = \frac{\pi}{4}.$$

Example 14.12 *Find the Fourier series of*

$$f(x) = \begin{cases} x/2 & \text{if } x \in (-\pi, \pi) \\ 0 & \text{if } x = \pi \end{cases}$$

for $x \in (-\pi, \pi]$ and extended by $2\pi-$periodicity. Compare Ff and f. Compute

$$\sum_{n=1}^{\infty} \frac{1}{n^2}.$$

Discussion (i) We first observe that, since the function is odd, all the coefficients $a_n = 0$. For $n \geq 1$, we get

$$b_n = \frac{1}{\pi} \int_{-\pi}^{\pi} \frac{x}{2} \sin(nx)\, dx = \frac{2}{\pi} \int_0^{\pi} \frac{x}{2} \sin(nx)\, dx$$

$$= \frac{2}{\pi} \left[\frac{-x}{2n} \cos(nx) \right]_0^{\pi} + \frac{2}{\pi} \int_0^{\pi} \frac{\cos(nx)}{2n}\, dx = \frac{(-1)^{n-1}}{n}.$$

Since the hypotheses of Theorem 14.3 are satisfied, we find

$$Ff(x) = f(x) = \sum_{n=1}^{\infty} \frac{(-1)^{n-1}}{n} \sin(nx), \ \forall\, x \in [-\pi, \pi].$$

(ii) Parseval identity gives

$$\sum_{n=1}^{\infty} \frac{1}{n^2} = \frac{2}{2\pi} \int_{-\pi}^{\pi} \frac{x^2}{4}\, dx = \frac{1}{\pi} \left[\frac{x^3}{12} \right]_{-\pi}^{\pi} = \frac{\pi^2}{6}.$$

Example 14.13 *Let*

$$f(x) = \begin{cases} x & \text{if } 0 \le x \le \frac{\pi}{2} \\ \pi - x & \text{if } \frac{\pi}{2} < x \le \pi. \end{cases}$$

Find its Fourier series in sines. Compare $F_s f$ and f. Can we differentiate the series term by term?

Discussion We apply Corollary 14.9 with $L = \pi$ and get, for $n = 1, 2, \cdots$,

$$b_n = \frac{2}{\pi} \int_0^\pi f(x) \sin(nx) \, dx = \frac{4}{n^2 \pi} \sin(n\pi/2).$$

We thus find $b_{2n} = 0$ and

$$b_{2n+1} = \frac{4(-1)^n}{(2n+1)^2 \pi}.$$

We therefore obtain

$$F_s f(x) = \sum_{n=0}^\infty \frac{4(-1)^n}{(2n+1)^2 \pi} \sin((2n+1)x).$$

The function f being continuous on $[0, \pi]$ and $f(0) = f(\pi) = 0$, we can extend it in an odd way to $[-\pi, 0]$ and then by 2π−periodicity to $\mathbb{R}$. The function that we obtain in this way is therefore continuous and its derivative is piecewise regular. We can therefore apply Theorem 14.5 and differentiate term by term its Fourier series. We thus deduce that, for every $x \in (0, \pi)$ except $x = \pi/2$,

$$f'(x) = \sum_{n=0}^\infty \frac{4(-1)^n}{(2n+1)\pi} \cos((2n+1)x).$$

Example 14.14 *Let f be the 4π−periodic function defined by*

$$f(x) = \begin{cases} 0 & \text{if } x = -2\pi \\ -(\pi + x)/2 & \text{if } -2\pi < x < 0 \\ 0 & \text{if } x = 0 \\ (\pi - x)/2 & \text{if } 0 < x < 2\pi \\ 0 & \text{if } x = 2\pi. \end{cases}$$

Find its Fourier series. Compare Ff and f. Can we differentiate the series term by term?

Discussion Since f is odd, we have that $a_n = 0$ and

$$b_n = \frac{2}{4\pi} \int_{-2\pi}^{2\pi} f(x) \sin(nx)\, dx = \frac{1}{\pi} \int_0^{2\pi} f(x) \sin(nx)\, dx$$

$$= \frac{1}{\pi} \int_0^{2\pi} \frac{\pi - x}{2} \sin(nx)\, dx = \frac{1}{n}.$$

We therefore get

$$Ff(x) = f(x) = \sum_{n=1}^{\infty} \frac{\sin(nx)}{n}, \quad \forall x \in [-2\pi, 2\pi].$$

Differentiating the series term by term, we find, for $x \in (-2\pi, 2\pi)$,

$$\frac{d}{dx} Ff(x) = \sum_{n=1}^{\infty} \cos(nx) \quad \text{although } f'(x) = \frac{-1}{2} \text{ if } x \neq 0.$$

The above series clearly diverges. We have, however, no contradiction with Theorem 14.5, since f is not continuous at $x = 2n\pi$, $n \in \mathbb{Z}$.

14.3   Exercises

Exercise 14.1 (i) Find the Fourier series of $f(x) = e^{(x-\pi)}$ on $[0, 2\pi)$ and extended by 2π−periodicity.

(ii) Invoking Dirichlet theorem compare Ff and f on $(0, 2\pi)$.

(iii) Using (i) and (ii) prove that

$$\sum_{n=2}^{\infty} \frac{(-1)^n}{1 + n^2} = \frac{\pi}{e^\pi - e^{-\pi}}.$$

Exercise 14.2 (i) Find the Fourier series of $f(x) = (x - \pi)^2$ on $[0, 2\pi)$ and extended by 2π−periodicity.

(ii) Appealing to Dirichlet theorem compare Ff and f on $[0, 2\pi]$.

(iii) Deduce that

$$\sum_{n=1}^{\infty} \frac{(-1)^n}{n^2} = -\frac{\pi^2}{12} \quad \text{and} \quad \sum_{n=1}^{\infty} \frac{1}{n^2} = \frac{\pi^2}{6}.$$

Exercise 14.3 Let $f : \mathbb{R} \to \mathbb{R}$ be the 2π−periodic function defined by

$$f(t) = \begin{cases} \sin t & \text{if } 0 \leq t \leq \dfrac{\pi}{2} \\ 0 & \text{if } \dfrac{\pi}{2} < t < \dfrac{3\pi}{2} \\ \sin t & \text{if } \dfrac{3\pi}{2} \leq t < 2\pi. \end{cases}$$

Find its Fourier series.

Exercise 14.4 Let $f : \mathbb{R} \to \mathbb{R}$ be the 2−periodic function defined by

$$f(t) = t \quad \text{if } t \in [0, 2).$$

Find its complex Fourier series.

Exercise 14.5 (i) Find the Fourier series for the 2π−periodic and odd function defined by

$$f(x) = x(\pi - x) \quad \text{if } x \in [0, \pi].$$

(ii) Invoking (i) and Parseval identity, evaluate

$$\sum_{n=1}^{\infty} \frac{1}{(2n - 1)^6}.$$

Exercise 14.6 Appealing to Parseval identity, prove that

$$\int_{-\pi}^{\pi} \cos^4 x \, dx = \frac{3}{4}\pi.$$

Exercise 14.7 (i) Write the Fourier series of the 2π−periodic function defined by

$$f(x) = |x| \quad \text{if } x \in [-\pi, \pi).$$

(ii) Evaluate the following two series

$$\sum_{n=1}^{\infty} \frac{1}{n^4} \quad \text{and} \quad \sum_{n=0}^{\infty} \frac{1}{(2n + 1)^4}.$$

Exercise 14.8 Find the Fourier series of $f(x) = |\cos x|$ and compute

$$\sum_{k=1}^{\infty} \frac{1}{4k^2 - 1}.$$

Exercise 14.9 Let f be the 2π−periodic and even function such that

$$f(t) = \sin(3t) \quad \text{if } t \in [0, \pi].$$

Find its Fourier series.

Exercise 14.10 (i) Let $\alpha \in \mathbb{R} \backslash \mathbb{Z}$ be given. Find the Fourier series of the 2π−periodic function defined by

$$f(x) = \cos(\alpha x) \quad \text{if } x \in [-\pi, \pi).$$

(ii) Deduce that

$$\sum_{n=1}^{\infty} \frac{1}{n^2 - \alpha^2} = \frac{1}{2\alpha^2} - \frac{\pi}{2\alpha \tan(\alpha \pi)}.$$

Exercise 14.11 (i) Find the complex Fourier series of the 2π−periodic and odd function given on $[0, \pi]$ by

$$f(x) = \begin{cases} x & \text{if } 0 \leq x \leq \frac{\pi}{2} \\ \pi - x & \text{if } \frac{\pi}{2} < x \leq \pi. \end{cases}$$

(ii) Deduce that for $a > 0$ the following result holds, for every $x \in [-a, a]$,

$$x = \frac{4}{\pi^2} ia \sum_{k=-\infty}^{+\infty} \frac{(-1)^k}{(2k-1)^2} e^{\frac{\pi i x (2k-1)}{2a}}.$$

(iii) Evaluate

$$\sum_{k=-\infty}^{+\infty} \frac{1}{(2k-1)^2}.$$

Exercise 14.12 Let f be the 2π−periodic function such that

$$f(x) = x + \pi \quad \text{if } -\pi \leq x < \pi.$$

(i) Compute

$$|f(x) - F_3 f(x)| \quad \text{for } x = -\pi, -\frac{\pi}{2}, 0, \frac{\pi}{2}, \pi.$$

(ii) Evaluate (with the help of Exercise* 14.13 below)

$$\int_0^{2\pi} |f(x) - F_3 f(x)|^2 \, dx.$$

Exercise* 14.13 Let $f : \mathbb{R} \to \mathbb{R}$ be a piecewise regular 2π−periodic function. Let a_n and b_n be its Fourier coefficients and

$$F_N f(x) = \frac{a_0}{2} + \sum_{n=1}^{N} (a_n \cos(nx) + b_n \sin(nx)).$$

(i) Prove that

$$\int_0^{2\pi} |f(x) - F_N f(x)|^2 dx = \int_0^{2\pi} (f(x))^2 dx - \pi \left[\frac{a_0^2}{2} + \sum_{n=1}^{N} (a_n^2 + b_n^2) \right].$$

(ii) Establish **Bessel inequality**

$$\frac{a_0^2}{2} + \sum_{n=1}^{\infty} (a_n^2 + b_n^2) \leq \frac{1}{\pi} \int_0^{2\pi} (f(x))^2 dx.$$

Hint: (i) Show that, for every $k = 0, 1, \cdots, N$,

$$\int_0^{2\pi} F_N f(x) \cos(kx)\, dx = \pi a_k$$

$$\int_0^{2\pi} F_N f(x) \sin(kx)\, dx = \pi b_k.$$

(ii) Observing that

$$\int_0^{2\pi} F_N f(x) \cos(kx)\, dx = \frac{a_0}{2} \int_0^{2\pi} \cos(kx)\, dx$$

$$+ \sum_{n=1}^{N} a_n \int_0^{2\pi} \cos(nx) \cos(kx)\, dx$$

$$+ \sum_{n=1}^{N} b_n \int_0^{2\pi} \sin(nx) \cos(kx)\, dx$$

prove that

$$\int_0^{2\pi} (F_N f(x))^2 \, dx = \pi \left\{ \frac{a_0^2}{2} + \sum_{n=1}^{N} (a_n^2 + b_n^2) \right\}.$$

(iii) Show that

$$\int_0^{2\pi} f(x) \, F_N f(x) \, dx = \pi \left\{ \frac{a_0^2}{2} + \sum_{n=1}^{N} \left(a_n^2 + b_n^2 \right) \right\}.$$

(iv) Deduce the result from the two previous observations.

Exercise* 14.14 Let $f : \mathbb{R} \to \mathbb{R}$ be a continuous T-periodic function. Let a_n and b_n be its Fourier coefficients.

(i) Prove that if $f \in C^1$, then there exists a constant $c > 0$ such that

$$|a_n|, \; |b_n| \leq \frac{c}{n}, \quad n = 1, 2, \cdots$$

(ii) More generally, if $k \geq 1$ and if f is C^k, show that there exists a constant $c > 0$ such that

$$|a_n|, \; |b_n| \leq \frac{c}{n^k}, \quad n = 1, 2, \cdots$$

Chapter 15

Fourier transform

15.1 Definitions and theoretical results

Definition 15.1 *Let* $f : \mathbb{R} \to \mathbb{R}$ *be a piecewise continuous function (cf. Definition 14.1) and such that* $\int_{-\infty}^{+\infty} |f(x)| \, dx < \infty$. *The **Fourier transform** of* f *is defined by*

$$\mathfrak{F}(f)(\alpha) = \widehat{f}(\alpha) = \frac{1}{\sqrt{2\pi}} \int_{-\infty}^{+\infty} f(y) \, e^{-i\alpha y} \, dy.$$

Remark Some authors define the Fourier transform by replacing the coefficient $1/\sqrt{2\pi}$ by 1 or by $1/2\pi$ and sometimes $e^{-i\alpha y}$ by $e^{-2\pi i \alpha y}$.

Theorem 15.2 *Let* $f, g : \mathbb{R} \to \mathbb{R}$ *be piecewise continuous functions such that*

$$\int_{-\infty}^{+\infty} |f(x)| \, dx < \infty \quad and \quad \int_{-\infty}^{+\infty} |g(x)| \, dx < \infty.$$

The following properties then hold.

*(i) **Continuity**. The function* $\widehat{f} : \mathbb{R} \to \mathbb{C}$ *is then continuous and*

$$\lim_{|\alpha| \to \infty} \left| \widehat{f}(\alpha) \right| = 0.$$

*(ii) **Linearity**. Let* $a, b \in \mathbb{R}$, *then*

$$\mathfrak{F}(af + bg) = a \, \mathfrak{F}(f) + b \, \mathfrak{F}(g).$$

121

(iii) **Differentiation.** *If, moreover, $f \in C^1(\mathbb{R})$ and $\int_{-\infty}^{+\infty} |f'(x)| dx < \infty$, then*

$$\mathfrak{F}(f')(\alpha) = i\alpha \, \mathfrak{F}(f)(\alpha), \quad \forall \alpha \in \mathbb{R}.$$

More generally, if $n \in \mathbb{N}$, $f \in C^n(\mathbb{R})$ and

$$\int_{-\infty}^{+\infty} |f^{(k)}(x)| dx < \infty \quad \text{for every } k = 1, 2, \cdots, n,$$

then

$$\boxed{\mathfrak{F}\left(f^{(n)}\right)(\alpha) = (i\alpha)^n \, \mathfrak{F}(f)(\alpha), \quad \forall \alpha \in \mathbb{R}.}$$

(iv) **Shift.** *If $a, b \in \mathbb{R}$, $a \neq 0$, and*

$$h(x) = e^{-ibx} f(ax)$$

then

$$\boxed{\mathfrak{F}(h)(\alpha) = \frac{1}{|a|} \mathfrak{F}(f)\left(\frac{\alpha + b}{a}\right).}$$

(v) **Convolution.** *Let the convolution product be defined by*

$$f * g(x) = \int_{-\infty}^{\infty} f(x - t)g(t) \, dt$$

then

$$\boxed{\mathfrak{F}(f * g) = \sqrt{2\pi}\, \mathfrak{F}(f)\, \mathfrak{F}(g).}$$

(vi) **Plancherel identity.** *If, in addition,*

$$\int_{-\infty}^{+\infty} (f(x))^2 \, dx < \infty,$$

then

$$\boxed{\int_{-\infty}^{+\infty} (f(x))^2 dx = \int_{-\infty}^{+\infty} \left|\widehat{f}(\alpha)\right|^2 d\alpha.}$$

Theorem 15.3 *Let $f : \mathbb{R} \to \mathbb{R}$ be a continuous function such that*

$$\int_{-\infty}^{+\infty} |f(x)|\,dx < \infty \quad and \quad \int_{-\infty}^{+\infty} \left|\widehat{f}(\alpha)\right| d\alpha < \infty.$$

(i) Inversion formula. The following formula holds

$$\boxed{f(x) = \frac{1}{\sqrt{2\pi}} \int_{-\infty}^{+\infty} \widehat{f}(\alpha)\, e^{i\alpha x} d\alpha.}$$

(ii) Cosine Fourier transform. If f is even, then

$$\mathfrak{F}(f)(\alpha) = \sqrt{\frac{2}{\pi}} \int_{0}^{+\infty} f(y) \cos(\alpha y)\, dy$$

and

$$f(x) = \sqrt{\frac{2}{\pi}} \int_{0}^{+\infty} \widehat{f}(\alpha) \cos(\alpha x)\, d\alpha.$$

(iii) Sine Fourier transform. If f is odd, then

$$\mathfrak{F}(f)(\alpha) = -i\sqrt{\frac{2}{\pi}} \int_{0}^{+\infty} f(y) \sin(\alpha y)\, dy$$

and

$$f(x) = i\sqrt{\frac{2}{\pi}} \int_{0}^{+\infty} \widehat{f}(\alpha) \sin(\alpha x)\, d\alpha.$$

(For more details, cf. [2] 79–89, [6] 467–484, 503, [10] 617–635, [12] 246–264, [13] 1–9, [14] Chapter 5, [16] 202–204.)

15.2 Examples

Example 15.4 *Let*

$$f(x) = \begin{cases} 1 & if\ |x| \leq 1 \\ 0 & otherwise. \end{cases}$$

Find the Fourier transform of f.

Discussion Observe first that f is piecewise regular and that

$$\int_{-\infty}^{+\infty} |f(x)|\,dx = \int_{-1}^{1} dx = 2 < \infty.$$

We have

$$\widehat{f}(\alpha) = \frac{1}{\sqrt{2\pi}} \int_{-\infty}^{+\infty} f(y) e^{-i\alpha y} dy = \frac{1}{\sqrt{2\pi}} \int_{-1}^{1} e^{-i\alpha y} dy$$

$$= \frac{1}{\sqrt{2\pi}} \left[-\frac{e^{-i\alpha y}}{i\alpha} \right]_{-1}^{1} = \frac{2}{\alpha\sqrt{2\pi}} \frac{e^{i\alpha} - e^{-i\alpha}}{2i}$$

and thus

$$\widehat{f}(\alpha) = \sqrt{\frac{2}{\pi}} \frac{\sin \alpha}{\alpha}.$$

Example 15.5 *Let $f(x) = e^{-|x|}$. Find the Fourier transform of f. Evaluate*

$$\int_{-\infty}^{+\infty} \frac{\cos x}{1 + x^2} dx.$$

Discussion We have that $f \in C^1$ except at 0. Indeed f is continuous at 0 and

$$f'(x) = \begin{cases} -e^{-x} & \text{if } x > 0 \\ e^{x} & \text{if } x < 0 \end{cases}$$

is piecewise continuous. We also have

$$\int_{-\infty}^{+\infty} |f(x)| dx = \int_{-\infty}^{+\infty} e^{-|x|} dx = 2 \int_{0}^{+\infty} e^{-x} dx = 2 < \infty.$$

We can therefore compute $\widehat{f}$ and find

$$\widehat{f}(\alpha) = \frac{1}{\sqrt{2\pi}} \int_{-\infty}^{+\infty} f(y) e^{-i\alpha y} dy = \frac{1}{\sqrt{2\pi}} \int_{-\infty}^{+\infty} e^{-|y|} e^{-i\alpha y} dy$$

$$= \frac{1}{\sqrt{2\pi}} \int_{-\infty}^{0} e^{y(1-i\alpha)} dy + \frac{1}{\sqrt{2\pi}} \int_{0}^{+\infty} e^{-y(1+i\alpha)} dy$$

$$= \frac{1}{\sqrt{2\pi}} \left\{ \left[\frac{e^{y(1-i\alpha)}}{(1-i\alpha)} \right]_{-\infty}^{0} - \left[\frac{e^{-y(1+i\alpha)}}{(1+i\alpha)} \right]_{0}^{+\infty} \right\}.$$

Since $\alpha \in \mathbb{R}$, we get

$$\lim_{y \to +\infty} \left| e^{-(1+i\alpha)y} \right| = \lim_{y \to +\infty} e^{-y} = 0$$

$$\lim_{y \to -\infty} \left| e^{(1-i\alpha)y} \right| = \lim_{y \to -\infty} e^{y} = 0$$

and thus

$$\hat{f}(\alpha) = \sqrt{\frac{2}{\pi}} \frac{1}{1+\alpha^2}.$$

Note that

$$\int_{-\infty}^{+\infty} \left| \hat{f}(\alpha) \right| d\alpha < \infty$$

and therefore Theorem 15.3 (i) leads to

$$e^{-|x|} = \frac{1}{\sqrt{2\pi}} \int_{-\infty}^{+\infty} \sqrt{\frac{2}{\pi}} \frac{e^{i\alpha x}}{1+\alpha^2} \, d\alpha = \frac{1}{\pi} \int_{-\infty}^{+\infty} \frac{e^{i\alpha x}}{1+\alpha^2} \, d\alpha.$$

In particular, if $x = 0$, we have $f(0) = 1$ and hence

$$1 = \frac{1}{\pi} \int_{-\infty}^{+\infty} \frac{d\alpha}{1+\alpha^2}.$$

If $x = 1$, we find

$$\frac{1}{e} = \frac{1}{\pi} \int_{-\infty}^{+\infty} \frac{e^{i\alpha}}{1+\alpha^2} \, d\alpha = \frac{1}{\pi} \int_{-\infty}^{+\infty} \frac{\cos\alpha}{1+\alpha^2} \, d\alpha + \frac{i}{\pi} \int_{-\infty}^{+\infty} \frac{\sin\alpha}{1+\alpha^2} \, d\alpha.$$

We therefore deduce that

$$\int_{-\infty}^{+\infty} \frac{\sin\alpha}{1+\alpha^2} \, d\alpha = 0 \quad \text{and} \quad \frac{\pi}{e} = \int_{-\infty}^{+\infty} \frac{\cos\alpha}{1+\alpha^2} \, d\alpha.$$

15.3   Exercises

Exercise 15.1 Find the Fourier transform of the function

$$f(x) = \begin{cases} 0 & \text{if } x \leq 0 \\ e^{-x} & \text{if } x > 0. \end{cases}$$

Exercise 15.2 Let $f : \mathbb{R}_+ \to \mathbb{R}$ be given by

$$f(x) = e^{-x} \cos x.$$

(i) Find the cosine Fourier transform of f (extended as an even function to $\mathbb{R}$).

(ii) Find the sine Fourier transform of f (extended as an odd function to $\mathbb{R}$).

Exercise 15.3 Prove (iii) of Theorem 15.2 under the further hypotheses

$$\lim_{|y|\to\infty} \left| f^{(k)}(y) \right| = 0, \quad k = 0, 1, \cdots, n-1.$$

Exercise 15.4 Show (iv) of Theorem 15.2.

Exercise 15.5 Prove (ii) and (iii) of Theorem 15.3.

Exercise 15.6 Under the hypotheses of Theorem 15.3, show that if f is continuous and even, then

$$\mathfrak{F}(\mathfrak{F}(f))(t) = \mathfrak{F}(\widehat{f})(t) = \widehat{\widehat{f}}(t) = f(t).$$

Hint: observe that if f is even, then $\widehat{f}$ is even.

Exercise 15.7 Let
$$g(x) = x^n f(x).$$

Assuming that all formal computations are allowed (in particular, we assume that the function g satisfy the hypotheses of Definition 15.1), prove that
$$\mathfrak{F}(g)(\alpha) = i^n \mathfrak{F}^{(n)}(f)(\alpha).$$

Exercise* 15.8 Show (v) of Theorem 15.2.

Chapter 16

Laplace transform

16.1 Definitions and theoretical results

Definition 16.1 *Let* $f : \mathbb{R}_+ = [0, +\infty) \to \mathbb{R}$ *be a piecewise continuous function (extended to* $\mathbb{R}$ *so that* $f(x) = 0$, $\forall x < 0$*) and* $\gamma_0 \in \mathbb{R}$ *be such that*

$$\int_0^{+\infty} |f(t)| e^{-\gamma_0 t} \, dt < \infty$$

(γ_0 *is called an **abscissa of convergence** of* f*). The **Laplace transform** of* f *is defined by*

$$\boxed{\mathfrak{L}(f)(z) = F(z) = \int_0^{+\infty} f(t) e^{-tz} \, dt, \quad \forall z \in \overline{O}}$$

where

$$O = \{z \in \mathbb{C} : \operatorname{Re} z > \gamma_0\} \quad \text{and} \quad \overline{O} = \{z \in \mathbb{C} : \operatorname{Re} z \geq \gamma_0\}.$$

Theorem 16.2 *Let* f, γ_0 *and* O *be as in the previous definition. Let* $g : \mathbb{R}_+ \to \mathbb{R}$ *be piecewise continuous (with* $g(x) = 0$, $\forall x < 0$*) such that*

$$\int_0^{+\infty} |g(t)| e^{-\gamma_0 t} \, dt < \infty.$$

Then the following properties hold.

*(i) **Holomorphy**. The Laplace transform* F *of* f *is holomorphic in* O

and

$$F'(z) = -\int_0^{+\infty} tf(t)\, e^{-tz}\, dt = -\mathfrak{L}(h)(z), \quad \forall\, z \in O$$

where $h(t) = tf(t)$.

    (ii) **Linearity.** *Let* $a, b \in \mathbb{R}$, *then*

$$\mathfrak{L}(af + bg) = a\,\mathfrak{L}(f) + b\,\mathfrak{L}(g).$$

    (iii) **Differentiation.** *If, moreover,* $f \in C^1(\mathbb{R}_+)$ *and*

$$\int_0^{+\infty} |f'(t)|\, e^{-\gamma_0 t}\, dt < \infty,$$

then

$$\mathfrak{L}(f')(z) = z\,\mathfrak{L}(f)(z) - f(0), \quad \forall\, z \in O.$$

More generally, if $n \in \mathbb{N}$, $f \in C^n(\mathbb{R}_+)$ *and*

$$\int_0^{+\infty} |f^{(k)}(t)|\, e^{-\gamma_0 t}\, dt < \infty, \quad \text{for every } k = 0, 1, 2, \cdots, n,$$

then

$$\mathfrak{L}(f^{(n)})(z) = z^n\,\mathfrak{L}(f)(z) - \sum_{k=0}^{n-1} z^k\, f^{(n-k-1)}(0), \quad \forall\, z \in O.$$

    (iv) **Integration.** *If* $f \in C(\mathbb{R})$, $\gamma_0 \geq 0$ *and*

$$\varphi(t) = \int_0^t f(s)\, ds,$$

then

$$\mathfrak{L}(\varphi)(z) = \frac{\mathfrak{L}(f)(z)}{z}, \quad \forall\, z \in O.$$

*(v) **Shift**. If $a > 0$, $b \in \mathbb{R}$ and*

$$\varphi(t) = e^{-bt} f(at),$$

then

$$\mathfrak{L}(\varphi)(z) = \frac{1}{a} \mathfrak{L}(f)\left(\frac{z+b}{a}\right), \quad \forall z \text{ such that } \operatorname{Re}\left(\frac{z+b}{a}\right) \geq \gamma_0.$$

*(vi) **Convolution**. Define the convolution product as*

$$f * g(t) = \int_{-\infty}^{+\infty} f(t-s)g(s)\,ds = \int_0^t f(t-s)g(s)\,ds.$$

Then

$$\mathfrak{L}(f * g)(z) = \mathfrak{L}(f)(z)\,\mathfrak{L}(g)(z), \quad \forall z \in O.$$

Theorem 16.3 (Inversion formula) *Let f be a continuous function such that $f(0) = 0$ ($f(t) \equiv 0$ if $t < 0$) and let $F(z) = \mathfrak{L}(f)(z)$ be its Laplace transform. If*

$$\int_0^{+\infty} |f(t)|e^{-\gamma t}\,dt < \infty \quad \text{and} \quad \int_{-\infty}^{+\infty} |F(\gamma + is)|\,ds < \infty$$

for a certain $\gamma \in \mathbb{R}$, then

$$\frac{1}{2\pi} \int_{-\infty}^{+\infty} F(\gamma + is)e^{(\gamma + is)t}\,ds = f(t), \quad \forall t \in \mathbb{R}.$$

(For more details, cf. [2] 91–109, [3] 75–83, [6] 529–549, [10] 242–300, [16] 35–70.)

16.2   Examples

Example 16.4 *Let $f(t) \equiv 1$ for $t \geq 0$. Find the Laplace transform of f.*

Discussion We first observe that any $\gamma_0 > 0$ is an abscissa of convergence since

$$\int_0^{+\infty} e^{-\gamma_0 t}\, dt = -\frac{e^{-\gamma_0 t}}{\gamma_0}\Big|_0^{+\infty} = \frac{1}{\gamma_0} < \infty.$$

We then find

$$\mathfrak{L}\left(f\right)(z) = \int_0^{+\infty} e^{-tz}\, dt = -\frac{e^{-tz}}{z}\Big|_0^{+\infty} = \frac{1}{z}, \quad \text{if } \operatorname{Re} z > 0.$$

Example 16.5 *Find the Laplace transform of $f(t) = e^{at}$ if $t \geq 0$ and $a \in \mathbb{R}$.*

Discussion We immediately find

$$\mathfrak{L}\left(f\right)(z) = \int_0^{+\infty} e^{(a-z)t}\, dt = \frac{e^{(a-z)t}}{a-z}\Big|_0^{+\infty} = \frac{1}{z-a}, \quad \text{if } \operatorname{Re} z > a.$$

Example 16.6 *Find the Laplace transform of $f(t) = t^2$ if $t \geq 0$.*

Discussion Note that since $f(0) = 0$, $f'(0) = 0$ and $f''(t) = 2$, then

$$\mathfrak{L}(f'')(z) = z^2 \mathfrak{L}(f)(z) - z f(0) - f'(0).$$

Invoking Example 16.4, we find

$$\mathfrak{L}(2)(z) = \mathfrak{L}(f'')(z) = \frac{2}{z},$$

and thus

$$\mathfrak{L}(f)(z) = \frac{2}{z^3}, \quad \text{if } \operatorname{Re} z > 0.$$

Example 16.7 *Find, appealing to the residue theorem (cf. Theorem 12.1), a function f whose Laplace transform is*

$$F(z) = \frac{1}{(z+1)(z+2)^2}.$$

Discussion This example could be treated in a more elementary way, but we want to present here a more powerful and more general method. We proceed into three steps.

   Step 1. Define, for $t > 0$ fixed, the function

$$\widetilde{F}(z) = F(z)\, e^{zt}.$$

We first find the singularities of $\widetilde{F}$ and calculate their residue. The singularities are in $z = -1$ (pole of order 1) and $z = -2$ (pole of order 2). We easily have

$$\text{Res}_{-1}(\widetilde{F}) = \lim_{z \to -1} (z+1)\widetilde{F}(z) = \lim_{z \to -1} \frac{e^{zt}}{(z+2)^2} = e^{-t}$$

$$\text{Res}_{-2}(\widetilde{F}) = \lim_{z \to -2} \frac{d}{dz}\left[(z+2)^2 \widetilde{F}(z)\right] = \lim_{z \to -2} \frac{d}{dz}\frac{e^{zt}}{(z+1)}$$

$$= \lim_{z \to -2} \frac{te^{zt}(z+1) - e^{zt}}{(z+1)^2} = -te^{-2t} - e^{-2t}.$$

Step 2. We choose $\gamma \in \mathbb{R}$ such that all singularities of F are to the left of the line $\text{Re}\, z = \gamma$. In the particular case that we consider we can take, for example, $\gamma = 0$. We next choose $r > 0$ sufficiently large so that all singularities of F (assumed to be a finite number) are in the interior of $\Gamma_{\gamma,r} = C_{\gamma,r} \cup L_{\gamma,r}$ (cf. Figure 16.1), where

$$C_{\gamma,r} = \{z \in \mathbb{C} : |z - \gamma| = r \text{ and } \text{Re}\, z < \gamma\}$$

$$L_{\gamma,r} = \{z \in \mathbb{C} : \text{Re}\, z = \gamma \text{ and } -r < \text{Im}\, z < r\}.$$

The residue theorem gives

$$\int_{\Gamma_{\gamma,r}} \widetilde{F}(z)\, dz = 2\pi i \sum \text{Res}_{z_k}(f) = 2\pi i \left(e^{-t} - te^{-2t} - e^{-2t}\right). \tag{16.1}$$

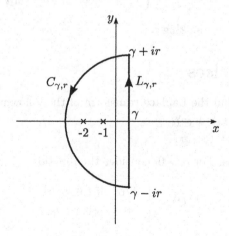

Figure 16.1: Example 16.7

Step 3. Since $t > 0$, it is not too difficult to show that

$$\lim_{r \to \infty} \int_{C_{\gamma,r}} \widetilde{F}(z)\, dz = 0. \tag{16.2}$$

This last result is true more generally and can be proved for every function F satisfying, for some constants $a > 0$ and $k > 1$, and for r sufficiently large (in our particular case $k = 3$)

$$|F(z)| \leq \frac{a}{r^k}, \quad \forall z \in C_{\gamma,r}.$$

Since $\Gamma_{\gamma,r} = C_{\gamma,r} \cup L_{\gamma,r}$ and

$$\int_{L_{\gamma,r}} \widetilde{F}(z)\, dz = i \int_{-r}^{+r} F(\gamma + is)\, e^{(\gamma+is)t}\, ds,$$

we obtain, combining (16.1) and (16.2), that (we consider here the particular case where $\gamma = 0$)

$$\frac{1}{2\pi} \int_{-\infty}^{+\infty} F(is)\, e^{ist}\, ds = \frac{1}{2\pi i} \lim_{r \to \infty} \int_{L_{\gamma,r}} \widetilde{F}(z)\, dz = \frac{1}{2\pi i} \int_{\Gamma_{\gamma,r}} \widetilde{F}(z)\, dz$$
$$= e^{-t} - te^{-2t} - e^{-2t}.$$

Appealing to Theorem 16.3, we find that

$$\mathcal{L}^{-1}(F)(t) = f(t) = \begin{cases} e^{-t} - te^{-2t} - e^{-2t} & \text{if } t \geq 0 \\ 0 & \text{if } t < 0 \end{cases}$$

is the function we are looking for.

16.3 Exercises

Exercise 16.1 Find the Laplace transform of the following functions.
  (i) $f(t) = \cos(kt)$, $k \in \mathbb{N}$.
  (ii) $f(t) = t\, e^{\alpha t}$, $\alpha \in \mathbb{R}$.
  (iii) Dirac mass. For $\alpha > 0$, consider the function

$$f_\alpha(t) = \begin{cases} 1/\alpha & \text{if } t \in [0, \alpha] \\ 0 & \text{otherwise.} \end{cases}$$

Evaluate $\lim_{\alpha \to 0} \mathcal{L}(f_\alpha)$.

Exercise 16.2 Find the function f such that its Laplace transform is

$$F(z) = \frac{1}{(z-a)(z-b)}, \quad a \neq b.$$

(One can here proceed directly without appealing to the residue theorem.)

Exercise 16.3 Find the Laplace transform of

$$f(t) = e^{-2t}\left(3\cos\left(6t\right) - 5\sin\left(6t\right)\right).$$

Exercise 16.4 Find the inverse Laplace transform (if possible without using the residue theorem) of

$$F(z) = \frac{4z}{z^2 + 64} \quad \text{and} \quad F(z) = \frac{z}{(z+1)(z+2)}.$$

Exercise 16.5 Find, invoking the residue theorem, the inverse Laplace transform of

$$F(z) = \frac{z^2}{(z^2+1)^2}.$$

Exercise 16.6 Prove (i) of Theorem 16.2.

Exercise 16.7 Show (iii) of Theorem 16.2, under the further hypotheses

$$\left|f^{(k)}(t)\right|e^{-\gamma_0 t} \leq c, \quad \forall t \geq 0 \text{ and } \forall k = 0, \cdots, n-1$$

for some constant $c > 0$.

Exercise 16.8 Establish (iv) of Theorem 16.2, under the further hypothesis

$$\int_0^{+\infty} |\varphi(t)|e^{-\gamma_0 t}\,dt < \infty.$$

Exercise 16.9 Prove (v) of Theorem 16.2.

Exercise 16.10 Let f be as in Theorem 16.2 and

$$|f(t)|e^{-\gamma_0 t} \leq c, \quad \forall t \geq 0$$

for some constant $c > 0$. If $\mathfrak{L}(f)(z) = F(z)$, prove that

$$\lim_{\operatorname{Re} z \to \infty} F(z) = 0.$$

Exercise 16.11 Let f satisfy the hypotheses of Theorem 16.2 (iii) and let $\mathcal{L}(f)(z) = F(z)$. Show the following properties.

(i) **Initial value.** If, moreover,

$$\left| f'(t) \right| e^{-\gamma_0 t} \le c, \quad \forall t \ge 0$$

for some constant $c > 0$, then

$$\lim_{\operatorname{Re} z \to \infty} z\, F(z) = f(0).$$

(ii) **Final value.** If, furthermore,

$$\int_0^{+\infty} \left| f'(t) \right| dt < \infty$$

and $\lim_{t \to \infty} f(t)$ exists, then

$$\lim_{z \to 0,\ \operatorname{Re} z \ge 0} z\, F(z) = \lim_{t \to \infty} f(t).$$

Exercise* 16.12 Establish (vi) of Theorem 16.2.

Exercise* 16.13 (i) Let f be as in Theorem 16.2. Assume that f is T–periodic. Prove that

$$\mathcal{L}(f)(z) = \frac{1}{1 - e^{-zT}} \int_0^T e^{-tz} f(t)\, dt, \quad \operatorname{Re} z > 0.$$

(ii) Using (i), find the Laplace transform of the function

$$f(t) = \begin{cases} \sin t & \text{if } 0 \le t < \pi \\ 0 & \text{if } \pi \le t < 2\pi \end{cases}$$

extended by 2π–periodicity to $\mathbb{R}_+$.

Hint: recall that if $|q| < 1$, then

$$\sum_{n=0}^{\infty} q^n = \frac{1}{1 - q}.$$

Exercise* 16.14 Using the formula for the inverse Fourier transform (cf. Theorem 15.3), establish Theorem 16.3.

Chapter 17

Applications to ordinary differential equations

In the present and the next chapters we will show how to apply the results of the three preceding chapters to differential equations. The way of finding the solutions is not rigorous, however, the result is true and can be justified without too many difficulties.

17.1 Cauchy problem

Example 17.1 *We wish to find a solution $y = y(t)$ to*

$$\begin{cases} a_2 y''(t) + a_1 y'(t) + a_0 y(t) = f(t) & t > 0 \\ y(0) = y_0 \quad and \quad y'(0) = y_1 \end{cases}$$

$$(17.1)$$

where a_0, a_1, a_2, y_0, $y_1 \in \mathbb{R}$ are given constants ($a_2 \neq 0$) and $f : \mathbb{R}_+ \to \mathbb{R}$ is given ($f(t) \equiv 0$, if $t < 0$) with an appropriate regularity. We will treat in particular the case where $a_0 = a_2 = 1$, $a_1 = 0$, $f(t) = \sin t$, $y_0 = y_1 = 1$, i.e. the problem

$$\begin{cases} y''(t) + y(t) = \sin t & t > 0 \\ y(0) = y'(0) = 1. \end{cases}$$

Discussion We proceed in three steps.

Step 1. We denote the Laplace transforms of f and of y respectively by

$$F(z) = \mathfrak{L}(f)(z) \quad and \quad Y(z) = \mathfrak{L}(y)(z).$$

We then obtain (cf. Theorem 16.2)

$$\mathcal{L}(y'')(z) = z^2 \mathcal{L}(y)(z) - zy(0) - y'(0) = z^2 Y - zy_0 - y_1$$
$$\mathcal{L}(y')(z) = z\mathcal{L}(y)(z) - y(0) = zY - y_0.$$

We therefore get

$$\begin{aligned}
\mathcal{L}(a_2 y'' + a_1 y' + a_0 y) &= a_2(z^2 Y - zy_0 - y_1) + a_1(zY - y_0) + a_0 Y \\
&= \mathcal{L}(f)(z) = F(z)
\end{aligned}$$

and hence

$$Y(z) = \frac{F(z) + a_2 y_0 z + a_2 y_1 + a_1 y_0}{a_2 z^2 + a_1 z + a_0} = H(z).$$

In the particular case $F(z) = \dfrac{1}{z^2 + 1}$ and thus

$$H(z) = \frac{1}{(z^2 + 1)^2} + \frac{z+1}{z^2 + 1}.$$

Step 2. The solution is then obtained by taking the inverse Laplace transform of both sides, i.e.

$$y(t) = \mathcal{L}^{-1}(Y)(t) = \mathcal{L}^{-1}(H)(t).$$

In the example we have

$$\begin{aligned}
y(t) &= \mathcal{L}^{-1}\left(\frac{1}{(z^2 + 1)^2} + \frac{z+1}{z^2 + 1}\right) \\
&= \mathcal{L}^{-1}\left(\frac{3}{2}\frac{1}{z^2 + 1} + \frac{z}{z^2 + 1} - \frac{1}{2}\frac{z^2 - 1}{(z^2 + 1)^2}\right) \\
&= \frac{3}{2}\sin t + \cos t - \frac{1}{2}t\cos t.
\end{aligned}$$

Step 3. The method is not rigorous. It therefore remains to check that the solution to (17.1) found heuristically is a true solution. This depends on the conditions on f and on the coefficients. It is worthwhile noticing that the Cauchy problem (17.1) has a unique solution.

Example 17.2 *Let $\lambda \in \mathbb{R}$, find all solutions to*

$$\begin{cases} y''(x) + \lambda y(x) = 0, & x > 0 \\ y(0) = y_0, & y'(0) = y_1. \end{cases}$$

Discussion We consider three cases.

Case 1: $\lambda = 0$. The problem then becomes

$$\begin{cases} y''(x) = 0 \\ y(0) = y_0, \ y'(0) = y_1. \end{cases}$$

The solution is trivially given by

$$y(x) = y_0 + y_1 x.$$

Case 2: $\lambda < 0$. From the previous example we get

$$Y(z) = \frac{y_0 z + y_1}{z^2 + \lambda} = y_0 \frac{z}{z^2 + \lambda} + \frac{y_1}{\sqrt{-\lambda}} \frac{\sqrt{-\lambda}}{z^2 + \lambda}$$

$$= y_0 \, \mathcal{L}\left(\cosh\left(x\sqrt{-\lambda}\right)\right)(z) + \frac{y_1}{\sqrt{-\lambda}} \, \mathcal{L}\left(\sinh\left(x\sqrt{-\lambda}\right)\right)(z)$$

and hence

$$y(x) = y_0 \cosh\left(x\sqrt{-\lambda}\right) + \frac{y_1}{\sqrt{-\lambda}} \sinh\left(x\sqrt{-\lambda}\right).$$

Case 3: $\lambda > 0$. Invoking again the previous example we obtain

$$Y(z) = \frac{y_0 z + y_1}{z^2 + \lambda} = y_0 \frac{z}{z^2 + \lambda} + \frac{y_1}{\sqrt{\lambda}} \frac{\sqrt{\lambda}}{z^2 + \lambda}$$

$$= y_0 \, \mathcal{L}\left(\cos\left(x\sqrt{\lambda}\right)\right)(z) + \frac{y_1}{\sqrt{\lambda}} \, \mathcal{L}\left(\sin\left(x\sqrt{\lambda}\right)\right)(z)$$

and thus

$$y(x) = y_0 \cos\left(x\sqrt{\lambda}\right) + \frac{y_1}{\sqrt{\lambda}} \sin\left(x\sqrt{\lambda}\right).$$

17.2 Sturm–Liouville problem

Example 17.3 *Let $L > 0$. Find $\lambda \in \mathbb{R}$ and $y = y(x)$, $y \not\equiv 0$, satisfying*

$$\begin{cases} y''(x) + \lambda y(x) = 0, & x \in (0, L) \\ y(0) = y(L) = 0. \end{cases}$$

$$(17.2)$$

(Note that $y \equiv 0$ is a trivial solution to (17.2) $\forall \lambda \in \mathbb{R}$.)

Discussion We start by studying Example 17.2 for fixed λ, namely

$$\begin{cases} y''(x) + \lambda y(x) = 0 \\ \quad\quad y(0) = 0. \end{cases} \tag{17.3}$$

To solve (17.2) we still have to impose $y(L) = 0$ and examine for which values of λ we can find non-trivial solutions. We consider three cases.

Case 1: $\lambda = 0$. Observe that every solution to (17.3) is of the form, y_1 being arbitrary,

$$y(x) = y_1 x.$$

Since we also want $y(L) = 0$, we find $y_1 L = 0$ and thus $y_1 = 0$. Therefore, only the trivial solution $y \equiv 0$ satisfies (17.2) with $\lambda = 0$.

Case 2: $\lambda < 0$. Every solution to (17.3) is of the form

$$y(x) = \frac{y_1}{\sqrt{-\lambda}} \sinh\left(\sqrt{-\lambda} x\right).$$

If we also impose $y(L) = 0$, we must have

$$\frac{y_1}{\sqrt{-\lambda}} \sinh\left(\sqrt{-\lambda} L\right) = 0$$

which implies $y_1 = 0$. Thus, if $\lambda < 0$, there is no non-trivial solution to (17.2).

Case 3: $\lambda > 0$. The solutions to (17.3) are of the form

$$y(x) = \frac{y_1}{\sqrt{\lambda}} \sin\left(\sqrt{\lambda} x\right).$$

Since we also want $y(L) = 0$ and a non-trivial solution ($\Rightarrow y_1 \neq 0$), we get

$$\frac{y_1}{\sqrt{\lambda}} \sin\left(\sqrt{\lambda} L\right) = 0 \quad\Leftrightarrow\quad \sin\left(\sqrt{\lambda} L\right) = 0$$
$$\Leftrightarrow \quad \sqrt{\lambda} L = n\pi, \text{ with } n \in \mathbb{N}$$
$$\Leftrightarrow \quad \lambda = \left(\frac{n\pi}{L}\right)^2.$$

Therefore, for $\lambda > 0$ there do exist non-trivial solutions to (17.2) provided $\lambda = (n\pi/L)^2$, for any $n \in \mathbb{Z}$, $n \neq 0$. Moreover, the solutions are given by

$$y(x) = \alpha_n \sin\left(\frac{n\pi}{L} x\right)$$

where α_n is an arbitrary constant.

Example 17.4 *Let $L > 0$. Find $\lambda \in \mathbb{R}$ and $y = y(x)$, $y \not\equiv 0$, that verify*

$$\begin{cases} y''(x) + \lambda y(x) = 0, & x \in (0, L) \\ \qquad y'(0) = y'(L) = 0. \end{cases}$$

$$(17.4)$$

Discussion This problem is very similar to the previous one (we also use Example 17.2). We start by considering the system

$$\begin{cases} y''(x) + \lambda y(x) = 0, & x \in (0, L) \\ \qquad\quad y'(0) = 0. \end{cases}$$

$$(17.5)$$

We have to discuss three different cases.

Case 1: $\lambda = 0$. We immediately find that any solution to (17.5) is of the form

$$y(x) = y_0.$$

In particular, for any $y_0 \neq 0$ we have found a non-trivial solution to (17.4) with $\lambda = 0$ (contrary to the previous example).

Case 2: $\lambda < 0$. The general solution to (17.5) is then given by

$$y(x) = y_0 \cosh\left(\sqrt{-\lambda}x\right).$$

Since $y'(L) = y_0\sqrt{-\lambda}\sinh\left(\sqrt{-\lambda}L\right) = 0$ can be true only if $y_0 = 0$, we deduce that for $\lambda < 0$ there is no non-trivial solution to (17.4).

Case 3: $\lambda > 0$. The solutions are of the form

$$y(x) = y_0 \cos\left(\sqrt{\lambda}x\right).$$

Since we also want $y'(L) = 0$, we have non-trivial solutions ($\Rightarrow y_0 \neq 0$) provided

$$y_0 \sin\left(\sqrt{\lambda}L\right) = 0 \quad \Leftrightarrow \quad \sin\left(\sqrt{\lambda}L\right) = 0$$
$$\Leftrightarrow \quad \sqrt{\lambda}L = n\pi, \text{ with } n \in \mathbb{N}$$
$$\Leftrightarrow \quad \lambda = \left(\tfrac{n\pi}{L}\right)^2.$$

For $\lambda > 0$ we have, therefore, found non-trivial solutions to (17.4) whenever $\lambda = (n\pi/L)^2$, with $n \in \mathbb{Z}$, and the solutions are then given by

$$y(x) = \beta_n \cos\left(\frac{n\pi}{L}x\right)$$

where β_n is an arbitrary constant.

17.3   Some other examples solved by Fourier analysis

Example 17.5 *Let f be a $2\pi-$periodic and C^1 function. Let $m, k \in \mathbb{R}$, $m \neq 0$. Find a solution $y = y(t)$ to the problem*

$$\begin{cases} my''(t) + ky(t) = f(t), & t \in (0, 2\pi) \\ y(0) = y(2\pi) & and \quad y'(0) = y'(2\pi). \end{cases}$$

Consider, in particular, the case where $f(t) = \cos t$.

Discussion We first write the Fourier series of f, namely

$$f(t) = \frac{\alpha_0}{2} + \sum_{n=1}^{\infty} (\alpha_n \cos(nt) + \beta_n \sin(nt)).$$

(In the case $f(t) = \cos t$, we obviously have $\beta_n = 0 \ \forall n$ and $\alpha_n = 0 \ \forall n \neq 1$ and $\alpha_1 = 1$.) It is reasonable to find y as a $2\pi-$periodic function and we therefore search for solutions of the form

$$y(t) = \frac{a_0}{2} + \sum_{n=1}^{\infty} (a_n \cos(nt) + b_n \sin(nt)).$$

We therefore need to find the coefficients a_n and b_n (in terms of the known coefficients α_n and β_n). Differentiating twice the function y, we formally obtain

$$y''(t) = \sum_{n=1}^{\infty} \left[(-n^2 a_n) \cos(nt) + (-n^2 b_n) \sin(nt) \right].$$

The differential equation therefore leads to

$$my'' + ky = \frac{ka_0}{2} + \sum_{n=1}^{\infty} \left[(k - mn^2) a_n \cos(nt) + (k - mn^2) b_n \sin(nt) \right]$$

$$= f = \frac{\alpha_0}{2} + \sum_{n=1}^{\infty} (\alpha_n \cos(nt) + \beta_n \sin(nt)).$$

Identifying the coefficients we find

$$a_0 = \frac{\alpha_0}{k}, \quad a_n = \frac{\alpha_n}{k - mn^2}, \quad b_n = \frac{\beta_n}{k - mn^2}.$$

In the case $f(t) = \cos t$ we obtain $a_n = b_n = 0$ except for

$$a_1 = \frac{\alpha_1}{k - m} = \frac{1}{k - m}.$$

and thus the solution is given by

$$y(t) = \frac{\cos t}{k - m}.$$

(Clearly, in the above analysis we have assumed that $k/m \neq n^2$ for every $n \in \mathbb{N}$.)

Example 17.6 *Let $\alpha \neq \pm 1$ and f be a 2π−periodic and C^1 function. Find a solution $x = x(t)$ to*

$$\begin{cases} x(t) + \alpha x(t - \pi) = f(t), & t \in (0, 2\pi) \\ x(0) = x(2\pi). \end{cases}$$

Consider the particular case $f(t) = \cos t + 3\sin(2t) + 4\cos(5t)$.

Discussion We start by writing the Fourier series of f, namely

$$f(t) = \frac{\alpha_0}{2} + \sum_{n=1}^{\infty} (\alpha_n \cos(nt) + \beta_n \sin(nt)).$$

(When $f(t) = \cos t + 3\sin(2t) + 4\cos(5t)$, we find $\alpha_1 = 1$, $\alpha_5 = 4$, $\beta_2 = 3$ and all the others vanish.) We search for 2π−periodic solutions and we therefore write

$$x(t) = \frac{a_0}{2} + \sum_{n=1}^{\infty} (a_n \cos(nt) + b_n \sin(nt)).$$

We have to find the coefficients a_n and b_n (in terms of the known coefficients α_n and β_n). We thus have

$$x(t - \pi) = \frac{a_0}{2} + \sum_{n=1}^{\infty} (a_n \cos(nt - n\pi) + b_n \sin(nt - n\pi))$$

$$= \frac{a_0}{2} + \sum_{n=1}^{\infty} [(-1)^n a_n \cos(nt) + (-1)^n b_n \sin(nt)].$$

Inserting the functions in the equation, we get

$$x(t) + \alpha x(t - \pi) = \frac{a_0}{2}(1 + \alpha)$$

$$+ \sum_{n=1}^{\infty} [(1 + \alpha(-1)^n) a_n \cos(nt) + (1 + \alpha(-1)^n) b_n \sin(nt)]$$

$$= f(t) = \frac{\alpha_0}{2} + \sum_{n=1}^{\infty} (\alpha_n \cos(nt) + \beta_n \sin(nt))$$

and thus (recalling that $\alpha \neq \pm 1$)

$$a_0 = \frac{\alpha_0}{1+\alpha}, \quad a_n = \frac{\alpha_n}{1+\alpha\,(-1)^n}, \quad b_n = \frac{\beta_n}{1+\alpha\,(-1)^n}.$$

In the example, we get

$$a_1 = \frac{1}{1-\alpha}, \quad a_5 = \frac{4}{1-\alpha}, \quad b_2 = \frac{3}{1+\alpha}$$

and all the other coefficients are 0. The solution is then given by

$$x(t) = \frac{\cos t}{1-\alpha} + \frac{4\cos(5t)}{1-\alpha} + \frac{3\sin(2t)}{1+\alpha}.$$

Example 17.7 *Find a solution to*

$$x(t) + 3\int_{-\infty}^{+\infty} e^{-|\tau|}x(t-\tau)d\tau = e^{-|t|}, \quad t \in \mathbb{R}.$$

Discussion We divide the discussion into two steps.

Step 1. For $\alpha \in \mathbb{R}$ we compute the Fourier transform of $f(t) = e^{-|t|}$, i.e.

$$\mathfrak{F}(f)(\alpha) = \widehat{f}(\alpha) = \frac{1}{\sqrt{2\pi}}\int_{-\infty}^{+\infty} f(t)\,e^{-i\alpha t}dt$$

$$= \frac{1}{\sqrt{2\pi}}\int_{-\infty}^{0} e^{(1-i\alpha)t}dt + \frac{1}{\sqrt{2\pi}}\int_{0}^{+\infty} e^{-(1+i\alpha)t}dt$$

$$= \frac{1}{\sqrt{2\pi}}\left[\frac{1}{1+i\alpha} + \frac{1}{1-i\alpha}\right] = \frac{1}{\sqrt{2\pi}}\frac{2}{1+\alpha^2}.$$

Step 2. The equation can be rewritten as ($f * x$ denoting the convolution product)

$$x(t) + 3(f * x)(t) = f(t).$$

We apply Fourier transform on both sides of the equation, letting $\mathfrak{F}(x)(\alpha) = \widehat{x}(\alpha)$, we get

$$\widehat{x}(\alpha) + 3\sqrt{2\pi}\,\widehat{f}(\alpha)\,\widehat{x}(\alpha) = \widehat{f}(\alpha)$$

and thus

$$\widehat{x}(\alpha) = \frac{\widehat{f}(\alpha)}{1+3\sqrt{2\pi}\,\widehat{f}(\alpha)} = \frac{1}{\sqrt{2\pi}}\frac{2}{\alpha^2+7} = \sqrt{\frac{2}{\pi}}\frac{1}{\alpha^2+7}.$$

Applying the inverse Fourier transform we obtain

$$x(t) = \mathfrak{F}^{-1}(x)(t) = \mathfrak{F}^{-1}\left(\sqrt{\frac{2}{\pi}}\frac{1}{\alpha^2+7}\right)(t) = \frac{1}{\sqrt{7}}e^{-\sqrt{7}|t|}.$$

17.4   Exercises

Exercise 17.1 Let $n \geq 1$, $a_0, a_1, \cdots, a_n$, $y_0, y_1, \cdots, y_{n-1} \in \mathbb{R}$, $a_n \neq 0$, and $f : \mathbb{R}_+ \to \mathbb{R}$ a sufficiently regular function. Solve

$$\begin{cases} a_n y^{(n)}(t) + \cdots + a_1 y'(t) + a_0 y(t) = f(t), & t > 0 \\ y^{(k)}(0) = y_k, & k = 0, 1, \cdots, n - 1. \end{cases}$$

Exercise 17.2 Find non-trivial solutions to

$$\begin{cases} y''(x) + \lambda y(x) = 0, & x \in (0, 2\pi) \\ y(0) = y(2\pi) & \text{and} \quad y'(0) = y'(2\pi). \end{cases}$$

Exercise 17.3 Find $\mu \in \mathbb{R}$ and $v = v(x)$, $v \not\equiv 0$, solution to

$$\begin{cases} v''(x) + 2v'(x) + (1 + \mu)v(x) = 0 \\ v(0) = v(\pi) = 0. \end{cases}$$

Exercise 17.4 Verify that the solutions to the equation

$$r^2 f''(r) + r f'(r) - n^2 f(r) = 0, \; r > 0$$

are given by

$$f(r) = \begin{cases} a_n r^n + a_{-n} r^{-n} & \text{if } n > 0 \\ a_0 + b_0 \log r & \text{if } n = 0 \end{cases}$$

where a_n and b_0 are constants.

Exercise 17.5 Let $f : \mathbb{R} \to \mathbb{R}$ be a 2π–periodic function such that

$$f(x) = |x|, \quad \text{if } x \in [-\pi, \pi].$$

(i) Find its Fourier series.

(ii) Find a solution (discuss for which values of α the method applies) to

$$\begin{cases} \dfrac{d^4 y(x)}{dx^4} - \alpha y(x) = f(x) \\ y(x + 2\pi) = y(x). \end{cases}$$

Exercise 17.6 Find a solution to

$$\begin{cases} x'\left(t - \dfrac{\pi}{2}\right) + x(t) = 1 + 2\cos t + \sin(2t) \\ x(t + 2\pi) = x(t). \end{cases}$$

Hint: write $\sin\left(n\left(t - \dfrac{\pi}{2}\right)\right)$, $\cos\left(n\left(t - \dfrac{\pi}{2}\right)\right)$ when $n = 4k$, $4k - 1$, $4k - 2$, $4k - 3$ where $k \geq 1$ is an integer.

Exercise 17.7 Find a solution $x = x(t)$ to

$$\begin{cases} x''(t) + 5x'(t - \pi) - x(t) = \cos t - 3\sin(2t) + 2 \\ \qquad\qquad x(t + 2\pi) = x(t). \end{cases}$$

Exercise 17.8 (i) Let $f : \mathbb{R} \to \mathbb{R}$ be a 2π−periodic and even function such that

$$f(t) = \sin(3t) \quad \text{if } t \in [0, \pi].$$

Find its Fourier series.

(ii) Find a 2π−periodic and even function such that

$$x(t) - 2x(t - \pi) = \sin(3t) \quad \text{if } t \in [0, \pi].$$

Exercise 17.9 (i) Let f be a 2π−periodic and C^1 function. Find a 2π−periodic solution $x = x(t)$ (in terms of f) to

$$x'(t) + 2x(t - \pi) = f(t).$$

(ii) Write explicitly $x(t)$ when $f(t) = 1 + 4\sin(6t)$.

Exercise 17.10 Let $f : \mathbb{R} \to \mathbb{R}$ be a T−periodic and C^1 function. Let $\alpha \neq 0$ and consider the differential equation

$$x'(t) + \alpha x(t) = f(t).$$

Find a T−periodic solution to the equation by writing both $x(t)$ and $f(t)$ in Fourier series. Apply the result to the case where $\alpha = 1$, $T = 2\pi$ and f is the following function (just write the first terms)

$$f(t) = \begin{cases} \left(t - \frac{\pi}{2}\right)^2 & \text{if } 0 \leq t < \pi \\ -\left(t - \frac{3\pi}{2}\right)^2 + \frac{\pi^2}{2} & \text{if } \pi \leq t < 2\pi. \end{cases}$$

Exercise 17.11 Let

$$f(t) = e^{-|t|} \quad \text{and} \quad g(t) = te^{-t^2}.$$

Solve

$$3y(t) + \int_{-\infty}^{\infty} [y''(\tau) - y(\tau)] f(t - \tau)d\tau = g(t).$$

Chapter 18

Applications to partial differential equations

We explain in several examples the method of *separation of variables*. As in the previous chapter the method is heuristic. However, with appropriate hypotheses on the given data, it is not too difficult to justify (but we will not do it here) that the solution found by formal computations is, in fact, a true solution. An important point to note is that, essentially, all the equations we consider in this chapter have a *unique* solution. Our method, therefore, finds *the* solution, even though in some cases, notably for the wave equation, other methods seem to provide a different one.

18.1 Heat equation

Example 18.1 (Bar of finite length) *Let $a \neq 0$, $L > 0$ and f be a $C^1([0,L])$ function such that $f(0) = f(L) = 0$. Find $u = u(x,t)$ a solution to*

$$\begin{cases} \dfrac{\partial u}{\partial t} = a^2 \dfrac{\partial^2 u}{\partial x^2} & \text{if } x \in (0,L), \ t > 0 \\ u(0,t) = u(L,t) = 0 & \text{if } t > 0 \\ u(x,0) = f(x) & \text{if } x \in (0,L). \end{cases} \tag{18.1}$$

Consider the particular case where $f(x) = 2\sin(\pi x/L) - \sin(3\pi x/L)$.

Remark The boundary conditions

$$u(0,t) = u(L,t) = 0$$

are called *Dirichlet* conditions. The same method allows us to treat, in a completely analogous way, other boundary conditions (cf. exercises below) notably *Neumann* conditions, namely

$$\frac{\partial u}{\partial x}(0,t) = \frac{\partial u}{\partial x}(L,t) = 0.$$

Discussion We divide the discussion into two steps.

Step 1 (Separation of variables). We start by solving the problem ignoring the initial data $(u(x,0) = f(x))$, namely

$$\begin{cases} \dfrac{\partial u}{\partial t} = a^2 \dfrac{\partial^2 u}{\partial x^2} & \text{if } x \in (0,L),\ t > 0 \\ u(0,t) = u(L,t) = 0 & \text{if } t > 0. \end{cases} \tag{18.2}$$

Step 1.1. We search for solutions of the form

$$u(x,t) = v(x)w(t).$$

In order to satisfy the boundary conditions $(u(0,t) = u(L,t) = 0)$ for every $t > 0$ we should have

$$\left. \begin{aligned} u(0,t) &= v(0)w(t) = 0 \\ u(L,t) &= v(L)w(t) = 0 \end{aligned} \right\} \Rightarrow v(0) = v(L) = 0.$$

Since

$$\frac{\partial u}{\partial t} = v(x)w'(t) \quad \text{and} \quad \frac{\partial^2 u}{\partial x^2} = v''(x)w(t)$$

we find that

$$\frac{\partial u}{\partial t} = a^2 \frac{\partial^2 u}{\partial x^2} \Rightarrow v(x)w'(t) = a^2 v''(x)w(t)$$

and thus

$$\frac{v''(x)}{v(x)} = \frac{w'(t)}{a^2 w(t)}.$$

Since we want this equation to be satisfied for every t and every x, we must have both sides equal to a constant, denoted $-\lambda$. We therefore have to solve the two problems

$$\begin{cases} v''(x) + \lambda v(x) = 0 & \text{if } x \in (0,L) \\ v(0) = v(L) = 0 \end{cases} \tag{18.3}$$

and

$$w'(t) + a^2 \lambda w(t) = 0. \tag{18.4}$$

We know (cf. Example 17.3) that the non-trivial solutions to (18.3) are given by

$$\lambda = \left(\frac{n\pi}{L}\right)^2 \quad \text{and} \quad v_n(x) = \sin\left(\frac{n\pi}{L}x\right).$$

The solutions to (18.4), for $\lambda = (n\pi/L)^2$, are easily obtained by straight integration, namely

$$w_n(t) = e^{-a^2\left(\frac{n\pi}{L}\right)^2 t}.$$

We have therefore found that

$$u_n(x,t) = \sin\left(\frac{n\pi}{L}x\right) e^{-a^2\left(\frac{n\pi}{L}\right)^2 t}$$

is a solution to (18.2) $\forall n = 1, 2, 3, \cdots$.

Step 1.2. We now observe that if φ_1 and φ_2 are two solutions to (18.2) and if $\alpha_1, \alpha_2 \in \mathbb{R}$, then $\alpha_1\varphi_1 + \alpha_2\varphi_2$ is also a solution to (18.2). By iterating this observation a finite number of times, we find that, for every $\alpha_n \in \mathbb{R}$, the function

$$u(x,t) = \sum_{n=1}^{N} \alpha_n \sin\left(\frac{n\pi}{L}x\right) e^{-a^2\left(\frac{n\pi}{L}\right)^2 t}$$

is still a solution to (18.2). Until now, the reasoning is rigorous and it is at this stage that we proceed formally in allowing N to be infinite. We therefore find that the general solution to (18.2) is given by

$$u(x,t) = \sum_{n=1}^{\infty} \alpha_n \sin\left(\frac{n\pi}{L}x\right) e^{-a^2\left(\frac{n\pi}{L}\right)^2 t}$$

where α_n are "arbitrary" constants.

Step 2 (Initial condition). We now go back to our original problem (18.1). In order to solve (18.1) it remains to choose the constants α_n so that the initial condition, $u(x,0) = f(x)$, is satisfied. We should therefore have

$$u(x,0) = \sum_{n=1}^{\infty} \alpha_n \sin\left(\frac{n\pi}{L}x\right) = f(x).$$

We hence choose the α_n as the Fourier coefficients of the sine series, namely

$$\alpha_n = \frac{2}{L}\int_0^L f(y) \sin\left(\frac{n\pi}{L}y\right) dy.$$

The solution to (18.1) is therefore

$$\boxed{u(x,t) = \sum_{n=1}^{\infty} \alpha_n \sin\left(\frac{n\pi}{L}x\right) e^{-a^2\left(\frac{n\pi}{L}\right)^2 t}}$$

with

$$\alpha_n = \frac{2}{L} \int_0^L f(y) \sin\left(\frac{n\pi}{L}y\right) \, dy.$$

It can be shown that as soon as $t > 0$ the presence of the term $e^{-a^2\left(\frac{n\pi}{L}\right)^2 t}$ and the fact that f is C^1 ensure that the series and all its derivatives, in t and in x, converge. The function u is hence C^∞ (as soon as $t > 0$). It can also be proved that $u(x,t) \to 0$ if $t \to \infty$.

In the particular case, the coefficients α_n are given by

$$\alpha_1 = 2, \quad \alpha_3 = -1, \quad \alpha_n = 0 \text{ otherwise}$$

and, thus, the solution is

$$u(x,t) = 2\sin\left(\frac{\pi}{L}x\right) e^{-a^2\left(\frac{\pi}{L}\right)^2 t} - \sin\left(\frac{3\pi}{L}x\right) e^{-a^2\left(\frac{3\pi}{L}\right)^2 t}.$$

Example 18.2 (Bar of infinite length) *Let $a \neq 0$ and $f : \mathbb{R} \to \mathbb{R}$ be a C^1 function such that*

$$\int_{-\infty}^{+\infty} |f(x)| \, dx < \infty \quad and \quad \int_{-\infty}^{+\infty} \left|\widehat{f}(\alpha)\right| \, d\alpha < \infty.$$

Find $u = u(x,t)$ a solution to

$$\begin{cases} \dfrac{\partial u}{\partial t} = a^2 \dfrac{\partial^2 u}{\partial x^2} & \text{if } t > 0, \ x \in \mathbb{R} \\ u(x,0) = f(x) & \text{if } x \in \mathbb{R}. \end{cases}$$

(18.5)

Discuss the particular case where $f(x) = e^{-x^2}$.

Discussion Before starting we should point out that, contrary to the previous case, the problem under consideration has other solutions than the one that we will find. The one that we provide here is the "best" both from a mathematical and physical point of view. We proceed in two steps.

Step 1 (Fourier transform in x). We define the function v as

$$v(\xi, t) = \mathfrak{F}(u)(\xi, t) = \frac{1}{\sqrt{2\pi}} \int_{-\infty}^{+\infty} u(y, t) e^{-i\xi y} \, dy$$

(it, therefore, is the Fourier transform in x, considering t as a parameter, of the function u). Using the properties of the Fourier transform we find

$$\mathfrak{F}\left(\frac{\partial^2 u}{\partial x^2}\right)(\xi,t) = (i\xi)^2\mathfrak{F}(u)(\xi,t) = -\xi^2 v(\xi,t)$$

$$\frac{\partial v}{\partial t}(\xi,t) = \frac{1}{\sqrt{2\pi}}\int_{-\infty}^{+\infty}\frac{\partial u}{\partial t}(y,t)\,e^{-i\xi y}\,dy = \mathfrak{F}\left(\frac{\partial u}{\partial t}\right)(\xi,t).$$

We next let

$$\widehat{f}(\xi) = \mathfrak{F}(f)(\xi) = \frac{1}{\sqrt{2\pi}}\int_{-\infty}^{+\infty} f(y)\,e^{-i\xi y}\,dy.$$

Applying Fourier transform (in x) to both sides of the differential equation in (18.5) we find that (ξ being considered as a parameter)

$$\begin{cases} \dfrac{\partial v}{\partial t}(\xi,t) = -a^2\xi^2 v(\xi,t), & t > 0 \\[2mm] v(\xi,0) = \widehat{f}(\xi). \end{cases} \tag{18.6}$$

The problem (18.6) has as an obvious solution

$$v(\xi,t) = \widehat{f}(\xi)\,e^{-a^2\xi^2 t}.$$

Step 2 (Solution to (18.5)). Applying to v the inverse Fourier transform (in ξ), we find that a solution to (18.5) is given by

$$\boxed{u(x,t) = \frac{1}{\sqrt{2\pi}}\int_{-\infty}^{+\infty}\widehat{f}(\xi)\,e^{i\xi x - a^2\xi^2 t}\,d\xi.}$$

In the particular case we have $\widehat{f}(\xi) = (1/\sqrt{2})e^{-\xi^2/4}$. Therefore, the solution is given by

$$u(x,t) = \frac{1}{\sqrt{2\pi}}\int_{-\infty}^{+\infty}\frac{1}{\sqrt{2}}e^{i\xi x - \xi^2(1/4 + a^2 t)}\,d\xi$$

$$= \frac{1}{\sqrt{2}}\mathfrak{F}(e^{-\xi^2(1/4 + a^2 t)})(-x).$$

An elementary computation leads to

$$u(x,t) = \frac{e^{-\frac{x^2}{1+4a^2 t}}}{\sqrt{1+4a^2 t}}, \quad t \geq 0.$$

18.2   Wave equation

Example 18.3 *Let* $c, L > 0$, *and* $f, g \in C^3\left([0, L]\right)$ *be such that*

$$f(0) = f(L) = g(0) = g(L) = 0.$$

Find $u = u(x, t)$ *a solution to*

$$
\begin{cases}
\dfrac{\partial^2 u}{\partial t^2} = c^2 \dfrac{\partial^2 u}{\partial x^2} & \text{if } x \in (0, L),\ t > 0 \\[2mm]
u(0, t) = u(L, t) = 0 & \text{if } t > 0 \\[2mm]
u(x, 0) = f(x) & \text{if } x \in (0, L) \\[2mm]
\dfrac{\partial u}{\partial t}(x, 0) = g(x) & \text{if } x \in (0, L).
\end{cases}
$$

$$(18.7)$$

Consider the special case where $f = 0$ *and* $g(x) = \sin(\pi x/L) - \sin(2\pi x/L)$.

Discussion We proceed by separation of variables (for another method, see Exercise 18.17).

Step 1 (Separation of variables). We first solve the problem ignoring the initial conditions ($u(x, 0) = f(x)$ and $\dfrac{\partial u}{\partial t}(x, 0) = g(x)$), namely

$$
\begin{cases}
\dfrac{\partial^2 u}{\partial t^2} = c^2 \dfrac{\partial^2 u}{\partial x^2} & \text{if } x \in (0, L),\ t > 0 \\[2mm]
u(0, t) = u(L, t) = 0 & \text{if } t > 0.
\end{cases}
$$

$$(18.8)$$

Step 1.1. We look for solutions of the form $u(x, t) = v(x) w(t)$. As for the heat equation, we are lead to (λ being a constant)

$$
\begin{cases}
\dfrac{v''(x)}{v(x)} = -\lambda = \dfrac{w''(t)}{c^2 w(t)} \\[2mm]
v(0) w(t) = v(L) w(t) = 0
\end{cases}
$$

and, hence,

$$
\begin{cases}
v''(x) + \lambda v(x) = 0 & \text{if } x \in (0, L) \\[2mm]
v(0) = v(L) = 0
\end{cases}
$$

$$(18.9)$$

$$w''(t) + c^2 \lambda w(t) = 0. \qquad (18.10)$$

We have seen (cf. Example 17.3) that the non-trivial solutions to (18.9) are given by $\lambda = (n\pi/L)^2$ and $v_n = \sin\left(\frac{n\pi}{L} x/L\right)$ where $n = 1, 2, \cdots$. Moreover, the solutions to (18.10), for $\lambda = (n\pi/L)^2$, turn out to be (cf. Example 17.2)

$$w_n(t) = a_n \cos\left(\frac{n\pi c}{L}t\right) + b_n \sin\left(\frac{n\pi c}{L}t\right).$$

Step 1.2. As for the heat equation, any finite combination of solutions found above is still a solution to (18.8). By abuse of reasoning, we deduce that the general solution to (18.8) is then given by

$$u(x,t) = \sum_{n=1}^{\infty} \left[a_n \cos\left(\frac{n\pi c}{L}t\right) + b_n \sin\left(\frac{n\pi c}{L}t\right) \right] \sin\left(\frac{n\pi}{L}x\right).$$

$$(18.11)$$

Step 2 (Initial conditions). We now have to determine the constants a_n and b_n so as to satisfy the initial conditions $u(x,0) = f(x)$ and $\frac{\partial u}{\partial t}(x,0) = g(x)$. For the first one we have

$$u(x,0) = \sum_{n=1}^{\infty} a_n \sin\left(\frac{n\pi}{L}x\right) = f(x)$$

and, thus, we choose a_n to be the coefficients of the sine Fourier series of f, i.e.

$$a_n = \frac{2}{L} \int_0^L f(y) \sin\left(\frac{n\pi}{L}y\right) dy.$$

$$(18.12)$$

The second initial condition (differentiating the series with respect to t and then setting $t = 0$) leads us to

$$\frac{\partial u}{\partial t}(x,0) = \sum_{n=1}^{\infty} b_n \frac{n\pi c}{L} \sin\left(\frac{n\pi}{L}x\right) = g(x).$$

We therefore choose b_n in the following manner

$$b_n \frac{n\pi c}{L} = \frac{2}{L} \int_0^L g(y) \sin\left(\frac{n\pi}{L}y\right) dy,$$

i.e.

$$b_n = \frac{2}{n\pi c} \int_0^L g(y) \sin\left(\frac{n\pi}{L}y\right) dy.$$

$$(18.13)$$

The solution to (18.7) is therefore given by (18.11) with a_n and b_n satisfying (18.12) and (18.13).

It can be shown that the hypotheses on f and g ensure the convergence of the series (18.11) as well as of its derivatives $\partial u/\partial x$, $\partial u/\partial t$, $\partial^2 u/\partial x^2$, $\partial^2 u/\partial x \partial t$ and $\partial^2 u/\partial t^2$. The function u is thus C^2. However, contrary to the heat equation, unless f and g are more regular, the formal series expressing the higher order derivatives do not, in general, converge.

In the particular case we find

$$a_n = 0 \ \forall n, \quad b_n = 0 \ \forall n \geq 3, \quad b_1 \frac{\pi c}{L} = 1 \quad \text{and} \quad b_2 \frac{2\pi c}{L} = -1$$

and the solution is

$$u(x,t) = \frac{L}{\pi c} \sin\left(\frac{\pi c}{L}t\right) \sin\left(\frac{\pi}{L}x\right) - \frac{L}{2\pi c} \sin\left(\frac{2\pi c}{L}t\right) \sin\left(\frac{2\pi}{L}x\right).$$

18.3 Laplace equation in a rectangle

Example 18.4 *Let $L, M > 0$. Let*

$$\alpha, \beta : [0, L] \to \mathbb{R} \quad \text{and} \quad \gamma, \delta : [0, M] \to \mathbb{R}$$

be C^1 functions satisfying

$$\alpha(0) = \beta(0) = \gamma(0) = \delta(0) = \alpha(L) = \beta(L) = \gamma(M) = \delta(M) = 0.$$

Find $u = u(x, y)$ a solution to

$$\begin{cases} \Delta u = \dfrac{\partial^2 u}{\partial x^2} + \dfrac{\partial^2 u}{\partial y^2} = 0 & \text{if } x \in (0, L), \ y \in (0, M) \\ u(x, 0) = \alpha(x), \ u(x, M) = \beta(x) & \text{if } x \in (0, L) \\ u(0, y) = \gamma(y), \ u(L, y) = \delta(y) & \text{if } y \in (0, M). \end{cases}$$

(18.14)

Discuss the particular case where $\gamma = \delta \equiv 0$,

$$\alpha(x) = 4\sin(\pi x/L) \quad \text{and} \quad \beta(x) = -\sin(2\pi x/L).$$

Discussion In fact, it is enough to solve the two problems

$$\begin{cases} \Delta v = 0 & \text{if } x \in (0, L), \ y \in (0, M) \\ v(x, 0) = 0, \quad v(x, M) = 0 & \text{if } x \in (0, L) \\ v(0, y) = \gamma(y), \ v(L, y) = \delta(y) & \text{if } y \in (0, M) \end{cases}$$

and

$$\begin{cases} \quad\quad\quad \Delta w = 0 & \text{if } x \in (0, L), \ y \in (0, M) \\ w(x, 0) = \alpha(x), \ w(x, M) = \beta(x) & \text{if } x \in (0, L) \\ \quad w(0, y) = 0, \quad w(L, y) = 0 & \text{if } y \in (0, M). \end{cases}$$

The solution to problem (18.14) is then given by

$$u = v + w.$$

Since both problems are solved in the same manner, we will treat only the second one (this means that we solve (18.14) with $\gamma = \delta \equiv 0$).

Step 1 (Separation of variables). We first solve the problem (ignoring the conditions $w(x, 0) = \alpha(x)$, $w(x, M) = \beta(x)$)

$$\begin{cases} \quad\quad \Delta w = 0 & \text{if } x \in (0, L), \ y \in (0, M) \\ w(0, y) = w(L, y) = 0 & \text{if } y \in (0, M). \end{cases} \tag{18.15}$$

Step 1.1. We look for solutions of the form

$$w(x, y) = f(x) g(y).$$

As before, by separating the variables x and y, (λ being a constant) we find

$$\begin{cases} f''(x) g(y) + f(x) g''(y) = 0 \\ f(0) = f(L) = 0 \end{cases} \Leftrightarrow \begin{cases} \frac{f''(x)}{f(x)} = -\lambda = -\frac{g''(y)}{g(y)} \\ f(0) = f(L) = 0. \end{cases}$$

We have, therefore, to solve the two following problems

$$\begin{cases} f''(x) + \lambda f(x) = 0 & \text{if } x \in (0, L) \\ f(0) = f(L) = 0 \end{cases} \tag{18.16}$$

and

$$g''(y) - \lambda g(y) = 0. \tag{18.17}$$

In Example 17.3 we saw that the non-trivial solutions to (18.16) are of the form

$$\lambda = \left(\frac{n\pi}{L}\right)^2 \quad \text{and} \quad f_n(x) = \sin\left(\frac{n\pi}{L} x\right)$$

while (cf. Example 17.2) the solutions to (18.17) are given by

$$g_n(y) = a_n \cosh\left(\frac{n\pi}{L} y\right) + b_n \sinh\left(\frac{n\pi}{L} y\right).$$

Step 1.2. Since the problem (18.15) is linear, we have, by the usual abuse of reasoning, that the general solution is given by

$$w\left(x,y\right) = \sum_{n=1}^{\infty} \left[a_n \cosh\left(\frac{n\pi}{L}y\right) + b_n \sinh\left(\frac{n\pi}{L}y\right)\right] \sin\left(\frac{n\pi}{L}x\right).$$

Step 2 (Boundary conditions). To solve (18.14) it remains to fix the constants a_n and b_n so as to verify

$$w\left(x,0\right) = \alpha\left(x\right) \quad \text{and} \quad w\left(x,M\right) = \beta\left(x\right).$$

Since we want that

$$\alpha\left(x\right) = \sum_{n=1}^{\infty} a_n \sin\left(\frac{n\pi}{L}x\right)$$

we find

$$a_n = \frac{2}{L}\int_0^L \alpha\left(t\right)\sin\left(\frac{n\pi}{L}t\right) dt.$$

The coefficients b_n are obtained by solving

$$\beta\left(x\right) = \sum_{n=1}^{\infty} \left[a_n \cosh\left(\frac{n\pi}{L}M\right) + b_n \sinh\left(\frac{n\pi}{L}M\right)\right] \sin\left(\frac{n\pi}{L}x\right)$$

$$= \sum_{n=1}^{\infty} c_n \sin\left(\frac{n\pi}{L}x\right)$$

where we have set

$$c_n = a_n \cosh\left(\frac{n\pi}{L}M\right) + b_n \sinh\left(\frac{n\pi}{L}M\right).$$

We therefore choose

$$c_n = \frac{2}{L}\int_0^L \beta\left(t\right)\sin\left(\frac{n\pi}{L}t\right) dt$$

and thus b_n are found through the equations

$$b_n \sinh\left(\frac{n\pi}{L}M\right) = c_n - a_n \cosh\left(\frac{n\pi}{L}M\right).$$

It can be shown that if $0 < y < M$ and under our hypotheses, then the series, as well as all series obtained by differentiating the original series, converges. We therefore have the same phenomenon as for the heat equation, namely the function $w \in C^\infty\left((0, L) \times (0, M)\right)$.

In the particular case, we find $a_1 = 4$, $a_n = 0$, $\forall n \geq 2$ and $c_2 = -1$, $c_n = 0$, $\forall n \neq 2$, i.e.

$$b_1 = -4\coth\left(\frac{\pi}{L}M\right) \quad \text{and} \quad b_2 = -1/\sinh\left(\frac{2\pi}{L}M\right).$$

Thus the solution is given by

$$w\left(x, y\right) = 4\left[\cosh\left(\frac{\pi}{L}y\right) - \coth\left(\frac{\pi}{L}M\right)\sinh\left(\frac{\pi}{L}y\right)\right]\sin\left(\frac{\pi}{L}x\right)$$
$$- \frac{1}{\sinh\left(2\pi M/L\right)}\sinh\left(\frac{2\pi}{L}y\right)\sin\left(\frac{2\pi}{L}x\right).$$

18.4 Laplace equation in a disk

Example 18.5 *Find $u\left(x, y\right)$ a solution to*

$$
\begin{cases}
\Delta u = \dfrac{\partial^2 u}{\partial x^2} + \dfrac{\partial^2 u}{\partial y^2} = 0 & \text{if } (x, y) \in \Omega \\
u\left(x, y\right) = \varphi\left(x, y\right) & \text{if } (x, y) \in \partial\Omega
\end{cases}
$$

(18.18)

where $\Omega = \left\{(x, y) \in \mathbb{R}^2 : x^2 + y^2 < R^2\right\}$ and φ is a C^1 function. Consider the special case where $\varphi = x^2 + y$.

Discussion We use the separation of variables method again, but after passing to polar coordinates.

Step 1 (Polar coordinates). We write

$$x = r\cos\theta \quad \text{and} \quad y = r\sin\theta$$

$$v\left(r, \theta\right) = u\left(x, y\right) = u\left(r\cos\theta, r\sin\theta\right)$$

and therefore find

$$\Delta u = \frac{\partial^2 v}{\partial r^2} + \frac{1}{r}\frac{\partial v}{\partial r} + \frac{1}{r^2}\frac{\partial^2 v}{\partial \theta^2}.$$

Letting

$$\psi\left(\theta\right) = \varphi(R\cos\theta, R\sin\theta)$$

the problem to be solved becomes

$$
\begin{cases}
\dfrac{\partial^2 v}{\partial r^2} + \dfrac{1}{r}\dfrac{\partial v}{\partial r} + \dfrac{1}{r^2}\dfrac{\partial^2 v}{\partial \theta^2} = 0 & \text{if } r \in (0,R),\ \theta \in (0,2\pi) \\
\quad v(R,\theta) = \psi(\theta) & \text{if } \theta \in (0,2\pi).
\end{cases}
$$

$$(18.19)$$

Note that, since we are dealing with polar coordinates, we also have to impose

$$
v(r,0) = v(r,2\pi) \quad \text{and} \quad \frac{\partial v}{\partial \theta}(r,0) = \frac{\partial v}{\partial \theta}(r,2\pi).
$$

In the particular case we have

$$
\psi(\theta) = \varphi(R,\theta) = R^2\cos^2\theta + R\sin\theta = \frac{R^2}{2} + \frac{R^2}{2}\cos(2\theta) + R\sin\theta.
$$

Step 2 (Separation of variables). We start, as usual, by ignoring the boundary condition $(v(R,\theta) = \psi(\theta))$ and we solve the equation

$$
\begin{cases}
\dfrac{\partial^2 v}{\partial r^2} + \dfrac{1}{r}\dfrac{\partial v}{\partial r} + \dfrac{1}{r^2}\dfrac{\partial^2 v}{\partial \theta^2} = 0 & \text{if } r \in (0,R),\ \theta \in (0,2\pi) \\
\quad v(r,0) = v(r,2\pi) & \text{if } r \in (0,R) \\
\quad \dfrac{\partial v}{\partial \theta}(r,0) = \dfrac{\partial v}{\partial \theta}(r,2\pi) & \text{if } r \in (0,R).
\end{cases}
$$

We search for a function $v(r,\theta)$ under the special form

$$
v(r,\theta) = f(r)\,g(\theta).
$$

We therefore get, by separating the variables r and θ, (λ being a constant),

$$
f''(r)\,g(\theta) + \frac{1}{r}f'(r)\,g(\theta) + \frac{1}{r^2}f(r)\,g''(\theta) = 0
$$

and thus

$$
\frac{r^2 f''(r) + r f'(r)}{f(r)} = \lambda = -\frac{g''(\theta)}{g(\theta)}
$$

and from the periodicity conditions

$$
g(0) = g(2\pi) \quad \text{and} \quad g'(0) = g'(2\pi).
$$

We hence obtain

$$
\begin{cases}
\quad g''(\theta) + \lambda g(\theta) = 0 \\
g(0) = g(2\pi) \quad \text{and} \quad g'(0) = g'(2\pi)
\end{cases}
$$

$$(18.20)$$

and

$$r^2 f''(r) + r f'(r) - \lambda f(r) = 0. \tag{18.21}$$

The non-trivial solutions to (18.20) (cf. Exercise 17.2) are therefore given, for $n \in \mathbb{Z}$, by

$$\lambda = n^2 \quad \text{and} \quad g_n(\theta) = \alpha_n \cos(n\theta) + \beta_n \sin(n\theta).$$

The solutions to (18.21) with $\lambda = n^2$ are then (cf. Exercise 17.4)

$$f_n(r) = \begin{cases} \gamma_n r^n & \text{if } n \neq 0, \ n \in \mathbb{Z} \\ \gamma_0 + \delta_0 \log r & \text{if } n = 0. \end{cases}$$

The general solution is then of the form

$$v(r, \theta) = \alpha_0 (\gamma_0 + \delta_0 \log r) + \sum_{n \in \mathbb{Z}, n \neq 0} \gamma_n((\alpha_n \cos(n\theta) + \beta_n \sin(n\theta)) r^n.$$

Relabeling the constants

$$a_0 = 2\alpha_0 \gamma_0, \quad b_0 = \alpha_0 \delta_0, \quad a_n = \alpha_n \gamma_n \text{ if } n \neq 0,$$

$$b_n = \begin{cases} \beta_n \gamma_n & \text{if } n > 0 \\ -\beta_n \gamma_n & \text{if } n < 0 \end{cases}$$

we can rewrite the general solution as

$$v(r, \theta) = \frac{a_0}{2} + b_0 \log r + \sum_{n=1}^{\infty} [(a_n \cos(n\theta) + b_n \sin(n\theta)) r^n]$$

$$+ \sum_{n=1}^{\infty} [(a_{-n} \cos(n\theta) + b_{-n} \sin(n\theta)) r^{-n}].$$

$$\tag{18.22}$$

Since we are interested in solutions in the disk centered at 0, the solutions of the form r^{-n} and $\log r$ are not allowed, as they are singular at $r = 0$. We therefore deduce that

$$b_0 = b_{-n} = a_{-n} = 0$$

and, consequently, the general solution v to our problem is

$$v(r, \theta) = \frac{a_0}{2} + \sum_{n=1}^{\infty} (a_n \cos(n\theta) + b_n \sin(n\theta)) r^n.$$

(Note that if the domain Ω is an annulus instead of a disk, then the general solution is indeed of the form (18.22).)

Step 3 (Boundary conditions). It remains to identify the coefficients a_n and b_n in order to have

$$v(R,\theta) = \varphi(R\cos\theta, R\sin\theta) = \psi(\theta) = \frac{a_0}{2} + \sum_{n=1}^{\infty}(a_n\cos(n\theta) + b_n\sin(n\theta))R^n.$$

This leads to

$$a_n R^n = \frac{1}{\pi}\int_0^{2\pi}\psi(\theta)\cos(n\theta)\,d\theta$$

and

$$b_n R^n = \frac{1}{\pi}\int_0^{2\pi}\psi(\theta)\sin(n\theta)\,d\theta.$$

One can also show that, under our hypotheses, the function v that we obtained is C^∞ whenever $r < R$.

In the particular case

$$\psi(\theta) = \frac{R^2}{2} + \frac{R^2}{2}\cos(2\theta) + R\sin\theta,$$

we get

$$a_0 = R^2, \quad a_2 = \frac{1}{2}, \quad a_n = 0 \text{ otherwise}$$

$$b_1 = 1, \quad b_n = 0 \text{ otherwise}$$

and thus

$$v(r,\theta) = \frac{R^2}{2} + \frac{r^2}{2}\cos(2\theta) + r\sin\theta = \frac{R^2}{2} + \frac{r^2}{2}\cos^2\theta - \frac{r^2}{2}\sin^2\theta + r\sin\theta.$$

Returning to the Cartesian coordinates, we have found the desired solution, namely

$$u(x,y) = \frac{R^2}{2} + \frac{x^2 - y^2}{2} + y.$$

18.5 Laplace equation in a simply connected domain

Example 18.6 *Find* $u = u(x, y)$ *a solution to*

$$
\begin{cases}
\Delta u = \dfrac{\partial^2 u}{\partial x^2} + \dfrac{\partial^2 u}{\partial y^2} = 0 & \text{if } (x, y) \in \Omega \\[2mm]
u(x, y) = \varphi(x, y) & \text{if } (x, y) \in \partial\Omega
\end{cases}
$$

where $\Omega \neq \mathbb{R}^2$ *is a simply connected domain whose boundary is sufficiently regular and* φ *is a* C^1 *function. Solve the special case where*

$$
\begin{cases}
\Delta u = \dfrac{\partial^2 u}{\partial x^2} + \dfrac{\partial^2 u}{\partial y^2} = 0 & \text{if } x > 1,\ y \in \mathbb{R} \\[2mm]
u(1, y) = \varphi(1, y) = \dfrac{2}{1 + y^2} & \text{if } y \in \mathbb{R}.
\end{cases}
$$

Discussion The idea is to get, via a conformal mapping, to the case of the disk. We divide the discussion into three steps.

Step 1. It follows from Riemann theorem (cf. Theorem 13.5), that there exists a conformal mapping

$$
f = \alpha + i\beta : \Omega \to D
$$

where $D = \{z \in \mathbb{C} : |z| < 1\}$. We write the inverse of the function f as

$$
f^{-1} = a + ib : D \to \Omega.
$$

In the particular case

$$
\Omega = \{z \in \mathbb{C} : \operatorname{Re} z > 1\}
$$

we easily find that

$$
f(z) = \frac{2}{z} - 1 \quad \text{and} \quad f^{-1}(w) = \frac{2}{1 + w}.
$$

In terms of coordinates this leads to

$$
f(x, y) = \alpha(x, y) + i\beta(x, y) = \frac{2}{x + iy} - 1
$$

$$
= \frac{2x}{x^2 + y^2} - 1 + i\frac{-2y}{x^2 + y^2}
$$

and

$$\begin{aligned} f^{-1}(\alpha,\beta) &= a(\alpha,\beta)+ib(\alpha,\beta) = \frac{2}{1+\alpha+i\beta} \\ &= \frac{2(1+\alpha)}{(1+\alpha)^2+\beta^2} + i\frac{-2\beta}{(1+\alpha)^2+\beta^2}. \end{aligned}$$

Step 2. We set

$$\psi(\alpha,\beta) = (\varphi\circ f^{-1})(\alpha,\beta) = \varphi(a(\alpha,\beta),b(\alpha,\beta))$$

and we solve, as in Section 18.4, the problem

$$\begin{cases} \Delta v = \dfrac{\partial^2 v}{\partial\alpha^2} + \dfrac{\partial^2 v}{\partial\beta^2} = 0 & \text{if } \alpha^2+\beta^2 < 1 \\ v(\alpha,\beta) = \psi(\alpha,\beta) & \text{if } \alpha^2+\beta^2 = 1. \end{cases}$$

In the particular case we have, if $\alpha^2+\beta^2 = 1$ and hence

$$(1+\alpha)^2 + \beta^2 = 2(1+\alpha),$$

that

$$\begin{aligned} \psi(\alpha,\beta) &= \frac{2}{1+\left(\frac{-2\beta}{(1+\alpha)^2+\beta^2}\right)^2} = \frac{2}{1+\left(\frac{-2\beta}{2(1+\alpha)}\right)^2} \\ &= \frac{2(1+\alpha)^2}{(1+\alpha)^2+\beta^2} = 1+\alpha. \end{aligned}$$

The solution to the problem is, trivially, given by

$$v(\alpha,\beta) = 1+\alpha.$$

Step 3. The solution is obtained by setting $u = v\circ f$, namely

$$u(x,y) = v(\alpha(x,y),\beta(x,y)).$$

Indeed, from Exercise 9.10 we have $\Delta u = 0$ while since $\psi = \varphi\circ f^{-1}$, we clearly have $u = \varphi$ on $\partial\Omega$.

In the particular case we get

$$u(x,y) = v(\alpha(x,y),\beta(x,y)) = v\left(\frac{2x}{x^2+y^2}-1, \frac{-2y}{x^2+y^2}\right) = \frac{2x}{x^2+y^2}.$$

Example 18.7 *Find* $u = u(x, y)$*, a solution to*

$$\begin{cases} \Delta u = \dfrac{\partial^2 u}{\partial x^2} + \dfrac{\partial^2 u}{\partial y^2} = 0 & \text{if } (x, y) \in \Omega \\[3mm] u(x, 0) = \dfrac{8x^2}{(1 + x^2)^2} & \text{if } x \in \mathbb{R} \\[3mm] u(x, y) \to 0 & \text{if } y \to +\infty \end{cases}$$

where $\Omega = \{(x, y) \in \mathbb{R}^2 : y > 0\}$.

Discussion This problem can be solved as in the above example (cf. Exercise 18.15 below). We propose here another way of solving the problem similar to the one in Example 18.2. As usual, it is only *a posteriori* that the procedure can be justified.

Step 1 (Fourier transform). We write $v(\alpha, y)$ for the Fourier transform in x (y being considered as a parameter) of $u(x, y)$, by abuse of notations we write $\mathfrak{F}(u)$, i.e.

$$v(\alpha, y) = \mathfrak{F}(u)(\alpha, y) = \frac{1}{\sqrt{2\pi}} \int_{-\infty}^{+\infty} u(x, y) \, e^{-i\alpha x} \, dx.$$

We easily find that

$$v(\alpha, 0) = \mathfrak{F}(f)(\alpha) = \widehat{f}(\alpha) = 2\sqrt{2\pi} \, (1 - |\alpha|) \, e^{-|\alpha|}.$$

The properties of the Fourier transform lead us to

$$\begin{cases} \mathfrak{F}\left(\dfrac{\partial^2 u}{\partial x^2}\right)(\alpha, y) = (i\alpha)^2 \mathfrak{F}(u)(\alpha, y) = -\alpha^2 v(\alpha, y) \\[3mm] \mathfrak{F}\left(\dfrac{\partial^2 u}{\partial y^2}\right)(\alpha, y) = \dfrac{\partial^2 v}{\partial y^2}(\alpha, y). \end{cases}$$

Returning to our problem, we apply Fourier transform (in x) to both sides of the equation and we are driven to the following problem

$$\begin{cases} \dfrac{\partial^2 v}{\partial y^2}(\alpha, y) - \alpha^2 v(\alpha, y) = 0 & \text{if } y > 0 \\[3mm] v(\alpha, 0) = \widehat{f}(\alpha) \\[3mm] v(\alpha, y) \to 0 & \text{if } y \to +\infty. \end{cases}$$

Considering α as a parameter (here we suppose that $\alpha \neq 0$) and appealing to Case 2 of Example 17.2, we find

$$v(\alpha, y) = \widehat{f}(\alpha) \cosh(|\alpha| y) + \frac{\gamma}{|\alpha|} \sinh(|\alpha| y)$$

$$= \widehat{f}(\alpha) \frac{e^{|\alpha| y} + e^{-|\alpha| y}}{2} + \frac{\gamma}{|\alpha|} \frac{e^{|\alpha| y} - e^{-|\alpha| y}}{2},$$

where γ is a constant to be determined so as to satisfy the condition $v(\alpha, y) \to 0$ if $y \to +\infty$. We therefore choose

$$\gamma = -|\alpha| \widehat{f}(\alpha).$$

The solution is then

$$v(\alpha, y) = \widehat{f}(\alpha) e^{-|\alpha|y} = 2\sqrt{2\pi} (1 - |\alpha|) e^{-|\alpha|(1+y)}.$$

Step 2. The solution to the problem is obtained by applying the inverse Fourier transform, namely

$$u(x, y) = \mathfrak{F}^{-1}((v(\alpha, y)))(x) = \frac{1}{\sqrt{2\pi}} \int_{-\infty}^{+\infty} \widehat{f}(\alpha) e^{-|\alpha|y} e^{i\alpha x} d\alpha$$

$$= \mathfrak{F}^{-1}\left(2\sqrt{2\pi}\left(\frac{1}{1+y} - |\alpha|\right) e^{-|\alpha|(1+y)}\right)(x)$$

$$+ \mathfrak{F}^{-1}\left(2\sqrt{2\pi}\left(1 - \frac{1}{1+y}\right) e^{-|\alpha|(1+y)}\right)(x)$$

which implies

$$u(x, y) = 2\mathfrak{F}^{-1}\left(\sqrt{2\pi}\left(\frac{1}{1+y} - |\alpha|\right) e^{-|\alpha|(1+y)}\right)(x)$$

$$+ 4y\mathfrak{F}^{-1}\left(\sqrt{\frac{\pi}{2}} \frac{e^{-|\alpha|(1+y)}}{1+y}\right)(x)$$

and thus

$$u(x, y) = 2\frac{4x^2}{((1+y)^2 + x^2)^2} + 4y\frac{1}{(1+y)^2 + x^2}$$

$$= 4\frac{y(1+y)^2 + x^2(2+y)}{((1+y)^2 + x^2)^2}.$$

18.6   Exercises

Exercise 18.1 Find a solution $u = u(x, t)$ to

$$\begin{cases} \dfrac{\partial u}{\partial t} = \dfrac{\partial^2 u}{\partial x^2} & \text{if } x \in (0, \pi),\ t > 0 \\ u(0, t) = u(\pi, t) = 0 & \text{if } t > 0 \\ u(x, 0) = \cos x - \cos(3x) & \text{if } x \in (0, \pi). \end{cases}$$

Exercise 18.2 Find $u = u(x, t)$ solving

$$
\begin{cases}
\dfrac{\partial u}{\partial t} = \dfrac{\partial^2 u}{\partial x^2} & \text{if } x \in (0, \pi),\ t > 0 \\
\dfrac{\partial u}{\partial x}(0, t) = \dfrac{\partial u}{\partial x}(\pi, t) = 0 & \text{if } t > 0 \\
u(x, 0) = \cos(2x) & \text{if } x \in (0, \pi).
\end{cases}
$$

Exercise 18.3 Find a function $u = u(x, t)$ satisfying

$$
\begin{cases}
(2 + t)\dfrac{\partial u}{\partial t} = \dfrac{\partial^2 u}{\partial x^2} & \text{if } x \in (0, \pi),\ t > 0 \\
u(0, t) = u(\pi, t) = 0 & \text{if } t > 0 \\
u(x, 0) = \sin x + 4\sin(2x) & \text{if } x \in (0, \pi).
\end{cases}
$$

Exercise 18.4 Let $u = u(x, t)$ and solve

$$
\begin{cases}
\dfrac{\partial^2 u}{\partial t^2} = \dfrac{\partial^2 u}{\partial x^2} & \text{if } x \in (0, 1),\ t > 0 \\
\dfrac{\partial u}{\partial x}(0, t) = \dfrac{\partial u}{\partial x}(1, t) = 0 & \text{if } t > 0 \\
u(x, 0) = 0 & \text{if } x \in (0, 1) \\
\dfrac{\partial u}{\partial t}(x, 0) = 1 + \cos^2(\pi x) & \text{if } x \in (0, 1).
\end{cases}
$$

Exercise 18.5 Find a solution $u = u(x, t)$ to

$$
\begin{cases}
\dfrac{\partial u}{\partial t} + \dfrac{\partial^4 u}{\partial x^4} = 0 & \text{if } x \in (0, \pi),\ t > 0 \\
u(0, t) = u(\pi, t) = 0 & \text{if } t > 0 \\
\dfrac{\partial^2 u}{\partial x^2}(0, t) = \dfrac{\partial^2 u}{\partial x^2}(\pi, t) = 0 & \text{if } t > 0 \\
u(x, 0) = 2\sin x + \sin(3x) & \text{if } x \in (0, \pi).
\end{cases}
$$

Exercise 18.6 Find $u = u(x, t)$ satisfying

$$
\begin{cases}
\dfrac{\partial u}{\partial t} = \dfrac{\partial^2 u}{\partial x^2} + 2\dfrac{\partial u}{\partial x} + 2u & \text{if } x \in (0, \pi),\ t > 0 \\
u(0, t) = u(\pi, t) = 0 & \text{if } t > 0 \\
u(x, 0) = e^{-x}\sin(2x) & \text{if } x \in (0, \pi).
\end{cases}
$$

Exercise 18.7 Find $u = u(x, y)$ solution to

$$\begin{cases} \Delta u = \dfrac{\partial^2 u}{\partial x^2} + \dfrac{\partial^2 u}{\partial y^2} = 0 & \text{if } x, y \in (0, \pi) \\[2mm] \dfrac{\partial u}{\partial y}(x, 0) = \dfrac{\partial u}{\partial y}(x, \pi) = 0 & \text{if } x \in (0, \pi) \\[2mm] u(0, y) = \cos(2y), \ u(\pi, y) = 0 & \text{if } y \in (0, \pi). \end{cases}$$

Exercise 18.8 Find $u = u(x, y)$ satisfying

$$\begin{cases} \Delta u = \dfrac{\partial^2 u}{\partial x^2} + \dfrac{\partial^2 u}{\partial y^2} = 0 & \text{if } x^2 + y^2 < 4 \\[2mm] u(x, y) = 1 + x^2 + y^3 & \text{if } x^2 + y^2 = 4. \end{cases}$$

Exercise 18.9 Let $\Omega = \{(x, y) \in \mathbb{R}^2 : x^2 + y^2 < 1\}$ and $u = u(x, y)$. Solve

$$\begin{cases} \Delta u = \dfrac{\partial^2 u}{\partial x^2} + \dfrac{\partial^2 u}{\partial y^2} = 0 & \text{if } x^2 + y^2 < 1 \\[2mm] x\dfrac{\partial u}{\partial x} + y\dfrac{\partial u}{\partial y} = x^2 - \dfrac{1}{2} & \text{if } x^2 + y^2 = 1. \end{cases}$$

First write $v(r, \theta) = u(r \cos \theta, r \sin \theta)$ and then compute $\dfrac{\partial v}{\partial r}$ as a function of $\dfrac{\partial u}{\partial x}$ and $\dfrac{\partial u}{\partial y}$.

Exercise 18.10 Let $\Omega = \{(x, y) \in \mathbb{R}^2 : 1 < x^2 + y^2 < 4\}$. Find $u = u(x, y)$ such that

$$\begin{cases} \Delta u = \dfrac{\partial^2 u}{\partial x^2} + \dfrac{\partial^2 u}{\partial y^2} = 0 & \text{if } (x, y) \in \Omega \\[2mm] u(x, y) = x & \text{if } x^2 + y^2 = 1 \\[2mm] u(x, y) = x - y & \text{if } x^2 + y^2 = 4. \end{cases}$$

Exercise 18.11 Let $u = u(x, y)$ and $v(r, \theta) = u(r \cos \theta, r \sin \theta)$.

(i) Compute $\dfrac{\partial v}{\partial r}$ as a function of $\dfrac{\partial u}{\partial x}$ and $\dfrac{\partial u}{\partial y}$.

(ii) Let $\Omega = \{(x, y) \in \mathbb{R}^2 : 1 < x^2 + y^2 < 4\}$ and solve

$$\begin{cases} \Delta u = \dfrac{\partial^2 u}{\partial x^2} + \dfrac{\partial^2 u}{\partial y^2} = 0 & \text{if } (x, y) \in \Omega \\[2mm] u(x, y) = 0 & \text{if } x^2 + y^2 = 1 \\[2mm] \dfrac{1}{2}\left(x\dfrac{\partial u}{\partial x} + y\dfrac{\partial u}{\partial y}\right) + u = \dfrac{1}{2} + \log 2 + x - \dfrac{y}{2} & \text{if } x^2 + y^2 = 4. \end{cases}$$

Exercise 18.12 Let $\Omega = \{(x,y) \in \mathbb{R}^2 : y > 0 \text{ and } x^2 + y^2 < 1\}$. Find a function $u = u(x,y)$ such that

$$\begin{cases} \Delta u = \dfrac{\partial^2 u}{\partial x^2} + \dfrac{\partial^2 u}{\partial y^2} = 0 & \text{if } (x,y) \in \Omega \\ u(x,0) = 0 & \text{if } -1 \le x \le 1 \\ u(x,y) = y^2 & \text{if } y > 0 \text{ and } x^2 + y^2 = 1. \end{cases}$$

Hint: use polar coordinates.

Exercise 18.13 Find, using Fourier transform, a function $u = u(x,y)$ satisfying

$$\begin{cases} \Delta u = \dfrac{\partial^2 u}{\partial x^2} + \dfrac{\partial^2 u}{\partial y^2} = 0 & \text{if } x \in \mathbb{R}, \ y > 0 \\ u(x,0) = e^{-x^2} & \text{if } x \in \mathbb{R} \\ u(x,y) \to 0 & \text{if } y \to +\infty. \end{cases}$$

Hint: (i) The result can be given in terms of integrals.

(ii) We recall that if $\alpha \in \mathbb{R}$ $(\alpha \neq 0)$, then the solution $\varphi = \varphi(y)$ to

$$\begin{cases} \varphi''(y) - \alpha^2 \varphi(y) = 0 & \text{if } y > 0 \\ \varphi(0) = \varphi_0 \\ \varphi(y) \to 0 & \text{if } y \to +\infty. \end{cases}$$

is given by $\varphi(y) = \varphi_0 \, e^{-|\alpha|y}$, $y > 0$.

Exercise 18.14 Let $u = u(x,y)$ and solve

$$\begin{cases} \Delta u = \dfrac{\partial^2 u}{\partial x^2} + \dfrac{\partial^2 u}{\partial y^2} = 0 & \text{if } y \in \mathbb{R}, \ x > 0 \\ u(0,y) = \dfrac{1}{(1+y^2)^2} & \text{if } y \in \mathbb{R}. \end{cases}$$

Exercise 18.15 Consider Example 18.7 (which was solved using Fourier transform) and solve it by the method that uses conformal mapping.

Exercise 18.16 Let $\Omega = \{(x,y) \in \mathbb{R}^2 : (x-2)^2 + (y-2)^2 > 4\}$. Find a solution $u = u(x,y)$ to

$$\begin{cases} \Delta u = \dfrac{\partial^2 u}{\partial x^2} + \dfrac{\partial^2 u}{\partial y^2} = 0 & \text{if } (x,y) \in \Omega \\ u(x,y) = x^2 + 2y^2 & \text{if } (x,y) \in \partial\Omega. \end{cases}$$

Exercise 18.17 (D'Alembert method) Let $c \in \mathbb{R}$, $f \in C^2(\mathbb{R})$ and $g \in C^1(\mathbb{R})$. Verify that

$$u(x,t) = \frac{1}{2}\left[f(x+ct) + f(x-ct)\right] + \frac{1}{2c}\int_{x-ct}^{x+ct} g(s)\,ds$$

is a solution to the wave equation

$$\begin{cases} \dfrac{\partial^2 u}{\partial t^2} = c^2 \dfrac{\partial^2 u}{\partial x^2} & \text{if } x \in \mathbb{R}, \ t > 0 \\[2mm] u(x,0) = f(x) & \text{if } x \in \mathbb{R} \\[2mm] \dfrac{\partial u}{\partial t}(x,0) = g(x) & \text{if } x \in \mathbb{R}. \end{cases}$$

Part IV

Solutions to the exercises

Chapter 1

Differential operators of mathematical physics

Exercise 1.1 (i) We first compute the gradient of f_1, f_2, f_3 and find

$$\operatorname{grad} f_1 = \begin{pmatrix} y^2 z \cos\left(xz\right) \\ 2y \sin\left(xz\right) \\ xy^2 \cos\left(xz\right) \end{pmatrix}, \quad \operatorname{grad} f_2 = \begin{pmatrix} -2xe^y \sin\left(x^2 + z\right) \\ e^y \cos\left(x^2 + z\right) \\ -e^y \sin\left(x^2 + z\right) \end{pmatrix}$$

$$\operatorname{grad} f_3 = \begin{pmatrix} -\dfrac{y \sin\left(xy\right)}{2 + \cos\left(xy\right)} \\ -\dfrac{x \sin\left(xy\right)}{2 + \cos\left(xy\right)} \\ 0 \end{pmatrix}.$$

(ii) We have

$$\operatorname{div} F = \frac{\partial f_1}{\partial x} + \frac{\partial f_2}{\partial y} + \frac{\partial f_3}{\partial z} = y^2 z \cos\left(xz\right) + e^y \cos\left(x^2 + z\right).$$

(iii) Finally, we get

$$\operatorname{curl} F = \begin{pmatrix} \dfrac{\partial f_3}{\partial y} - \dfrac{\partial f_2}{\partial z} \\ \dfrac{\partial f_1}{\partial z} - \dfrac{\partial f_3}{\partial x} \\ \dfrac{\partial f_2}{\partial x} - \dfrac{\partial f_1}{\partial y} \end{pmatrix} = \begin{pmatrix} -\dfrac{x \sin\left(xy\right)}{2 + \cos\left(xy\right)} + e^y \sin\left(x^2 + z\right) \\ xy^2 \cos\left(xz\right) + \dfrac{y \sin\left(xy\right)}{2 + \cos\left(xy\right)} \\ -2(xe^y \sin\left(x^2 + z\right) + y \sin\left(xz\right)) \end{pmatrix}.$$

Exercise 1.2 (i) The following expressions make sense: (i); (ii); (iii); (v); (vi) and (viii).

(ii) The following expressions are not well defined: (iv); (vii) and (ix).

Exercise 1.3 We first deduce from Example 1.3 (by taking $a = 0$) that

$$\frac{\partial r}{\partial x_i} = \frac{x_i}{r}.$$

(i) We begin by writing

$$\frac{\partial f}{\partial x_i} = \psi'(r)\frac{\partial r}{\partial x_i} = \psi'(r)\frac{x_i}{r}.$$

We then get

$$\frac{\partial^2 f}{\partial x_i^2} = \psi''(r)\frac{\partial r}{\partial x_i}\frac{x_i}{r} + \psi'(r)\frac{\partial}{\partial x_i}\left(x_i r^{-1}\right)$$

$$= \psi''(r)\frac{x_i^2}{r^2} + \psi'(r)\left[\frac{1}{r} - x_i\frac{1}{r^2}\frac{\partial r}{\partial x_i}\right]$$

$$= \psi''(r)\frac{x_i^2}{r^2} + \psi'(r)\left[\frac{1}{r} - \frac{x_i^2}{r^3}\right].$$

We finally have

$$\Delta f = \sum_{i=1}^{n}\frac{\partial^2 f}{\partial x_i^2} = \frac{\psi''(r)}{r^2}\sum_{i=1}^{n}x_i^2 + \psi'(r)\left[\frac{1}{r}\sum_{i=1}^{n}1 - \frac{1}{r^3}\sum_{i=1}^{n}x_i^2\right]$$

$$= \psi''(r) + \psi'(r)\left[\frac{n}{r} - \frac{1}{r}\right] = \psi''(r) + \frac{n-1}{r}\psi'(r).$$

(ii) If $\Delta f = 0$, we then find

$$\psi''(r) = \frac{1-n}{r}\psi'(r) \;\Rightarrow\; \frac{\psi''(r)}{\psi'(r)} = \frac{1-n}{r}$$

and thus

$$\log \psi'(r) = (1-n)\log r \;\Rightarrow\; \psi'(r) = r^{1-n}.$$

We easily get

$$\psi(r) = \begin{cases} \log r & \text{if } n = 2 \\ \dfrac{r^{2-n}}{2-n} & \text{if } n \geq 3. \end{cases}$$

Exercise 1.4 (i) Since $x, y > 0$, we have

$$r = \left(x^2 + y^2\right)^{1/2} \quad \text{and} \quad \tan\theta = \frac{y}{x} \;\Rightarrow\; \theta = \arctan\frac{y}{x}.$$

Moreover, we saw (cf. Example 1.3) that

$$\frac{\partial r}{\partial x} = \frac{x}{r} \quad \text{and} \quad \frac{\partial r}{\partial y} = \frac{y}{r}.$$

We thus get

$$\frac{\partial \theta}{\partial x} = \frac{\dfrac{-y}{x^2}}{1 + \dfrac{y^2}{x^2}} = \frac{-y}{x^2 + y^2} = \frac{-y}{r^2}$$

and

$$\frac{\partial \theta}{\partial y} = \frac{\dfrac{1}{x}}{1 + \dfrac{y^2}{x^2}} = \frac{x}{r^2}.$$

(ii) We write

$$\frac{\partial f}{\partial x} = \frac{\partial g}{\partial r}\frac{\partial r}{\partial x} + \frac{\partial g}{\partial \theta}\frac{\partial \theta}{\partial x} = \frac{x}{r}\frac{\partial g}{\partial r} - \frac{y}{r^2}\frac{\partial g}{\partial \theta}$$

and

$$\frac{\partial f}{\partial y} = \frac{\partial g}{\partial r}\frac{\partial r}{\partial y} + \frac{\partial g}{\partial \theta}\frac{\partial \theta}{\partial y} = \frac{y}{r}\frac{\partial g}{\partial r} + \frac{x}{r^2}\frac{\partial g}{\partial \theta}.$$

(iii) We have

$$\frac{\partial^2 f}{\partial x^2} = \frac{\partial}{\partial x}\left(\frac{x}{r}\right)\frac{\partial g}{\partial r} + \frac{x}{r}\frac{\partial}{\partial x}\left(\frac{\partial g}{\partial r}\right)$$
$$-y\frac{\partial}{\partial x}\left(r^{-2}\right)\frac{\partial g}{\partial \theta} - \frac{y}{r^2}\frac{\partial}{\partial x}\left(\frac{\partial g}{\partial \theta}\right)$$

and thus

$$\frac{\partial^2 f}{\partial x^2} = \frac{r - x\frac{x}{r}}{r^2}\frac{\partial g}{\partial r} + \frac{x}{r}\left(\frac{x}{r}\frac{\partial^2 g}{\partial r^2} - \frac{y}{r^2}\frac{\partial^2 g}{\partial r\partial \theta}\right)$$
$$+ \frac{2xy}{r^4}\frac{\partial g}{\partial \theta} - \frac{y}{r^2}\left(\frac{x}{r}\frac{\partial^2 g}{\partial r\partial \theta} - \frac{y}{r^2}\frac{\partial^2 g}{\partial \theta^2}\right).$$

We therefore find

$$\frac{\partial^2 f}{\partial x^2} = \frac{y^2}{r^3}\frac{\partial g}{\partial r} + \frac{x^2}{r^2}\frac{\partial^2 g}{\partial r^2} - \frac{2xy}{r^3}\frac{\partial^2 g}{\partial r\partial \theta} + \frac{2xy}{r^4}\frac{\partial g}{\partial \theta} + \frac{y^2}{r^4}\frac{\partial^2 g}{\partial \theta^2}.$$

A similar calculation gives

$$\frac{\partial^2 f}{\partial y^2} = \frac{x^2}{r^3}\frac{\partial g}{\partial r} + \frac{y^2}{r^2}\frac{\partial^2 g}{\partial r^2} + \frac{2xy}{r^3}\frac{\partial^2 g}{\partial r\partial \theta} - \frac{2xy}{r^4}\frac{\partial g}{\partial \theta} + \frac{x^2}{r^4}\frac{\partial^2 g}{\partial \theta^2}$$

and so

$$\Delta f = \frac{1}{r}\frac{\partial g}{\partial r} + \frac{\partial^2 g}{\partial r^2} + \frac{1}{r^2}\frac{\partial^2 g}{\partial \theta^2}.$$

(iv) Writing the function in polar coordinates, we obtain

$$f(x,y) = r + \theta^2$$

$$\Delta f = \frac{1}{r} + \frac{2}{r^2} = \frac{2 + \sqrt{x^2 + y^2}}{x^2 + y^2}.$$

Exercise 1.5 (i) We first show that $\operatorname{div}\operatorname{grad} f = \Delta f$. We have

$$\operatorname{div}\operatorname{grad} f = \operatorname{div}\left(\frac{\partial f}{\partial x_1}, \frac{\partial f}{\partial x_2}, \cdots, \frac{\partial f}{\partial x_n}\right)$$

$$= \frac{\partial}{\partial x_1}\left(\frac{\partial f}{\partial x_1}\right) + \frac{\partial}{\partial x_2}\left(\frac{\partial f}{\partial x_2}\right) + \cdots + \frac{\partial}{\partial x_n}\left(\frac{\partial f}{\partial x_n}\right)$$

$$= \frac{\partial^2 f}{\partial x_1^2} + \frac{\partial^2 f}{\partial x_2^2} + \cdots + \frac{\partial^2 f}{\partial x_n^2} = \Delta f.$$

(ii) We now verify that $\operatorname{curl}\operatorname{grad} f = 0$. We indeed have

$$\operatorname{curl}\operatorname{grad} f = \begin{vmatrix} e_1 & e_2 & e_3 \\ \dfrac{\partial}{\partial x} & \dfrac{\partial}{\partial y} & \dfrac{\partial}{\partial z} \\ \dfrac{\partial f}{\partial x} & \dfrac{\partial f}{\partial y} & \dfrac{\partial f}{\partial z} \end{vmatrix} = \begin{pmatrix} \dfrac{\partial^2 f}{\partial y \partial z} - \dfrac{\partial^2 f}{\partial z \partial y} \\ \dfrac{\partial^2 f}{\partial z \partial x} - \dfrac{\partial^2 f}{\partial x \partial z} \\ \dfrac{\partial^2 f}{\partial x \partial y} - \dfrac{\partial^2 f}{\partial y \partial x} \end{pmatrix} = \begin{pmatrix} 0 \\ 0 \\ 0 \end{pmatrix}.$$

(iii) Since the following equalities hold

$$\operatorname{curl} F = \begin{vmatrix} e_1 & e_2 & e_3 \\ \dfrac{\partial}{\partial x} & \dfrac{\partial}{\partial y} & \dfrac{\partial}{\partial z} \\ F_1 & F_2 & F_3 \end{vmatrix} = \begin{pmatrix} \dfrac{\partial F_3}{\partial y} - \dfrac{\partial F_2}{\partial z} \\ \dfrac{\partial F_1}{\partial z} - \dfrac{\partial F_3}{\partial x} \\ \dfrac{\partial F_2}{\partial x} - \dfrac{\partial F_1}{\partial y} \end{pmatrix},$$

we get

$$\operatorname{div}\operatorname{curl} F = \frac{\partial}{\partial x}\left(\frac{\partial F_3}{\partial y} - \frac{\partial F_2}{\partial z}\right) + \frac{\partial}{\partial y}\left(\frac{\partial F_1}{\partial z} - \frac{\partial F_3}{\partial x}\right)$$

$$+ \frac{\partial}{\partial z}\left(\frac{\partial F_2}{\partial x} - \frac{\partial F_1}{\partial y}\right)$$

and thus

$$\operatorname{div}\operatorname{curl} F = \frac{\partial^2 F_3}{\partial x \partial y} - \frac{\partial^2 F_2}{\partial x \partial z} + \frac{\partial^2 F_1}{\partial y \partial z} - \frac{\partial^2 F_3}{\partial y \partial x} + \frac{\partial^2 F_2}{\partial z \partial x} - \frac{\partial^2 F_1}{\partial z \partial y} = 0.$$

Exercise 1.6 (i) We have

$$\operatorname{div}\left(f \operatorname{grad} g\right) = \operatorname{div}\left(f\frac{\partial g}{\partial x_1}, f\frac{\partial g}{\partial x_2}, \cdots, f\frac{\partial g}{\partial x_n}\right)$$

$$= \frac{\partial}{\partial x_1}\left(f\frac{\partial g}{\partial x_1}\right) + \frac{\partial}{\partial x_2}\left(f\frac{\partial g}{\partial x_2}\right) + \cdots + \frac{\partial}{\partial x_n}\left(f\frac{\partial g}{\partial x_n}\right)$$

$$= \sum_{i=1}^{n} \frac{\partial}{\partial x_i}\left(f\frac{\partial g}{\partial x_i}\right) = \sum_{i=1}^{n}\left[\frac{\partial f}{\partial x_i}\frac{\partial g}{\partial x_i} + f\frac{\partial^2 g}{\partial x_i^2}\right]$$

and then

$$\operatorname{div}\left(f\operatorname{grad} g\right) = \operatorname{grad} f \cdot \operatorname{grad} g + f\,\Delta g.$$

(ii) Similar calculations give

$$\operatorname{grad}\left(fg\right) = \left(\frac{\partial f}{\partial x_1}g + \frac{\partial g}{\partial x_1}f, \frac{\partial f}{\partial x_2}g + \frac{\partial g}{\partial x_2}f, \cdots, \frac{\partial f}{\partial x_n}g + \frac{\partial g}{\partial x_n}f\right)$$

$$= g\left(\frac{\partial f}{\partial x_1}, \frac{\partial f}{\partial x_2}, \cdots, \frac{\partial f}{\partial x_n}\right) + f\left(\frac{\partial g}{\partial x_1}, \frac{\partial g}{\partial x_2}, \cdots, \frac{\partial g}{\partial x_n}\right)$$

$$= g\operatorname{grad} f + f\operatorname{grad} g.$$

Exercise 1.7 (i) Since $F = (F_1, \cdots, F_n)$, we deduce

$$\operatorname{div}\left(fF\right) = \operatorname{div}\left(fF_1, fF_2, \cdots, fF_n\right)$$

$$= \sum_{i=1}^{n} \frac{\partial}{\partial x_i}\left(fF_i\right) = \sum_{i=1}^{n}\left[\frac{\partial f}{\partial x_i}F_i + f\frac{\partial F_i}{\partial x_i}\right]$$

$$= F \cdot \operatorname{grad} f + f\operatorname{div} F.$$

(ii) We now want to show that $\operatorname{curl}\operatorname{curl} F = -\Delta F + \operatorname{grad}\operatorname{div} F$. We have

$$\operatorname{curl} F = \left(\frac{\partial F_3}{\partial y} - \frac{\partial F_2}{\partial z}, \frac{\partial F_1}{\partial z} - \frac{\partial F_3}{\partial x}, \frac{\partial F_2}{\partial x} - \frac{\partial F_1}{\partial y}\right).$$

We therefore get

$$\operatorname{curl}\operatorname{curl} F = \begin{vmatrix} e_1 & e_2 & e_3 \\[2mm] \dfrac{\partial}{\partial x} & \dfrac{\partial}{\partial y} & \dfrac{\partial}{\partial z} \\[3mm] \dfrac{\partial F_3}{\partial y} - \dfrac{\partial F_2}{\partial z} & \dfrac{\partial F_1}{\partial z} - \dfrac{\partial F_3}{\partial x} & \dfrac{\partial F_2}{\partial x} - \dfrac{\partial F_1}{\partial y} \end{vmatrix},$$

i.e.

$$\operatorname{curl} \operatorname{curl} F = \begin{pmatrix} \dfrac{\partial^2 F_2}{\partial x \partial y} - \dfrac{\partial^2 F_1}{\partial y^2} - \dfrac{\partial^2 F_1}{\partial z^2} + \dfrac{\partial^2 F_3}{\partial x \partial z} \\[2mm] \dfrac{\partial^2 F_3}{\partial y \partial z} - \dfrac{\partial^2 F_2}{\partial z^2} - \dfrac{\partial^2 F_2}{\partial x^2} + \dfrac{\partial^2 F_1}{\partial x \partial y} \\[2mm] \dfrac{\partial^2 F_1}{\partial x \partial z} - \dfrac{\partial^2 F_3}{\partial x^2} - \dfrac{\partial^2 F_3}{\partial y^2} + \dfrac{\partial^2 F_2}{\partial y \partial z} \end{pmatrix}.$$

On the other hand, since we have

$$\Delta F = \begin{pmatrix} \dfrac{\partial^2 F_1}{\partial x^2} + \dfrac{\partial^2 F_1}{\partial y^2} + \dfrac{\partial^2 F_1}{\partial z^2} \\[2mm] \dfrac{\partial^2 F_2}{\partial x^2} + \dfrac{\partial^2 F_2}{\partial y^2} + \dfrac{\partial^2 F_2}{\partial z^2} \\[2mm] \dfrac{\partial^2 F_3}{\partial x^2} + \dfrac{\partial^2 F_3}{\partial y^2} + \dfrac{\partial^2 F_3}{\partial z^2} \end{pmatrix},$$

we deduce that

$$\operatorname{curl} \operatorname{curl} F + \Delta F = \begin{pmatrix} \dfrac{\partial^2 F_2}{\partial x \partial y} + \dfrac{\partial^2 F_3}{\partial x \partial z} + \dfrac{\partial^2 F_1}{\partial x^2} \\[2mm] \dfrac{\partial^2 F_3}{\partial y \partial z} + \dfrac{\partial^2 F_1}{\partial x \partial y} + \dfrac{\partial^2 F_2}{\partial y^2} \\[2mm] \dfrac{\partial^2 F_1}{\partial x \partial z} + \dfrac{\partial^2 F_2}{\partial y \partial z} + \dfrac{\partial^2 F_3}{\partial z^2} \end{pmatrix}.$$

The following equalities

$$\operatorname{grad} \operatorname{div} F = \operatorname{grad} \left(\frac{\partial F_1}{\partial x} + \frac{\partial F_2}{\partial y} + \frac{\partial F_3}{\partial z} \right)$$

$$= \begin{pmatrix} \dfrac{\partial^2 F_1}{\partial x^2} + \dfrac{\partial^2 F_2}{\partial x \partial y} + \dfrac{\partial^2 F_3}{\partial x \partial z} \\[2mm] \dfrac{\partial^2 F_1}{\partial x \partial y} + \dfrac{\partial^2 F_2}{\partial y^2} + \dfrac{\partial^2 F_3}{\partial y \partial z} \\[2mm] \dfrac{\partial^2 F_1}{\partial x \partial z} + \dfrac{\partial^2 F_2}{\partial y \partial z} + \dfrac{\partial^2 F_3}{\partial z^2} \end{pmatrix}$$

imply that

$$\operatorname{curl} \operatorname{curl} F = -\Delta F + \operatorname{grad} \operatorname{div} F.$$

(iii) Finally, we want to show that

$$\operatorname{curl} (fF) = \operatorname{grad} f \wedge F + f \operatorname{curl} F.$$

We have

$$
\operatorname{curl}(fF) =
\begin{vmatrix}
e_1 & e_2 & e_3 \\
\dfrac{\partial}{\partial x} & \dfrac{\partial}{\partial y} & \dfrac{\partial}{\partial z} \\
fF_1 & fF_2 & fF_3
\end{vmatrix}
=
\begin{pmatrix}
\dfrac{\partial}{\partial y}(fF_3) - \dfrac{\partial}{\partial z}(fF_2) \\[2mm]
\dfrac{\partial}{\partial z}(fF_1) - \dfrac{\partial}{\partial x}(fF_3) \\[2mm]
\dfrac{\partial}{\partial x}(fF_2) - \dfrac{\partial}{\partial y}(fF_1)
\end{pmatrix}.
$$

We thus get that

$$
\operatorname{curl}(fF) =
\begin{pmatrix}
F_3\dfrac{\partial f}{\partial y} - F_2\dfrac{\partial f}{\partial z} + f\left(\dfrac{\partial F_3}{\partial y} - \dfrac{\partial F_2}{\partial z}\right) \\[3mm]
F_1\dfrac{\partial f}{\partial z} - F_3\dfrac{\partial f}{\partial x} + f\left(\dfrac{\partial F_1}{\partial z} - \dfrac{\partial F_3}{\partial x}\right) \\[3mm]
F_2\dfrac{\partial f}{\partial x} - F_1\dfrac{\partial f}{\partial y} + f\left(\dfrac{\partial F_2}{\partial x} - \dfrac{\partial F_1}{\partial y}\right)
\end{pmatrix}.
$$

By definition, we have

$$
\operatorname{grad} f \wedge F =
\begin{vmatrix}
e_1 & e_2 & e_3 \\
\dfrac{\partial f}{\partial x} & \dfrac{\partial f}{\partial y} & \dfrac{\partial f}{\partial z} \\
F_1 & F_2 & F_3
\end{vmatrix}
=
\begin{pmatrix}
F_3\dfrac{\partial f}{\partial y} - F_2\dfrac{\partial f}{\partial z} \\[3mm]
F_1\dfrac{\partial f}{\partial z} - F_3\dfrac{\partial f}{\partial x} \\[3mm]
F_2\dfrac{\partial f}{\partial x} - F_1\dfrac{\partial f}{\partial y}
\end{pmatrix}.
$$

Gathering these results we have indeed obtained that

$$
\operatorname{curl}(fF) = \operatorname{grad} f \wedge F + f\operatorname{curl} F.
$$

Chapter 2

Line integrals

Exercise 2.1 (i) A parametrization of Γ is given by

$$x = t \quad \text{and} \quad y = f(t),$$

that is

$$\gamma(t) = (t, f(t)) \quad \Rightarrow \quad \gamma'(t) = (1, f'(t)).$$

We therefore get

$$\text{length}(\Gamma) = \int_a^b \sqrt{1 + (f'(t))^2}\, dt.$$

(ii) We now have $f(t) = \cosh t$ and so

$$\sqrt{1 + (f'(t))^2} = \cosh t.$$

We, hence, obtain that

$$\text{length}(\Gamma) = \sinh 1 - \sinh 0 = \frac{e - e^{-1}}{2}.$$

(iii) We let $\gamma(t) = (r(t)\cos t, \ r(t)\sin t)$ and we deduce that

$$\gamma'(t) = (r'(t)\cos t - r(t)\sin t, r'(t)\sin t + r(t)\cos t).$$

We then have

$$\|\gamma'(t)\|^2 = r'^2 + r^2$$

and finally

$$\text{length}(\Gamma) = \int_a^b \sqrt{r'(t)^2 + r(t)^2}\, dt.$$

Exercise 2.2 (i) Since

$$\gamma_1(t) = (t, t) \ \Rightarrow \ \gamma_1'(t) = (1, 1),$$

we find that

$$\int_{\Gamma_1} F \cdot dl = \int_0^1 \left(t^2, t^2 - t\right) \cdot (1, 1) \, dt = \int_0^1 \left(2t^2 - t\right) dt = \frac{1}{6}.$$

(ii) We choose

$$\gamma_2(t) = (t, e^t) \ \Rightarrow \ \gamma_2'(t) = (1, e^t)$$

and we get

$$\int_{\Gamma_2} F \cdot dl = \int_0^1 \left(te^t, e^{2t} - t\right) \cdot (1, e^t) \, dt = \int_0^1 e^{3t} dt = \frac{1}{3}(e^3 - 1).$$

(iii) In this case, we write

$$\gamma_3(t) = \left(\sqrt{t}, t^2\right) \ \Rightarrow \ \gamma_3'(t) = \left(\frac{1}{2\sqrt{t}}, 2t\right)$$

and we find

$$\int_{\Gamma_3} F \cdot dl = \int_1^2 \left(t^2\sqrt{t}, t^4 - \sqrt{t}\right) \cdot \left(\frac{1}{2\sqrt{t}}, 2t\right) dt$$

$$= \int_1^2 \left(\frac{1}{2} t^2 + 2t^5 - 2t\sqrt{t}\right) dt = \frac{689}{30} - \frac{16}{5}\sqrt{2}.$$

Exercise 2.3 (i) We choose the following parametrization

$$\gamma(t) = (\cos t, \sin t, 0) \ \Rightarrow \ \gamma'(t) = (-\sin t, \cos t, 0) \ \text{ with } t \in [0, 2\pi].$$

We therefore have that

$$\int_\Gamma F \cdot dl = \int_0^{2\pi} (\cos t, 0, \sin t) \cdot (-\sin t, \cos t, 0) \, dt$$

$$= \int_0^{2\pi} -\sin t \cos t \, dt = 0.$$

(ii) We set $\gamma(t) = (t, e^t, t)$ and thus

$$\gamma'(t) = (1, e^t, 1) \quad \text{with } t \in [0, 1].$$

We then infer that

$$\int_\Gamma F \cdot dl = \int_0^1 (t, e^t, t) \cdot (1, e^t, 1)\, dt = \int_0^1 (2t + e^{2t})\, dt$$

$$= t^2 + \frac{1}{2}e^{2t}\Big|_0^1 = \frac{1+e^2}{2}.$$

(Note that in (i), unlike in (ii), the orientation of Γ is not explicitly given; the choice of the opposite orientation would have given the same result, namely 0.)

Exercise 2.4 From the given parametrization, we deduce that

$$\gamma'(t) = (-\sin t, \cos t, t)$$

and thus

$$\|\gamma'(t)\| = \sqrt{1+t^2} \quad \text{and} \quad f(\gamma(t)) = 1 + t.$$

We therefore have

$$\int_\Gamma f\, dl = \int_0^1 (1+t)\sqrt{1+t^2}\, dt = \int_0^1 \sqrt{1+t^2}\, dt + \int_0^1 t\sqrt{1+t^2}\, dt$$

$$= \left[\frac{t}{2}\sqrt{t^2+1} + \frac{1}{2}\log\left(t + \sqrt{t^2+1}\right) + \frac{1}{3}(1+t^2)\sqrt{1+t^2} \right]\Big|_0^1$$

$$= \frac{-1}{3} + \frac{7\sqrt{2}}{6} + \frac{1}{2}\log\left(1 + \sqrt{2}\right).$$

Exercise 2.5 Let $\gamma : [a, b] \to \Gamma$ be a parametrization of Γ. We have $\gamma(a) = A$, $\gamma(b) = B$. We also have $F(\gamma(t)) = m\gamma''(t)$. We thus find that

$$\int_\Gamma F \cdot dl = \int_a^b m\gamma''(t) \cdot \gamma'(t)\, dt = m \int_a^b \frac{d}{dt}\left[\frac{1}{2}\|\gamma'(t)\|^2 \right] dt$$

$$= \frac{m}{2}\|\gamma'(b)\|^2 - \frac{m}{2}\|\gamma'(a)\|^2,$$

which is the variation in the kinetic energy.

Exercise 2.6 The curve is given by $y = \pm 2x\sqrt{1 - x^2}$. We can choose the following parametrization

$$\Gamma = \{\gamma(t) = (\sin t, \sin(2t)),\ t \in [0, \pi]\}.$$

Note that the orientation of Γ has not been explicitly given in the problem; we decide to choose the counterclockwise direction. We then find

$$\int_\Gamma F \cdot dl = \int_0^\pi (\sin t + \sin(2t), -\sin t) \cdot (\cos t, 2\cos(2t))\, dt$$

$$= \int_0^\pi \{-2\cos(2t)\sin t + \cos t\, [\sin t + \sin(2t)]\}\, dt = \frac{8}{3}.$$

Chapter 3

Gradient vector fields

Exercise 3.1 (i) Since

$$\operatorname{curl} F_1 = \frac{\partial}{\partial x}\left(xy - x\right) - \frac{\partial}{\partial y}\left(y\right) = y - 1 - 1 \neq 0,$$

the field F_1 is not a gradient vector field. We choose $\Gamma = \Gamma_1 \cup \Gamma_2$ (cf. Figure 3.1), where

$$\Gamma_1 = \left\{ \left(t, t^2\right) : t \in [0, 1] \right\} \quad \text{and} \quad \Gamma_2 = \left\{ (2 - t,\ 2 - t) : t \in [1, 2] \right\}.$$

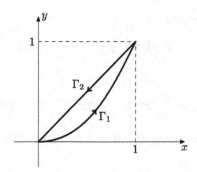

Figure 3.1: Exercise 3.1

We then infer that

$$\int_{\Gamma} F_1 \cdot dl = \int_{\Gamma_1} F_1 \cdot dl + \int_{\Gamma_2} F_1 \cdot dl$$

and thus

$$\int_{\Gamma} F_1 \cdot dl = \int_0^1 \left(t^2, t^3 - t\right) \cdot (1, 2t)\, dt$$

$$+ \int_1^2 \left(2 - t, (2 - t)^2 - (2 - t)\right) \cdot (-1, -1)\, dt$$

$$= \frac{-4}{15} \neq 0.$$

(ii) Since

$$\operatorname{curl} F_2 = \frac{\partial}{\partial x}\left(x^3\right) - \frac{\partial}{\partial y}\left(3x^2 y + 2x\right) = 3x^2 - 3x^2 = 0,$$

and the domain of F_2 is $\mathbb{R}^2$, which is convex, we deduce that F_2 is the gradient of a potential f. This potential satisfies

$$\operatorname{grad} f = \left(\begin{array}{c} \partial f / \partial x \\ \partial f / \partial y \end{array}\right) = \left(\begin{array}{c} 3x^2 y + 2x \\ x^3 \end{array}\right)$$

and thus

$$f(x, y) = x^3 y + h(x).$$

Differentiating this equation with respect to x and comparing to the first component of the gradient, we get

$$3x^2 y + h'(x) = 3x^2 y + 2x \implies h'(x) = 2x \implies h(x) = x^2 + \text{constant}.$$

The potential is therefore

$$f(x, y) = x^3 y + x^2 + \text{constant}.$$

(iii) Since

$$\operatorname{curl} F_3 = \frac{\partial}{\partial x}\left(x^2\right) - \frac{\partial}{\partial y}\left(3x^2 y\right) = 2x - 3x^2 \neq 0,$$

we find that F_3 is not the gradient of a potential. We choose $\Gamma = \Gamma_1 \cup \Gamma_2$ as in (i). We have

$$\int_{\Gamma} F_3 \cdot dl = \int_0^1 \left(3t^4, t^2\right) \cdot (1, 2t)\, dt$$

$$+ \int_1^2 \left(3(2 - t)^3, (2 - t)^2\right) \cdot (-1, -1)\, dt$$

$$= \frac{1}{60}.$$

Exercise 3.2 (i) Case $n = 2$. We first observe that

$$\frac{\partial}{\partial t} [tf(tx, ty)] = f(tx, ty) + t \left[x \frac{\partial f(tx, ty)}{\partial u} + y \frac{\partial f(tx, ty)}{\partial v} \right]$$

$$\frac{\partial}{\partial x} [xf(tx, ty) + yg(tx, ty)] = f(tx, ty) + tx \frac{\partial f(tx, ty)}{\partial u} + ty \frac{\partial g(tx, ty)}{\partial u}.$$

Since $\partial f / \partial v = \partial g / \partial u$, we deduce that these two quantities are equal. Returning to the definition

$$\varphi(x, y) = \int_0^1 [xf(tx, ty) + yg(tx, ty)] \, dt$$

and using the previous observation, we find

$$\frac{\partial \varphi}{\partial x} = \int_0^1 \frac{\partial}{\partial x} [xf(tx, ty) + yg(tx, ty)] \, dt$$

$$= \int_0^1 \frac{\partial}{\partial t} [tf(tx, ty)] \, dt = tf(tx, ty)|_0^1 = f(x, y).$$

In a similar way, we obtain $\dfrac{\partial \varphi}{\partial y} = g(x, y)$. In the example, we first check
that

$$\frac{\partial}{\partial y} (2xy) = \frac{\partial}{\partial x} (x^2 + y).$$

We then compute

$$\varphi(x, y) = \int_0^1 (2t^2 xy, t^2 x^2 + ty) \cdot (x, y) \, dt$$

$$= \int_0^1 (3t^2 x^2 y + ty^2) \, dt = x^2 y + \frac{y^2}{2}.$$

(ii) Case $n \geq 2$. As above, we observe that for every $j = 1, \cdots, n$,

$$\frac{\partial}{\partial t} [tF_j(tx)] = F_j(tx) + t \sum_{i=1}^n x_i \frac{\partial F_j}{\partial u_i}(tx)$$

$$\frac{\partial}{\partial x_j} \left[\sum_{i=1}^n x_i F_i(tx) \right] = F_j(tx) + t \sum_{i=1}^n x_i \frac{\partial F_i}{\partial u_j}(tx).$$

Since $\partial F_i / \partial u_j = \partial F_j / \partial u_i$, we find that these two quantities are equal.
From the definition

$$\varphi(x) = \int_0^1 F(tx) \cdot x \, dt = \int_0^1 \sum_{i=1}^n x_i F_i(tx) \, dt$$

and the previous identities, we finally get

$$\frac{\partial \varphi}{\partial x_j} = \int_0^1 \frac{\partial}{\partial x_j} \left[\sum_{i=1}^n x_i F_i(tx) \right] dt = \int_0^1 \frac{\partial}{\partial t} \left[t F_j(tx) \right] dt$$
$$= t F_j(tx) \big|_0^1 = F_j(x).$$

Exercise 3.3 We easily check that $\operatorname{curl} F = 0$. If $F = \operatorname{grad} f$ we must have

$$\frac{\partial f}{\partial x} = 2xy + \frac{z}{1+x^2}, \quad \frac{\partial f}{\partial y} = x^2 + 2yz, \quad \frac{\partial f}{\partial z} = y^2 + \arctan x.$$

Integrating the first equation with respect to x leads to

$$f(x, y, z) = x^2 y + z \arctan x + \varphi(y, z).$$

Differentiating f with respect to y and z, we find

$$\frac{\partial f}{\partial y} = x^2 + \frac{\partial \varphi}{\partial y} \quad \text{and} \quad \frac{\partial f}{\partial z} = \arctan x + \frac{\partial \varphi}{\partial z}$$

and therefore

$$\frac{\partial \varphi}{\partial y} = 2yz \quad \text{and} \quad \frac{\partial \varphi}{\partial z} = y^2.$$

Integrating the first equation with respect to y gives us

$$\varphi(y, z) = y^2 z + \phi(z) \quad \Rightarrow \quad \frac{\partial \varphi}{\partial z} = y^2 + \phi'(z).$$

From the second equation we get

$$\phi'(z) = 0 \quad \Rightarrow \quad \phi(z) = c \quad \Rightarrow \quad \varphi(y, z) = y^2 z + c.$$

We finally have

$$f(x, y, z) = x^2 y + z \arctan x + y^2 z + c.$$

Exercise 3.4 We first check that

$$\operatorname{curl} F = \frac{\partial}{\partial x} \left(\frac{x}{x^2 + y^2} \right) - \frac{\partial}{\partial y} \left(\frac{-y}{x^2 + y^2} \right) = 0.$$

We now want to find a potential. Recall that

$$\Omega = \mathbb{R}^2 \setminus \left\{ (0, y) \in \mathbb{R}^2 : y \geq 0 \right\}.$$

Case 1. We find a potential in $\Omega_1 = \left\{ (x, y) \in \mathbb{R}^2 : x > 0 \right\}$. We have

$$\left. \begin{array}{l} \dfrac{\partial f}{\partial x} = \dfrac{-y}{x^2 + y^2} \\[3mm] \dfrac{\partial f}{\partial y} = \dfrac{x}{x^2 + y^2} \end{array} \right\} \;\Rightarrow\; f(x, y) = \arctan \dfrac{y}{x} + \alpha_+ \quad (\alpha_+ \in \mathbb{R}).$$

Case 2. Reasoning as above we obtain

$$f(x, y) = \arctan \frac{y}{x} + \alpha_- \quad (\alpha_- \in \mathbb{R})$$

in $\Omega_2 = \left\{ (x, y) \in \mathbb{R}^2 : x < 0 \right\}$.

If the potential exists in Ω, it must therefore be given by

$$f(x, y) = \begin{cases} \arctan \dfrac{y}{x} + \alpha_+ & \text{if } x > 0 \\[3mm] \arctan \dfrac{y}{x} + \alpha_- & \text{if } x < 0. \end{cases}$$

We must now choose α_+ and α_- so that the potential is well defined on $(0, y)$ with $y < 0$. Since

$$\lim_{x \to 0^+} f(x, y) = -\frac{\pi}{2} + \alpha_+ \quad \text{and} \quad \lim_{x \to 0^-} f(x, y) = \frac{\pi}{2} + \alpha_-,$$

we have to choose $\alpha_+ = \alpha_- + \pi$. We then easily check that the required potential in Ω is given by

$$f(x, y) = \begin{cases} \arctan \dfrac{y}{x} + \alpha_- + \pi & \text{if } x > 0 \\[3mm] \alpha_- + \dfrac{\pi}{2} & \text{if } x = 0 \\[3mm] \arctan \dfrac{y}{x} + \alpha_- & \text{if } x < 0. \end{cases}$$

Exercise 3.5 (i) A necessary condition is $\operatorname{curl} F = 0$, that is

$$\frac{\partial}{\partial x} \left(\beta(r) y \right) - \frac{\partial}{\partial y} \left(\alpha(r) x \right) = y \beta'(r) \frac{x}{r} - x \alpha'(r) \frac{y}{r} = 0.$$

We, hence, deduce that we must have $\beta'(r) = \alpha'(r)$ which implies $\beta(r) = \alpha(r) + \alpha_0$.

(ii) Reasoning as in Exercise 3.2, we find that a candidate for the potential is

$$f(x,y) = \int_0^1 [F(tx,ty) \cdot (x,y)]\, dt = \int_0^1 (\alpha(tr)tx, \beta(tr)ty) \cdot (x,y)\, dt$$

$$= \int_0^1 t\left[\alpha(tr)x^2 + \beta(tr)y^2\right] dt = \int_0^1 t\left[\alpha(tr)r^2 + \alpha_0 y^2\right] dt$$

$$= \frac{1}{2}\alpha_0 y^2 + \int_0^1 t\, r^2 \alpha(tr)\, dt.$$

Setting $tr = s$ $(ds = r\, dt)$, we get

$$f(x,y) = \frac{1}{2}\alpha_0 y^2 + \int_0^r s\alpha(s)\, ds.$$

Since $\lim_{s \to 0} (s\alpha(s)) = 0$, the integral is well defined and so f is a candidate for the potential. It is easily seen that, indeed, $f \in C^1(\Omega)$ and $\operatorname{grad} f = F$.

Exercise 3.6 (i) Since $F = \operatorname{grad} f = \left(\dfrac{\partial f}{\partial x}, \dfrac{\partial f}{\partial y}\right)$, we have

$$0 = \frac{d}{dt} f(t, u(t)) = \frac{\partial f}{\partial x}(t, u(t)) + u'(t)\frac{\partial f}{\partial y}(t, u(t))$$

$$= f_1(t, u(t)) + u'(t) f_2(t, u(t)).$$

(ii) In this case, we have $F(x,y) = (\sin x, y^2)$ and thus $\dfrac{\partial f_2}{\partial x} - \dfrac{\partial f_1}{\partial y} = 0$. Since F is defined on $\mathbb{R}^2$, we can find a potential, namely

$$f(x,y) = -\cos x + \frac{y^3}{3}.$$

A solution is thus given by

$$\frac{1}{3} u^3(t) - \cos t = k.$$

Since $u(0) = 3$, we find that $k = 8$ and so

$$u(t) = (3\cos t + 24)^{1/3}.$$

Exercise 3.7 (i) Reasoning as in the previous exercise, we find that since

$$WF = \operatorname{grad} \Phi,$$

we have

$$0 = \frac{d}{dt}\,\Phi\left(t, u\left(t\right)\right) = \frac{\partial \Phi}{\partial x} + u'\left(t\right)\frac{\partial \Phi}{\partial y} = W f_1 + u'\left(t\right) W f_2$$
$$= W(f_1 + u'\left(t\right) f_2).$$

Since $W \neq 0$, we deduce that every solution to $\Phi\left(t, u\left(t\right)\right) = c$ satisfies

$$f_1\left(t, u\left(t\right)\right) + u'\left(t\right) f_2\left(t, u\left(t\right)\right) = 0.$$

(ii) We now have

$$F\left(x, y\right) = \left(4x \sin\left(xy\right) + y(x^2 + 1)\cos\left(xy\right), (x^2 + 1)x\cos\left(xy\right)\right).$$

Since

$$\frac{\partial f_2}{\partial x} - \frac{\partial f_1}{\partial y} \neq 0,$$

we cannot use the previous exercise. However, if we consider

$$WF = \left((x^2 + 1)\left[4x \sin\left(xy\right) + y(x^2 + 1)\cos\left(xy\right)\right], (x^2 + 1)^2 x\cos\left(xy\right)\right),$$

we get

$$\frac{\partial(W f_2)}{\partial x} = \frac{\partial(W f_1)}{\partial y}$$
$$= (1 + x^2)\left[\cos\left(xy\right) + 5x^2\cos\left(xy\right) - xy\sin\left(xy\right) - x^3 y\sin\left(xy\right)\right].$$

We find the potential

$$\Phi\left(x, y\right) = (1 + x^2)^2 \sin\left(xy\right).$$

We deduce that a solution is given in implicit form by

$$(1 + t^2)^2 \sin\left[t u\left(t\right)\right] = k = \text{constant}.$$

Exercise 3.8 (i) We set $F = (\alpha, \beta)$ and we find

$$\frac{\partial \alpha}{\partial y} = -x\left[-2(2y)(x^2 + y^2)^{-3}\right] = 4xy(x^2 + y^2)^{-3}$$
$$\frac{\partial \beta}{\partial x} = -y\left[-2(2x)(x^2 + y^2)^{-3}\right] = 4xy(x^2 + y^2)^{-3}$$

and therefore

$$\text{curl } F = 0.$$

We have

$$\frac{\partial f}{\partial x} = \alpha(x,y) = -x(x^2+y^2)^{-2}$$

$$\frac{\partial f}{\partial y} = \beta(x,y) = -y(x^2+y^2)^{-2}.$$

Integrating the first equation we get

$$f = \frac{(x^2+y^2)^{-1}}{2} + a(y)$$

and therefore, using the second equation, we obtain

$$a'(y) - \frac{2y}{2}(x^2+y^2)^{-2} = -y(x^2+y^2)^{-2}.$$

Thus, the potential is given by

$$f(x,y) = \frac{1}{2(x^2+y^2)} + \text{constant}$$

which belongs to $C^1(\Omega)$.

(ii) We set $G = (a,b)$ and we find

$$\frac{\partial a}{\partial y} = 3y^2(x^2+y^2)^{-2} + y^3[-4y(x^2+y^2)^{-3}]$$

$$= (x^2+y^2)^{-3}(3y^2x^2 - y^4)$$

$$\frac{\partial b}{\partial x} = -y^2(x^2+y^2)^{-2} - xy^2[-4x(x^2+y^2)^{-3}]$$

$$= (x^2+y^2)^{-3}(3y^2x^2 - y^4)$$

and thus

$$\operatorname{curl} G = 0.$$

However, if we take $\Gamma = \{(\cos t, \sin t), \ t \in (0, 2\pi)\}$, we find

$$\int_\Gamma G \cdot dl = \int_0^{2\pi} (\sin^3 t, -\cos t \sin^2 t) \cdot (-\sin t, \cos t)\, dt$$

$$= -\int_0^{2\pi} \sin^2 t\, dt = -\pi \neq 0.$$

Therefore, this field is not the gradient of a potential in $\Omega = \mathbb{R}^2 \backslash \{(0,0)\}$.

Chapter 4

Green theorem

Exercise 4.1 (i) Evaluation of $\iint_\Omega \text{curl} F\, dx\, dy$. We first compute the curl of F and find

$$\text{curl} F = \frac{\partial F_2}{\partial x} - \frac{\partial F_1}{\partial y} = -x.$$

Using polar coordinates, $x = r \cos t$, $y = r \sin t$ and since the Jacobian determinant is r, we have

$$\iint_\Omega \text{curl} F\, dx\, dy = \int_0^1 \int_0^{2\pi} (-r \cos t)\, r\, dr\, dt = 0.$$

(ii) Evaluation of $\int_{\partial\Omega} F \cdot dl$. Let $\gamma(t) = (\cos t, \sin t)$ be a parametrization of $\partial\Omega$, we then find

$$\int_{\partial\Omega} F \cdot dl = \int_0^{2\pi} (\cos t \sin t, \sin^2 t) \cdot (-\sin t, \cos t)\, dt$$

$$= \int_0^{2\pi} (-\cos t \sin^2 t + \cos t \sin^2 t)\, dt = 0.$$

Exercise 4.2 (i) Evaluation of $\iint_\Omega \text{curl} F\, dx\, dy$. We have

$$\text{curl} F = \frac{\partial F_2}{\partial x} - \frac{\partial F_1}{\partial y} = -1.$$

Using polar coordinates, we obtain

$$\iint_\Omega \text{curl} F\, dx\, dy = \int_1^2 \int_0^{2\pi} (-1)\, r\, dr\, d\theta = -3\pi.$$

(ii) Evaluation of $\int_{\partial\Omega} F \cdot dl$. We have

$$\int_{\partial\Omega} F \cdot dl = \int_{\Gamma_0} F \cdot dl - \int_{\Gamma_1} F \cdot dl,$$

189

where we have set

$$\Gamma_0 = \{\gamma_0\,(\theta) = (2\cos\theta, 2\sin\theta) : \theta \in [0, 2\pi]\}$$
$$\Gamma_1 = \{\gamma_1\,(\theta) = (\cos\theta, \sin\theta) : \theta \in [0, 2\pi]\}.$$

We find

$$\int_{\Gamma_0} F \cdot dl = \int_0^{2\pi} (2\cos\theta + 2\sin\theta, 4\sin^2\theta) \cdot (-2\sin\theta, 2\cos\theta)\,d\theta = -4\pi$$

$$\int_{\Gamma_1} F \cdot dl = \int_0^{2\pi} (\cos\theta + \sin\theta, \sin^2\theta) \cdot (-\sin\theta, \cos\theta)\,d\theta = -\pi.$$

We therefore get

$$\int_{\partial\Omega} F \cdot dl = -3\pi.$$

Exercise 4.3 (i) We first observe that $\Delta u\,(x, y) = e^x$. We have

$$\Omega = \left\{(x, y) \in \mathbb{R}^2 : 0 \le x \le 1 \text{ and } 0 \le y \le 1 - x\right\}$$

and, hence,

$$\int_\Omega \Delta u\, dx\, dy = \int_0^1 \int_0^{1-x} e^x\, dy\, dx = \int_0^1 (1 - x)\, e^x\, dx = [(2 - x)\, e^x]_0^1 = e - 2.$$

(ii) Note that $\partial\Omega = \Gamma_1 \cup \Gamma_2 \cup \Gamma_3$ where

$$\Gamma_1 = \{\gamma_1\,(t) = (t, 0) : t \in [0, 1]\} \;\Rightarrow\; \gamma_1'\,(t) = (1, 0) \;\Rightarrow\; \nu = (0, -1)$$
$$\Gamma_2 = \{\gamma_2\,(t) = (1 - t, t) : t \in [0, 1]\} \;\Rightarrow\; \gamma_2'\,(t) = (-1, 1) \;\Rightarrow\; \nu = \left(\frac{1}{\sqrt{2}}, \frac{1}{\sqrt{2}}\right)$$
$$\Gamma_3 = \{\gamma_3\,(t) = (0, 1 - t) : t \in [0, 1]\} \;\Rightarrow\; \gamma_3'\,(t) = (0, -1) \;\Rightarrow\; \nu = (-1, 0)\,.$$

Since $\operatorname{grad} u = (e^x, 1)$, we get (by setting $\operatorname{grad} u \cdot \nu = \partial u / \partial\nu$)

$$\int_{\Gamma_1} \frac{\partial u}{\partial\nu}\, dl = \int_0^1 (e^t, 1) \cdot (0, -1)\, dt = -1$$

$$\int_{\Gamma_2} \frac{\partial u}{\partial\nu}\, dl = \int_0^1 (e^{1-t}, 1) \cdot \left(\frac{1}{\sqrt{2}}, \frac{1}{\sqrt{2}}\right) \sqrt{2}\, dt = \int_0^1 (e^{1-t} + 1)\, dt = e$$

$$\int_{\Gamma_3} \frac{\partial u}{\partial\nu}\, dl = \int_0^1 (1, 1) \cdot (-1, 0)\, dt = -1\,.$$

We thus obtain

$$\int_{\partial\Omega} \frac{\partial u}{\partial\nu}\, dl = e - 2.$$

Note that the divergence theorem (which is a corollary of Green theorem) gives immediately

$$\int_\Omega \Delta u = \int_\Omega \operatorname{div}(\operatorname{grad} u) = \int_{\partial\Omega} (\operatorname{grad} u \cdot \nu) \, dl = \int_{\partial\Omega} \frac{\partial u}{\partial \nu} \, dl.$$

Exercise 4.4 (i) We first verify Green theorem for $F(x,y) = (-x^2 y, xy^2)$ and

$$\Omega = \left\{(x,y) \in \mathbb{R}^2 : x^2 + (y-1)^2 < 1\right\}.$$

Step 1. Evaluation of $\iint_\Omega \operatorname{curl} F \, dx \, dy$. We have

$$\operatorname{curl} F = \frac{\partial F_2}{\partial x} - \frac{\partial F_1}{\partial y} = y^2 - (-x^2) = x^2 + y^2.$$

Using polar coordinates, we write

$$\Omega = \left\{(x,y) \in \mathbb{R}^2 : x = r\cos\theta, \ y = 1 + r\sin\theta \text{ with } r \in [0,1), \ \theta \in (0, 2\pi)\right\}.$$

Since the Jacobian determinant is r, an easy calculation gives

$$\iint_\Omega \operatorname{curl} F \, dx \, dy = \int_0^1 \int_0^{2\pi} \left[(r\cos\theta)^2 + (1 + r\sin\theta)^2\right] r \, dr \, d\theta = \frac{3\pi}{2}.$$

Step 2. Evaluation of $\int_{\partial\Omega} F \cdot dl$. The curve is parameterized by

$$\partial\Omega = \left\{(x,y) \in \mathbb{R}^2 : \ x = \cos\theta, \ y = 1 + \sin\theta, \ \theta \in [0, 2\pi]\right\}$$

and thus

$$\int_{\partial\Omega} F \cdot dl = \int_0^{2\pi} \left(-(1+\sin\theta)\cos^2\theta, (1+\sin\theta)^2 \cos\theta\right) \cdot (-\sin\theta, \cos\theta) \, d\theta$$
$$= \frac{3\pi}{2}.$$

(ii) We verify Green theorem for

$$\Omega = \left\{(x,y) \in \mathbb{R}^2 : x > 0 \text{ and } x^2 + y^2 < 1\right\}$$

and

$$F(x,y) = \left(\frac{x}{2(1 + x^2 + y^2)}, \varphi(y)\right)$$

with $\varphi \in C^1(\mathbb{R})$.

Step 1. Evaluation of $\iint_\Omega \operatorname{curl} F \, dx \, dy$. We find

$$\operatorname{curl} F = \frac{\partial F_2}{\partial x} - \frac{\partial F_1}{\partial y} = 0 - \left(-\frac{xy}{(1 + x^2 + y^2)^2}\right) = \frac{xy}{(1 + x^2 + y^2)^2}$$

and (cf. Figure 4.1)

$$\Omega = \left\{ (x,y) \in \mathbb{R}^2 : x = r\cos\theta, \ y = r\sin\theta \text{ with } r \in [0,1), \ \theta \in \left(\frac{-\pi}{2}, \frac{\pi}{2} \right) \right\}.$$

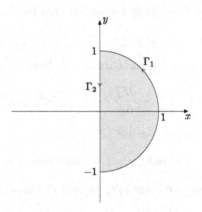

Figure 4.1: Exercise 4.4

We easily get

$$\iint_\Omega \text{curl} F \, dx \, dy = \int_0^1 \int_{-\frac{\pi}{2}}^{\frac{\pi}{2}} r \frac{r^2 \cos\theta \sin\theta}{(1+r^2)^2} \, dr \, d\theta = 0.$$

Step 2. Evaluation of $\int_{\partial\Omega} F \cdot dl$. We let

$$\Gamma = \partial\Omega = \Gamma_1 \cup \Gamma_2$$

where

$$\Gamma_1 = \left\{ (x,y) \in \mathbb{R}^2 : \ x = \cos\theta \text{ and } y = \sin\theta \text{ with } \theta \in \left[\frac{-\pi}{2}, \frac{\pi}{2} \right] \right\}$$
$$\Gamma_2 = \left\{ (x,y) \in \mathbb{R}^2 : \ x = 0 \text{ and } y = -t \text{ with } t \in [-1,1] \right\}.$$

We therefore obtain

$$\int_{\Gamma_1} F \cdot dl = \int_{-\frac{\pi}{2}}^{\frac{\pi}{2}} \left(\frac{\cos\theta}{4}, \varphi(\sin\theta) \right) \cdot (-\sin\theta, \cos\theta) \, d\theta$$
$$= \int_{-\frac{\pi}{2}}^{\frac{\pi}{2}} \varphi(\sin\theta) \cos\theta \, d\theta = \int_{-1}^{1} \varphi(t) \, dt$$

and

$$\int_{\Gamma_2} F \cdot dl = \int_{-1}^{1} (0, \varphi(-t)) \cdot (0, -1) \, dt = -\int_{-1}^{1} \varphi(-t) \, dt = -\int_{-1}^{1} \varphi(u) \, du.$$

We finally find

$$\int_{\partial \Omega} F \cdot dl = 0.$$

Exercise 4.5 (i) Evaluation of $\iint_{\Omega} \operatorname{curl} F \, dx \, dy$. We find $\operatorname{curl} F = -x$ and $\Omega = \Omega_1 \setminus \Omega_2$ where

$$\Omega_1 = \left\{ (x, y) \in \mathbb{R}^2 : x^2 - 4 < y < 2 \right\}$$
$$\Omega_2 = \left\{ (x, y) \in \mathbb{R}^2 : x^2 + y^2 < 1 \right\}.$$

We therefore have

$$\iint_{\Omega} \operatorname{curl} F \, dx \, dy = \iint_{\Omega_1} \operatorname{curl} F \, dx \, dy - \iint_{\Omega_2} \operatorname{curl} F \, dx \, dy.$$

We immediately find

$$-\iint_{\Omega_2} \operatorname{curl} F \, dx \, dy = -\int_{0}^{1} \int_{0}^{2\pi} -r \cos\theta \, r \, dr \, d\theta = 0$$

and

$$\iint_{\Omega_1} \operatorname{curl} F \, dx \, dy = \int_{-\sqrt{6}}^{\sqrt{6}} \int_{x^2-4}^{2} (-x) \, dx \, dy = -\int_{-\sqrt{6}}^{\sqrt{6}} x \left(2 - x^2 + 4 \right) dx$$

$$= -\int_{-\sqrt{6}}^{\sqrt{6}} \left(6x - x^3 \right) dx = -\left[3x^2 - \frac{x^4}{4} \right]_{-\sqrt{6}}^{\sqrt{6}} = 0.$$

We, hence, have

$$\iint_{\Omega} \operatorname{curl} F \, dx \, dy = 0.$$

(ii) Evaluation of $\int_{\partial \Omega} F \cdot dl$. We find (cf. Figure 4.2) that

$$\partial \Omega = \Gamma_0 \cup \Gamma_1 \cup (-\Gamma_2)$$

where

$$\Gamma_0 = \left\{ \alpha(t) = (-t, 2), \ t \in (-\sqrt{6}, \sqrt{6}) \right\}$$
$$\Gamma_1 = \left\{ \beta(t) = (t, t^2 - 4), \ t \in (-\sqrt{6}, \sqrt{6}) \right\}$$
$$\Gamma_2 = \left\{ \gamma(t) = (\cos t, \sin t), \ t \in (0, 2\pi) \right\}.$$

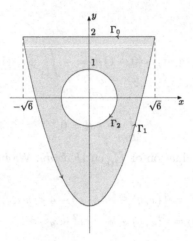

Figure 4.2: Exercise 4.5

The computation of the line integrals gives

$$\int_{\Gamma_0} F \cdot dl = \int_{-\sqrt{6}}^{\sqrt{6}} (-2t, 2) \cdot (-1, 0)\, dt = 0$$

$$\int_{\Gamma_1} F \cdot dl = \int_{-\sqrt{6}}^{\sqrt{6}} \left(t(t^2 - 4), t^2 - 4 \right) \cdot (1, 2t)\, dt = 0$$

$$\int_{-\Gamma_2} F \cdot dl = -\int_0^{2\pi} (\cos t \sin t, \sin t) \cdot (-\sin t, \cos t)\, dt = 0.$$

Therefore, the final result is

$$\int_{\partial \Omega} F \cdot dl = \int_{\Gamma_0} F \cdot dl + \int_{\Gamma_1} F \cdot dl + \int_{-\Gamma_2} F \cdot dl = 0.$$

Exercise 4.6 (i) Evaluation of $\iint_\Omega \operatorname{curl} F\, dx\, dy$. For

$$F = F(x, y) = (f(x, y), g(x, y))$$

we write

$$\iint_\Omega \frac{\partial g}{\partial x}\, dx\, dy = \int_{-1}^1 dy \int_{-\sqrt{1-y^2}}^{\sqrt{1-y^2}} \frac{\partial g}{\partial x}(x, y)\, dx$$

$$= \int_{-1}^1 \left[g\left(\sqrt{1-y^2}, y\right) - g\left(-\sqrt{1-y^2}, y\right) \right] dy$$

and

$$\iint_\Omega \left(\frac{-\partial f}{\partial y} \right) dx dy = - \int_{-1}^1 dx \int_{-\sqrt{1-x^2}}^{\sqrt{1-x^2}} \frac{\partial f}{\partial y}(x,y)\, dy$$

$$= \int_{-1}^1 \left[f\left(x, -\sqrt{1-x^2}\right) - f\left(x, \sqrt{1-x^2}\right) \right] dx.$$

We have thus found that

$$\iint_\Omega \operatorname{curl} F\, dx\, dy = \int_{-1}^1 \left[g\left(\sqrt{1-t^2}, t\right) + f\left(t, -\sqrt{1-t^2}\right) \right.$$
$$\left. - g\left(-\sqrt{1-t^2}, t\right) - f\left(t, \sqrt{1-t^2}\right) \right] dt.$$

(ii) Evaluation of $\int_{\partial \Omega} F \cdot dl$. We let $\partial\Omega = \Gamma_1 \cup \Gamma_2$ with

$$-\Gamma_1 = \left\{ \alpha\left(t\right) = (t, \sqrt{1-t^2}),\ t \in [-1,1] \right\}$$
$$\Gamma_2 = \left\{ \beta\left(t\right) = (t, -\sqrt{1-t^2}),\ t \in [-1,1] \right\}.$$

(The parametrizations $\alpha, \beta \notin C^1\left([-1,1]; \mathbb{R}^2\right)$. However, it is easy to make the following computation rigorous, but we will not go into detail.) The line integral is given by

$$\int_{\partial\Omega} F \cdot dl = -\int_{-1}^1 \left(f(t, \sqrt{1-t^2}),\ g(t, \sqrt{1-t^2}) \right) \cdot \left(1, \frac{-t}{\sqrt{1-t^2}} \right) dt$$
$$+ \int_{-1}^1 \left(f(t, -\sqrt{1-t^2}),\ g(t, -\sqrt{1-t^2}) \right) \cdot \left(1, \frac{t}{\sqrt{1-t^2}} \right) dt$$

and thus

$$\int_{\partial\Omega} F \cdot dl = \int_{-1}^1 \left[f(t, -\sqrt{1-t^2}) - f(t, \sqrt{1-t^2}) \right] dt$$
$$+ \int_{-1}^1 \left[g(t, -\sqrt{1-t^2}) + g(t, \sqrt{1-t^2}) \right] \frac{t}{\sqrt{1-t^2}}\, dt.$$

We make the substitution $s = -\sqrt{1-t^2}$ in the third integral and $s = \sqrt{1-t^2}$ in the fourth one. We then get

$$\int_0^1 g\left(t, -\sqrt{1-t^2}\right) \frac{t}{\sqrt{1-t^2}}\, dt = \int_{-1}^0 g\left(\sqrt{1-s^2}, s\right) ds$$

$$\int_{-1}^0 g\left(t, -\sqrt{1-t^2}\right) \frac{t}{\sqrt{1-t^2}}\, dt = -\int_{-1}^0 g\left(-\sqrt{1-s^2}, s\right) ds$$

and

$$\int_0^1 g\left(t, \sqrt{1-t^2}\right) \frac{t}{\sqrt{1-t^2}}\, dt = \int_0^1 g\left(\sqrt{1-s^2}, s\right) ds$$

$$\int_{-1}^0 g\left(t, \sqrt{1-t^2}\right) \frac{t}{\sqrt{1-t^2}}\, dt = -\int_0^1 g\left(-\sqrt{1-s^2}, s\right) ds.$$

Combining these identities, we obtain

$$\int_{-1}^1 \left[g(t, -\sqrt{1-t^2}) + g(t, \sqrt{1-t^2})\right] \frac{t}{\sqrt{1-t^2}}\, dt$$

$$= \int_{-1}^1 \left[g(\sqrt{1-s^2}, s) - g(-\sqrt{1-s^2}, s)\right] ds.$$

Green theorem is then verified.

Exercise 4.7 (i) Since the assumptions of Corollary 4.3 are satisfied, we write

$$\iint_\Omega \Delta u\, dx\, dy = \iint_\Omega \operatorname{div}(\operatorname{grad} u) dx\, dy = \int_{\partial\Omega} (\operatorname{grad} u \cdot \nu)\, dl.$$

(ii) Since (cf. Theorem 1.2 (iii))

$$\operatorname{div}(v \operatorname{grad} u) = v\Delta u + (\operatorname{grad} u \cdot \operatorname{grad} v),$$

Corollary 4.3 implies the first Green identity

$$\iint_\Omega [v\Delta u + (\operatorname{grad} u \cdot \operatorname{grad} v)]\, dx\, dy = \iint_\Omega \operatorname{div}(v \operatorname{grad} u)\, dx\, dy$$

$$= \int_{\partial\Omega} (v \operatorname{grad} u \cdot \nu)\, dl$$

$$= \int_{\partial\Omega} v\, (\operatorname{grad} u \cdot \nu)\, dl.$$

(iii) Theorem 1.2 (iii) implies that

$$u\Delta v - v\Delta u = \operatorname{div}(u \operatorname{grad} v) - \operatorname{div}(v \operatorname{grad} u).$$

Using Corollary 4.3, we therefore find the second Green identity, namely

$$\iint_\Omega (u\Delta v - v\Delta u)\, dx\, dy = \iint_\Omega [\operatorname{div}(u \operatorname{grad} v) - \operatorname{div}(v \operatorname{grad} u)]\, dx\, dy$$

$$= \int_{\partial\Omega} [u\, (\operatorname{grad} v \cdot \nu) - v\, (\operatorname{grad} u \cdot \nu)]\, dl.$$

Exercise 4.8 Assume that $\partial\Omega$ is composed of only one regular curve given by

$$\partial\Omega = \{\gamma(t) : t \in [a, b]\}.$$

We know (cf. Definition 8.9) that the exterior unit normal ν to Ω is such that

$$\nu = (\nu_1, \nu_2) = \frac{(\gamma_2'(t), -\gamma_1'(t))}{\|\gamma'(t)\|}$$

which is equivalent to

$$\gamma_1' = -\|\gamma'(t)\|\,\nu_2 \quad \text{and} \quad \gamma_2' = \|\gamma'(t)\|\,\nu_1.$$

We set $\Phi = (\Phi_1, \Phi_2)$ where

$$\Phi_1 = -F_2 \quad \text{and} \quad \Phi_2 = F_1.$$

We, hence, find

$$\operatorname{curl}\Phi = \frac{\partial\Phi_2}{\partial x} - \frac{\partial\Phi_1}{\partial y} = \frac{\partial F_1}{\partial x} + \frac{\partial F_2}{\partial y} = \operatorname{div} F$$

$$\Phi \cdot \gamma' = \Phi_1\gamma_1' + \Phi_2\gamma_2' = (F_1\nu_1 + F_2\nu_2)\,\|\gamma'(t)\| = (F \cdot \nu)\,\|\gamma'(t)\|.$$

Using Green theorem, we deduce the result, namely

$$\iint_\Omega \operatorname{div} F \, dx\, dy = \iint_\Omega \operatorname{curl}\Phi \, dx\, dy = \int_{\partial\Omega} \Phi \cdot dl = \int_a^b \Phi(\gamma(t)) \cdot \gamma'(t)\, dt$$

$$= \int_a^b (F \cdot \nu)\,\|\gamma'(t)\|\, dt = \int_{\partial\Omega} (F \cdot \nu)\, dl.$$

Exercise* 4.9 (i) We observe that

$$\operatorname{curl} F = 2 \quad \text{and} \quad \operatorname{curl} G_1 = \operatorname{curl} G_2 = 1.$$

We therefore find the result using Green theorem, since

$$\operatorname{Area}(\Omega) = \iint_\Omega dx\, dy = \frac{1}{2} \iint_\Omega \operatorname{curl} F \, dx\, dy = \frac{1}{2} \int_{\partial\Omega} F \cdot dl$$

$$= \iint_\Omega \operatorname{curl} G_1 \, dx\, dy = \int_{\partial\Omega} G_1 \cdot dl$$

$$= \iint_\Omega \operatorname{curl} G_2 \, dx\, dy = \int_{\partial\Omega} G_2 \cdot dl.$$

(ii) If $\partial\Omega = \{\gamma(t) = (\gamma_1(t), \gamma_2(t)), \ t \in [a, b]\}$, the above computation (i) implies that

$$\text{Area}(\Omega) = \frac{1}{2} \int_a^b (-\gamma_2(t), \gamma_1(t)) \cdot (\gamma_1'(t), \gamma_2'(t)) \, dt$$

$$= \frac{1}{2} \int_a^b (\gamma_1(t)\gamma_2'(t) - \gamma_1'(t)\gamma_2(t)) \, dt$$

as well as

$$\text{Area}(\Omega) = \int_a^b (0, \gamma_1(t)) \cdot (\gamma_1'(t), \gamma_2'(t)) \, dt = \int_a^b \gamma_1(t)\gamma_2'(t) \, dt$$

$$= \int_a^b (-\gamma_2(t), 0) \cdot (\gamma_1'(t), \gamma_2'(t)) \, dt = -\int_a^b \gamma_1'(t)\gamma_2(t) \, dt.$$

Exercise* 4.10 Let $u, v \in C^2(\overline{\Omega})$ be two solutions of (D) and let $w = u - v$. The function w therefore satisfies

$$\begin{cases} \Delta w(x, y) = 0 & (x, y) \in \Omega \\ w(x, y) = 0 & (x, y) \in \partial\Omega. \end{cases}$$

Using the first Green identity, we find

$$0 = \iint_\Omega w \Delta w \, dx dy = \int_{\partial\Omega} w(\text{grad } w \cdot \nu) dl - \iint_\Omega (w_x^2 + w_y^2) \, dx dy$$

$$= -\iint_\Omega \|\text{grad } w\|^2 \, dx dy.$$

This implies that $\text{grad } w \equiv 0$ and since Ω is a domain, we deduce that w is constant in Ω. Since $w \in C^2(\overline{\Omega})$ and $w = 0$ on $\partial\Omega$, we infer that $w \equiv 0$, that is $u = v$. We, hence, have shown that (D) has at most one solution.

Chapter 5

Surface integrals

Exercise 5.1 We let

$$\sigma\left(\theta, z\right) = \left(z\cos\theta, z\sin\theta, z\right), \quad \text{with } \left(\theta, z\right) \in A = (0, 2\pi) \times (0, 1).$$

Hence, the normal is given by

$$\sigma_\theta \wedge \sigma_z = \left(z\cos\theta, z\sin\theta, -z\right)$$

and thus

$$\|\sigma_\theta \wedge \sigma_z\| = \sqrt{2}\, z.$$

We therefore find

$$
\iint_\Sigma f\, ds = \int_0^1 \int_0^{2\pi} \sqrt{2}\, z\left(z^2\cos\theta\sin\theta + z^2\right) d\theta\, dz = 2\pi\sqrt{2}\int_0^1 z^3\, dz
$$
$$
= \frac{\pi}{\sqrt{2}}.
$$

Exercise 5.2 Let

$$\sigma\left(\theta, r\right) = \left(r\cos\theta, r\sin\theta, r\right), \quad \text{with } \left(\theta, r\right) \in A = (0, 2\pi) \times (0, 1)$$

and thus

$$
\sigma_\theta \wedge \sigma_r = \begin{vmatrix} e_1 & e_2 & e_3 \\ -r\sin\theta & r\cos\theta & 0 \\ \cos\theta & \sin\theta & 1 \end{vmatrix} = \left(r\cos\theta, r\sin\theta, -r\right).
$$

199

Note that the normal has a negative third component. We then have

$$\iint_{\Sigma} F \cdot ds = -\int_0^1 \int_0^{2\pi} \left(r^2 \cos^2 \theta, r^2 \sin^2 \theta, r^2\right) \cdot (r \cos \theta, r \sin \theta, -r)\, dr\, d\theta$$

$$= -\int_0^1 \int_0^{2\pi} \left(r^3 \cos^3 \theta + r^3 \sin^3 \theta - r^3\right) dr\, d\theta$$

$$= 2\pi \int_0^1 r^3\, dr = \frac{\pi}{2}.$$

Exercise 5.3 A parametrization of Σ is given by the previous exercise, namely

$$\Sigma = \{\sigma(\theta, r) = (r \cos \theta,\ r \sin \theta, r) : (\theta, r) \in A = (0, 2\pi) \times (0, 1)\}.$$

The normal is again given by

$$\sigma_\theta \wedge \sigma_r = (r \cos \theta, r \sin \theta, -r) \ \Rightarrow\ \|\sigma_\theta \wedge \sigma_r\| = \sqrt{2}\, r.$$

We find

$$\iint_{\Sigma} \sqrt{x^2 + y^2}\, ds = \sqrt{2} \int_0^1 \int_0^{2\pi} r^2\, dr\, d\theta = \frac{2\pi}{3} \sqrt{2}.$$

Exercise 5.4 We have

$$\Sigma = \left\{\sigma(x, y) = (x, y, 6 - 3x - 2y) : 0 \le x \le 2 \text{ and } 0 \le y \le \frac{6 - 3x}{2}\right\}$$

and the normal is (note that the normal computed below is pointing away from the origin)

$$\sigma_x \wedge \sigma_y = \begin{vmatrix} e_1 & e_2 & e_3 \\ 1 & 0 & -3 \\ 0 & 1 & -2 \end{vmatrix} = (3, 2, 1).$$

We thus find

$$\iint_{\Sigma} F \cdot ds = \int_0^2 \int_0^{(6-3x)/2} (0, 6 - 3x - 2y, 6 - 3x - 2y) \cdot (3, 2, 1)\, dx\, dy$$

$$= \int_0^2 \int_0^{(6-3x)/2} 3(6 - 3x - 2y)\, dy\, dx = 18.$$

Exercise 5.5 We write $\partial \Omega = \Sigma_1 \cup \Sigma_2$ where

$$\Sigma_1 = \left\{(x, y, z) \in \mathbb{R}^3 : z = x^2 + y^2 \text{ and } z \le 1\right\}$$

$$= \left\{\alpha(r, \theta) = (r \cos \theta, r \sin \theta, r^2) : 0 \le r \le 1,\ 0 \le \theta \le 2\pi\right\},$$

$$\Sigma_2 = \{(x, y, z) \in \mathbb{R}^3 : z = 1 \text{ and } x^2 + y^2 \leq 1\}$$
$$= \{\beta(r, \theta) = (r \cos\theta, r \sin\theta, 1) : 0 \leq r \leq 1, \ 0 \leq \theta \leq 2\pi\}.$$

We first compute Area (Σ_1). We have

$$\alpha_r \wedge \alpha_\theta = \begin{vmatrix} e_1 & e_2 & e_3 \\ \cos\theta & \sin\theta & 2r \\ -r\sin\theta & r\cos\theta & 0 \end{vmatrix} = (-2r^2 \cos\theta, -2r^2 \sin\theta, r)$$

and thus

$$\|\alpha_r \wedge \alpha_\theta\| = r\sqrt{1 + 4r^2}.$$

We therefore find

$$\text{Area}(\Sigma_1) = \int_0^1 \int_0^{2\pi} r\sqrt{1 + 4r^2} \, dr \, d\theta = 2\pi \int_0^1 r\sqrt{1 + 4r^2} \, dr$$
$$= 2\pi \left[\frac{1}{12}(1 + 4r^2)^{3/2} \right]_0^1 = \frac{\pi}{6}(5^{3/2} - 1).$$

We next compute Area (Σ_2) (one could observe that Σ_2 is the unit disk and deduce its area immediately). We get

$$\beta_r \wedge \beta_\theta = \begin{vmatrix} e_1 & e_2 & e_3 \\ \cos\theta & \sin\theta & 0 \\ -r\sin\theta & r\cos\theta & 0 \end{vmatrix} = (0, 0, r)$$

and thus

$$\text{Area}(\Sigma_2) = \int_0^1 \int_0^{2\pi} r \, dr \, d\theta = \pi.$$

The required result is

$$\text{Area}(\partial\Omega) = \text{Area}(\Sigma_1) + \text{Area}(\Sigma_2).$$

Exercise 5.6 (i) For this exercise, one could refer first to Example 8.19. We have

$$\begin{pmatrix} x \\ y \\ z \end{pmatrix} \xrightarrow{u} \begin{pmatrix} (R + r\cos\varphi)\cos\theta \\ (R + r\cos\varphi)\sin\theta \\ r\sin\varphi \end{pmatrix} = u(r, \theta, \varphi)$$

and thus

$$\nabla u = \begin{pmatrix} \cos\varphi\cos\theta & -(R + r\cos\varphi)\sin\theta & -r\sin\varphi\cos\theta \\ \cos\varphi\sin\theta & (R + r\cos\varphi)\cos\theta & -r\sin\varphi\sin\theta \\ \sin\varphi & 0 & r\cos\varphi \end{pmatrix}.$$

The Jacobian is

$$|\det \nabla u| = r(R + r\cos\varphi).$$

We thus deduce that

$$\mathrm{Vol}\,(\Omega) = \iiint_\Omega dx\,dy\,dz = \int_0^a \int_0^{2\pi} \int_0^{2\pi} r\,(R + r\cos\varphi)\,dr\,d\theta\,d\varphi$$

$$= 4\pi^2 R \int_0^a r\,dr = 2\pi^2 R a^2.$$

(ii) We have the following parametrization for $\partial\Omega$

$$\sigma\,(\theta,\varphi) = ((R + a\cos\varphi)\cos\theta, (R + a\cos\varphi)\sin\theta, a\sin\varphi)\ \text{ with } \theta,\varphi \in [0, 2\pi].$$

We find that

$$\sigma_\theta \wedge \sigma_\varphi = \begin{vmatrix} e_1 & e_2 & e_3 \\ -(R + a\cos\varphi)\sin\theta & (R + a\cos\varphi)\cos\theta & 0 \\ -a\sin\varphi\cos\theta & -a\sin\varphi\sin\theta & a\cos\varphi \end{vmatrix}$$

$$= \begin{pmatrix} a\,(R + a\cos\varphi)\cos\varphi\cos\theta \\ a\,(R + a\cos\varphi)\cos\varphi\sin\theta \\ a\,(R + a\cos\varphi)\sin\varphi \end{pmatrix}$$

and thus

$$\|\sigma_\theta \wedge \sigma_\varphi\| = a\,(R + a\cos\varphi).$$

(iii) The area is given by

$$\mathrm{Area}(\partial\Omega) = \iint_{\partial\Omega} ds = \int_0^{2\pi} \int_0^{2\pi} a\,(R + a\cos\varphi)\,d\theta\,d\varphi = 4\pi^2 R a.$$

(iv) We finally have

$$\iiint_\Omega z^2\,dx\,dy\,dz = \int_0^a \int_0^{2\pi} \int_0^{2\pi} r(r^2\sin^2\varphi)\,(R + r\cos\varphi)\,dr\,d\theta\,d\varphi$$

$$= 2\pi R \int_0^a r^3 dr \int_0^{2\pi} \sin^2\varphi\,d\varphi = \frac{\pi^2 R a^4}{2}.$$

Chapter 6

Divergence theorem

Exercise 6.1 (i) Evaluation of $\iiint_\Omega \operatorname{div} F \, dx \, dy \, dz$. We immediately find that

$$\operatorname{div} F = z + 1.$$

Using spherical coordinates

$$x = r \cos \theta \sin \varphi, \quad y = r \sin \theta \sin \varphi, \quad z = r \cos \varphi,$$

we find

$$\iiint_\Omega \operatorname{div} F \, dx \, dy \, dz = \int_0^1 \int_0^{2\pi} \int_0^\pi (r \cos \varphi + 1) \, r^2 \sin \varphi \, d\varphi \, d\theta \, dr$$

$$= 2\pi \int_0^1 \int_0^\pi (r^3 \cos \varphi \sin \varphi + r^2 \sin \varphi) \, d\varphi \, dr$$

and thus

$$\iiint_\Omega \operatorname{div} F \, dx \, dy \, dz = \frac{\pi}{2} \int_0^\pi \cos \varphi \sin \varphi \, d\varphi + \frac{2\pi}{3} \int_0^\pi \sin \varphi \, d\varphi = \frac{4\pi}{3}.$$

(ii) Evaluation of $\iint_{\partial\Omega} (F \cdot \nu) \, ds$. A parametrization of $\partial\Omega = \Sigma$ is given by

$$\sigma(\theta, \varphi) = (\cos \theta \sin \varphi, \sin \theta \sin \varphi, \cos \varphi) \quad (\theta, \varphi) \in [0, 2\pi] \times [0, \pi].$$

The normal is given by

$$\sigma_\theta \wedge \sigma_\varphi = -\sin \varphi \, (\cos \theta \sin \varphi, \sin \theta \sin \varphi, \cos \varphi)$$

which is an interior normal. We then get

$$\iint_{\partial\Omega} (F \cdot \nu) \, ds$$

$$= \int_0^{2\pi} \int_0^\pi (\cos^2 \theta \sin^3 \varphi \cos \varphi + \sin^2 \theta \sin^3 \varphi + \sin \theta \sin^2 \varphi \cos \varphi) \, d\varphi \, d\theta$$

and thus

$$\iint_{\partial\Omega} (F \cdot \nu)\, ds = \pi \int_0^\pi \left(\sin^3 \varphi \cos \varphi + \sin^3 \varphi \right) d\varphi$$

$$= \pi \int_0^\pi \left(\sin^3 \varphi \cos \varphi + \sin \varphi - \sin \varphi \cos^2 \varphi \right) d\varphi.$$

We therefore have

$$\iint_{\partial\Omega} (F \cdot \nu)\, ds = \pi \int_0^\pi \left(\frac{\sin^4 \varphi}{4} - \cos \varphi + \frac{\cos^3 \varphi}{3} \right)' d\varphi$$

$$= \pi \left[0 + 2 - \frac{2}{3} \right] = \frac{4\pi}{3}.$$

Exercise 6.2 (i) Evaluation of $\iiint_\Omega \operatorname{div} F \, dx\, dy\, dz$. We have that

$$\operatorname{div} F = xy.$$

Using cylindrical coordinates

$$x = r \cos \theta, \ y = r \sin \theta, \ z = z \quad \text{with} \quad (z, \theta, r) \in (0, 1) \times (0, 2\pi) \times (0, z)$$

we find, since the Jacobian is r,

$$\iiint_\Omega \operatorname{div} F \, dx\, dy\, dz = \int_0^1 \int_0^{2\pi} \int_0^z r^2 \cos \theta \sin \theta\, r \, dr\, d\theta\, dz$$

$$= \int_0^1 \int_0^z r^3 \left[-\frac{\cos(2\theta)}{4} \right]_0^{2\pi} dr\, dz = 0.$$

(ii) Evaluation of $\iint_{\partial\Omega} (F \cdot \nu)\, ds$. We set

$$\partial\Omega = \Sigma_1 \cup \Sigma_2$$

where

$$\Sigma_1 = \{(x, y, z) \in \mathbb{R}^3 : x^2 + y^2 \leq 1 \text{ and } z = 1\}$$
$$\Sigma_2 = \{(x, y, z) \in \mathbb{R}^3 : x^2 + y^2 = z^2 \text{ and } z \in [0, 1]\}.$$

A parametrization of Σ_1 is given by

$$\sigma^1 (r, \theta) = (r \cos \theta, r \sin \theta, 1) \quad \text{with} \ (r, \theta) \in [0, 1] \times [0, 2\pi].$$

We therefore find

$$\sigma_r^1 \wedge \sigma_\theta^1 = \begin{vmatrix} e_1 & e_2 & e_3 \\ \cos \theta & \sin \theta & 0 \\ -r \sin \theta & r \cos \theta & 0 \end{vmatrix} = (0, 0, r),$$

which is an exterior normal. We then get

$$\iint_{\Sigma_1} (F \cdot \nu) \, ds = \int_0^{2\pi} \int_0^1 (0, 0, r^2 \cos\theta \sin\theta) \cdot (0, 0, r) \, dr \, d\theta$$
$$= \int_0^{2\pi} \int_0^1 r^3 \cos\theta \sin\theta \, dr \, d\theta = 0.$$

We parametrize Σ_2 by

$$\sigma^2(r, \theta) = (r\cos\theta, r\sin\theta, r) \quad \text{with } (r, \theta) \in [0, 1] \times [0, 2\pi]$$

and we find

$$\sigma_r^2 \wedge \sigma_\theta^2 = \begin{vmatrix} e_1 & e_2 & e_3 \\ \cos\theta & \sin\theta & 1 \\ -r\sin\theta & r\cos\theta & 0 \end{vmatrix} = (-r\cos\theta, -r\sin\theta, r),$$

which is an interior normal. We get

$$\iint_{\Sigma_2} (F \cdot \nu) \, ds$$
$$= \int_0^{2\pi} \int_0^1 (0, 0, r^3 \cos\theta \sin\theta) \cdot (r\cos\theta, r\sin\theta, -r) \, dr \, d\theta$$
$$= \int_0^{2\pi} \int_0^1 (-r^4 \cos\theta \sin\theta) \, dr \, d\theta$$

and thus

$$\iint_{\partial\Omega} (F \cdot \nu) \, ds = 0.$$

Exercise 6.3 (i) Evaluation of $\iiint_\Omega \operatorname{div} F \, dx \, dy \, dz$. We immediately find

$$\operatorname{div} F = x + y + z.$$

We have

$$\Omega = \{(x, y, z) \in \mathbb{R}^3 : 0 < z < 1 - x - y, \ 0 < y < 1 - x, \ 0 < x < 1\}$$

and thus

$$\iiint_\Omega \operatorname{div} F \, dx \, dy \, dz = \int_0^1 \int_0^{1-x} \int_0^{1-x-y} (x + y + z) \, dz \, dy \, dx$$
$$= \frac{1}{2} \int_0^1 \int_0^{1-x} (1 - (x + y)^2) \, dy \, dx = \frac{1}{8}.$$

(ii) Evaluation of $\iint_{\partial \Omega} (F \cdot \nu)\, ds$. We have

$$\partial \Omega = \Sigma_1 \cup \Sigma_2 \cup \Sigma_3 \cup \Sigma_4$$

where

$$\Sigma_1 = \{\alpha\,(x,z) = (x,0,z) : 0 \leq z \leq 1-x,\ 0 \leq x \leq 1\}$$
$$\Sigma_2 = \{\beta\,(x,y) = (x,y,0) : 0 \leq y \leq 1-x,\ 0 \leq x \leq 1\}$$
$$\Sigma_3 = \{\gamma\,(y,z) = (0,y,z) : 0 \leq z \leq 1-y,\ 0 \leq y \leq 1\}$$
$$\Sigma_4 = \{\delta\,(x,y) = (x,y,1-x-y) : 0 \leq y \leq 1-x,\ 0 \leq x \leq 1\}.$$

The corresponding normals are

$$\alpha_x \wedge \alpha_z = \begin{vmatrix} e_1 & e_2 & e_3 \\ 1 & 0 & 0 \\ 0 & 0 & 1 \end{vmatrix} = (0,-1,0) \qquad \text{(exterior)}$$

$$\beta_x \wedge \beta_y = \begin{vmatrix} e_1 & e_2 & e_3 \\ 1 & 0 & 0 \\ 0 & 1 & 0 \end{vmatrix} = (0,0,1) \qquad \text{(interior)}$$

$$\gamma_y \wedge \gamma_z = \begin{vmatrix} e_1 & e_2 & e_3 \\ 0 & 1 & 0 \\ 0 & 0 & 1 \end{vmatrix} = (1,0,0) \qquad \text{(interior)}$$

$$\delta_x \wedge \delta_y = \begin{vmatrix} e_1 & e_2 & e_3 \\ 1 & 0 & -1 \\ 0 & 1 & -1 \end{vmatrix} = (1,1,1) \qquad \text{(exterior)}.$$

Observe that

$$\iint_{\Sigma_1} (F \cdot \nu)\,ds = \iint_{\Sigma_2} (F \cdot \nu)\,ds = \iint_{\Sigma_3} (F \cdot \nu)\,ds = 0$$

and thus

$$\iint_{\partial \Omega} (F \cdot \nu)\,ds = \iint_{\Sigma_4} (F \cdot \nu)\,ds$$
$$= \int_0^1 \int_0^{1-x} (xy, y\,(1-x-y), x\,(1-x-y)) \cdot (1,1,1)\, dy\, dx$$
$$= \frac{1}{8}.$$

Exercise 6.4 (i) Evaluation of $\iiint_\Omega \operatorname{div} F \, dx \, dy \, dz$. We have

$$\operatorname{div} F = 2(x + y + z).$$

We set

$$x = ar \cos \theta, \quad y = ar \sin \theta, \quad z = bt$$

with $0 < \theta < 2\pi$, $0 < r < t < 1$. We then find

$$\text{Jacobian} = \begin{vmatrix} a \cos \theta & -ar \sin \theta & 0 \\ a \sin \theta & ar \cos \theta & 0 \\ 0 & 0 & b \end{vmatrix} = a^2 br.$$

We finally infer that

$$\iiint_\Omega \operatorname{div} F \, dx \, dy \, dz = 2 \int_0^{2\pi} d\theta \int_0^1 dt \int_0^t a^2 br \, (ar \cos \theta + ar \sin \theta + bt) \, dr$$

$$= 4\pi a^2 b^2 \int_0^1 dt \int_0^t r \, t \, dr = \frac{\pi a^2 b^2}{2}.$$

(ii) Evaluation of $\iint_{\partial\Omega} (F \cdot \nu) \, ds$. We have

$$\partial\Omega = \Sigma_1 \cup \Sigma_2$$

where

$$\Sigma_1 = \left\{ (x, y, z) \in \mathbb{R}^3 : x^2 + y^2 \le a^2 \text{ and } z = b \right\}$$
$$= \left\{ \alpha(r, \theta) = (r \cos \theta, r \sin \theta, b) : 0 \le \theta \le 2\pi, \ 0 \le r \le a \right\}$$

$$\Sigma_2 = \left\{ (x, y, z) \in \mathbb{R}^3 : b^2 (x^2 + y^2) = a^2 z^2 \text{ and } 0 \le z \le b \right\}$$
$$= \left\{ \beta(\theta, t) = (at \cos \theta, at \sin \theta, bt) : 0 \le \theta \le 2\pi, \ 0 \le t \le 1 \right\}.$$

We find

$$\alpha_r \wedge \alpha_\theta = \begin{vmatrix} e_1 & e_2 & e_3 \\ \cos \theta & \sin \theta & 0 \\ -r \sin \theta & r \cos \theta & 0 \end{vmatrix} = (0, 0, r)$$

$$\beta_\theta \wedge \beta_t = \begin{vmatrix} e_1 & e_2 & e_3 \\ -at \sin \theta & at \cos \theta & 0 \\ a \cos \theta & a \sin \theta & b \end{vmatrix} = (abt \cos \theta, abt \sin \theta, -a^2 t)$$

which are both exterior normal. We then get

$$\iint_{\Sigma_1} (F \cdot \nu) ds = \int_0^{2\pi} \int_0^a (r^2 \cos^2 \theta, r^2 \sin^2 \theta, b^2) \cdot (0, 0, r) \, dr \, d\theta = \pi a^2 b^2$$

$$\iint_{\Sigma_2} (F \cdot \nu) ds$$

$$= \int_0^{2\pi} \int_0^1 (a^2 t^2 \cos^2 \theta, a^2 t^2 \sin^2 \theta, b^2 t^2) \cdot (abt \cos \theta, abt \sin \theta, -a^2 t) \, dt \, d\theta$$

$$= -2\pi a^2 b^2 \int_0^1 t^3 dt = -\frac{\pi}{2} a^2 b^2$$

and finally

$$\iint_{\partial \Omega} (F \cdot \nu) \, ds = \frac{\pi}{2} a^2 b^2.$$

Exercise 6.5 (i) Evaluation of $\iiint_\Omega \operatorname{div} F \, dx \, dy \, dz$. We get

$$\operatorname{div} F = 2x - 2y + 2z.$$

We set

$$x = r \cos \theta, \quad y = r \sin \theta, \quad z = z, \quad \text{with } 0 < r, z < 2, \ 0 < \theta < 2\pi.$$

We therefore have

$$\iiint_\Omega \operatorname{div} F \, dx \, dy \, dz = \int_0^{2\pi} \int_0^2 \int_0^2 (2r \cos \theta - 2r \sin \theta + 2z) \, r \, dr \, d\theta \, dz$$

$$= 16\pi.$$

(ii) Evaluation of $\iint_{\partial \Omega} (F \cdot \nu) \, ds$. We observe that

$$\partial \Omega = \Sigma_1 \cup \Sigma_2 \cup \Sigma_3$$

where

$$\Sigma_1 = \{(x, y, z) \in \mathbb{R}^3 : x^2 + y^2 < 4 \text{ and } z = 0\}$$
$$= \{\alpha(r, \theta) = (r \cos \theta, r \sin \theta, 0) : 0 \leq \theta \leq 2\pi \text{ and } 0 \leq r \leq 2\}$$

$$\Sigma_2 = \{(x, y, z) \in \mathbb{R}^3 : x^2 + y^2 = 4 \text{ and } 0 < z < 2\}$$
$$= \{\beta(\theta, z) = (2 \cos \theta, 2 \sin \theta, z) : 0 \leq \theta \leq 2\pi \text{ and } 0 \leq z \leq 2\}$$

$$\Sigma_3 = \{(x, y, z) \in \mathbb{R}^3 : x^2 + y^2 < 4 \text{ and } z = 2\}$$
$$= \{\gamma(r, \theta) = (r \cos \theta, r \sin \theta, 2) : 0 \leq \theta \leq 2\pi \text{ and } 0 \leq r \leq 2\}.$$

The normals are given by

$$\alpha_r \wedge \alpha_\theta = \begin{vmatrix} e_1 & e_2 & e_3 \\ \cos \theta & \sin \theta & 0 \\ -r \sin \theta & r \cos \theta & 0 \end{vmatrix} = (0, 0, r) \qquad \text{(interior)}$$

$$\beta_\theta \wedge \beta_z = \begin{vmatrix} e_1 & e_2 & e_3 \\ -2\sin\theta & 2\cos\theta & 0 \\ 0 & 0 & 1 \end{vmatrix} = (2\cos\theta, 2\sin\theta, 0) \qquad \text{(exterior)}$$

$$\gamma_r \wedge \gamma_\theta = \begin{vmatrix} e_1 & e_2 & e_3 \\ \cos\theta & \sin\theta & 0 \\ -r\sin\theta & r\cos\theta & 0 \end{vmatrix} = (0,0,r) \qquad \text{(exterior)}.$$

This gives

$$\iint_{\Sigma_1} (F \cdot \nu)\, ds = -\int_0^{2\pi}\int_0^2 (r^2\cos^2\theta, -r^2\sin^2\theta, 0) \cdot (0,0,r)\, dr\, d\theta = 0$$

$$\iint_{\Sigma_2} (F \cdot \nu)\, ds = \int_0^{2\pi}\int_0^2 (4\cos^2\theta, -4\sin^2\theta, z^2)\cdot(2\cos\theta, 2\sin\theta, 0)\, dz\, d\theta = 0$$

$$\iint_{\Sigma_3} (F \cdot \nu)\, ds = \int_0^{2\pi}\int_0^2 (r^2\cos^2\theta, -r^2\sin^2\theta, 4) \cdot (0,0,r)\, dr\, d\theta = 16\pi$$

and thus

$$\iint_{\partial\Omega} (F \cdot \nu)\, ds = 16\pi.$$

Exercise 6.6 (i) Evaluation of $\iiint_\Omega \operatorname{div} F\, dx\, dy\, dz$. We immediately find

$$\operatorname{div} F = 3.$$

Using cylindrical coordinates, we obtain (cf. Figure 6.1)

$$\Omega = \left\{ (r\cos\theta, r\sin\theta, z) : \theta \in (0, 2\pi),\ r \in \left(0, \sqrt{3}\right),\ \frac{r^2}{3} < z < \sqrt{4 - r^2} \right\}.$$

We therefore have

$$\iiint_\Omega \operatorname{div} F\, dx\, dy\, dz = 3\int_0^{\sqrt{3}}\int_{r^2/3}^{\sqrt{4-r^2}}\int_0^{2\pi} r\, d\theta\, dz\, dr$$

$$= 6\pi\int_0^{\sqrt{3}} \left(\sqrt{4 - r^2} - \frac{r^2}{3}\right) r\, dr = \frac{19\pi}{2}.$$

(ii) Evaluation of $\iint_{\partial\Omega} (F \cdot \nu)\, ds$. We note that $\partial\Omega = \Sigma_1 \cup \Sigma_2$, where

$$\Sigma_1 = \left\{ \alpha\,(\theta, \varphi) = (2\cos\theta\sin\varphi, 2\sin\theta\sin\varphi, 2\cos\varphi) : \begin{bmatrix} \theta \in [0, 2\pi] \\ \varphi \in \left[0, \dfrac{\pi}{3}\right] \end{bmatrix} \right\}$$

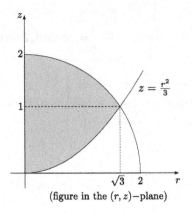

(figure in the (r, z)–plane)

Figure 6.1: Exercise 6.6

$$\Sigma_2 = \left\{ \beta(r, \theta) = \left(r \cos \theta, r \sin \theta, \frac{r^2}{3} \right) : \theta \in [0, 2\pi] \text{ and } r \in \left[0, \sqrt{3} \right] \right\}.$$

The normals are given by

$$\alpha_\theta \wedge \alpha_\varphi = \begin{vmatrix} e_1 & e_2 & e_3 \\ -2 \sin \theta \sin \varphi & 2 \cos \theta \sin \varphi & 0 \\ 2 \cos \theta \cos \varphi & 2 \sin \theta \cos \varphi & -2 \sin \varphi \end{vmatrix}$$

$$= -4 \sin \varphi (\cos \theta \sin \varphi, \sin \theta \sin \varphi, \cos \varphi)$$

$$\beta_r \wedge \beta_\theta = \begin{vmatrix} e_1 & e_2 & e_3 \\ \cos \theta & \sin \theta & \frac{2}{3} r \\ -r \sin \theta & r \cos \theta & 0 \end{vmatrix} = \left(-\frac{2}{3} r^2 \cos \theta, -\frac{2}{3} r^2 \sin \theta, r \right),$$

which are both interior normals. We next compute the surface integrals

$$\iint_{\Sigma_1} (F \cdot \nu) \, ds = \int_0^{2\pi} \int_0^{\pi/3} \left[(2 \cos \theta \sin \varphi, 2 \sin \theta \sin \varphi, 2 \cos \varphi) \cdot \right.$$

$$\left. \left(4 \cos \theta \sin^2 \varphi, 4 \sin \theta \sin^2 \varphi, 4 \sin \varphi \cos \varphi \right) \right] d\theta \, d\varphi$$

$$= 16\pi \int_0^{\pi/3} \sin \varphi \, d\varphi = 16\pi \left[-\cos \varphi \right]_0^{\pi/3} = 8\pi$$

$$\iint_{\Sigma_2} (F \cdot \nu)\, ds$$

$$= \int_0^{2\pi} \int_0^{\sqrt{3}} \left(r\cos\theta, r\sin\theta, \frac{r^2}{3} \right) \cdot \left(\frac{2}{3}r^2\cos\theta, \frac{2}{3}r^2\sin\theta, -r \right) dr\, d\theta$$

$$= 2\pi \int_0^{\sqrt{3}} \left(\frac{2}{3}r^3 - \frac{r^3}{3} \right) dr = 2\pi \left[\frac{r^4}{12} \right]_0^{\sqrt{3}} = \frac{9\pi}{6} = \frac{3\pi}{2}.$$

We finally obtain

$$\iint_{\partial\Omega} (F \cdot \nu)\, ds = 8\pi + \frac{3\pi}{2} = \frac{19\pi}{2}.$$

Exercise 6.7 (i) Evaluation of $\iiint_\Omega \operatorname{div} F\, dx\, dy\, dz$. We find

$$\operatorname{div} F = \frac{3}{1 + z^2}.$$

Using cylindrical coordinates (cf. Figure 6.2), we get

$$\Omega = \left\{ (r\cos\theta, r\sin\theta, z) : \tan r < z < 1,\ 0 < r < \frac{\pi}{4},\ \frac{\pi}{2} < \theta < \pi \right\}.$$

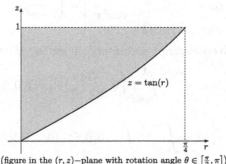

(figure in the (r, z)–plane with rotation angle $\theta \in \left[\frac{\pi}{2}, \pi \right]$)

Figure 6.2: Exercise 6.7

We, hence, deduce that

$$\iiint_\Omega \operatorname{div} F\, dx\, dy\, dz = \int_0^{\pi/4} \int_{\pi/2}^{\pi} \int_{\tan r}^{1} \frac{3}{1 + z^2}\, r\, dz\, d\theta\, dr$$

$$= \frac{3\pi}{2} \int_0^{\pi/4} [\arctan z]_{\tan r}^{1}\, r\, dr$$

$$= \frac{3\pi}{2} \int_0^{\pi/4} \left[\frac{\pi}{4} - r \right] r\, dr = \left(\frac{\pi}{4} \right)^4.$$

(ii) Evaluation of $\iint_{\partial\Omega} (F \cdot \nu)\, ds$. We have

$$\partial\Omega = \Sigma_1 \cup \Sigma_2 \cup \Sigma_3 \cup \Sigma_4$$

where

$$\Sigma_1 = \left\{ (0, y, z) : \tan y \leq z \leq 1 \text{ and } 0 \leq y \leq \frac{\pi}{4} \right\}$$

$$\Sigma_2 = \left\{ (x, 0, z) : \tan |x| \leq z \leq 1 \text{ and } -\frac{\pi}{4} \leq x \leq 0 \right\}$$

$$\Sigma_3 = \left\{ \alpha\, (r, \theta) = (r\cos\theta, r\sin\theta, 1),\ 0 \leq r \leq \frac{\pi}{4} \text{ and } \frac{\pi}{2} \leq \theta \leq \pi \right\}$$

$$\Sigma_4 = \left\{ \beta\, (r, \theta) = (r\cos\theta, r\sin\theta, \tan r) : 0 \leq r \leq \frac{\pi}{4} \text{ and } \frac{\pi}{2} \leq \theta \leq \pi \right\}.$$

The exterior normals to Σ_1 and Σ_2 are respectively $(1, 0, 0)$ and $(0, -1, 0)$. The two others normals are

$$\alpha_r \wedge \alpha_\theta = \begin{vmatrix} e_1 & e_2 & e_3 \\ \cos\theta & \sin\theta & 0 \\ -r\sin\theta & r\cos\theta & 0 \end{vmatrix} = (0, 0, r)$$

$$\beta_r \wedge \beta_\theta = \begin{vmatrix} e_1 & e_2 & e_3 \\ \cos\theta & \sin\theta & \dfrac{1}{\cos^2 r} \\ -r\sin\theta & r\cos\theta & 0 \end{vmatrix} = \left(\frac{-r\cos\theta}{\cos^2 r}, \frac{-r\sin\theta}{\cos^2 r}, r \right),$$

which are respectively exterior and interior normals. We therefore find

$$\iint_{\Sigma_1} (F \cdot \nu)\, ds = \int_0^{\pi/4} \int_{\tan y}^1 \left(0, \frac{3y}{1 + z^2}, 5 \right) \cdot (1, 0, 0)\, dy\, dz = 0$$

$$\iint_{\Sigma_2} (F \cdot \nu)\, ds = \int_{-\frac{\pi}{4}}^0 \int_{\tan(-x)}^1 (0, 0, 5) \cdot (0, -1, 0)\, dx\, dz = 0$$

$$\iint_{\Sigma_3} (F \cdot \nu)\, ds = \int_0^{\pi/4} \int_{\pi/2}^\pi \left(0, \frac{3r\sin\theta}{2}, 5 \right) \cdot (0, 0, r)\, d\theta\, dr = \frac{5\pi}{4} \left(\frac{\pi}{4} \right)^2$$

$$\iint_{\Sigma_4} (F \cdot \nu)\, ds = \int_0^{\pi/4} \int_{\pi/2}^\pi \left(0, \frac{3r\sin\theta}{1 + \tan^2 r}, 5 \right) \cdot \left(\frac{r\cos\theta}{\cos^2 r}, \frac{r\sin\theta}{\cos^2 r}, -r \right)\, d\theta\, dr$$

$$= \int_0^{\pi/4} \int_{\pi/2}^\pi (3r^2 \sin^2\theta - 5r)\, d\theta\, dr = \left(\frac{\pi}{4} \right)^4 - \frac{5\pi}{4} \left(\frac{\pi}{4} \right)^2$$

and thus

$$\iint_{\partial\Omega} (F \cdot \nu)\, ds = \left(\frac{\pi}{4} \right)^4.$$

Exercise 6.8 (i) Evaluation of $\iiint_\Omega \operatorname{div} F \, dx \, dy \, dz$. We find

$$\operatorname{div} F = 2 \left(x + y + z \right).$$

Using cylindrical coordinates (cf. Figure 6.3), we get

$$\Omega = \{(r \cos \theta, r \sin \theta, z) : 0 < z < r < 1 \text{ and } \theta \in (0, 2\pi)\}.$$

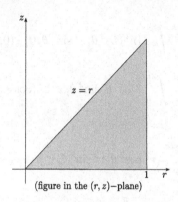

(figure in the (r, z)–plane)

Figure 6.3: Exercise 6.8

We, therefore, find

$$\iiint_\Omega \operatorname{div} F \, dx \, dy \, dz = \int_0^{2\pi} d\theta \int_0^1 dr \int_0^r 2 \left(r \cos \theta + r \sin \theta + z \right) r \, dz$$

$$= 2\pi \int_0^1 r \, dr \int_0^r 2z \, dz = 2\pi \int_0^1 r^3 \, dr = \frac{\pi}{2}.$$

(ii) Evaluation of $\iint_{\partial\Omega} \left(F \cdot \nu \right) ds$. We observe that $\partial\Omega = \Sigma_1 \cup \Sigma_2 \cup \Sigma_3$, where

$$\Sigma_1 = \{\alpha \left(r, \theta \right) = (r \cos \theta, r \sin \theta, 0) : 0 \le r \le 1 \text{ and } 0 \le \theta \le 2\pi\}$$
$$\Sigma_2 = \{\beta \left(\theta, z \right) = (\cos \theta, \sin \theta, z) : 0 \le z \le 1 \text{ and } 0 \le \theta \le 2\pi\}$$
$$\Sigma_3 = \{\gamma \left(r, \theta \right) = (r \cos \theta, r \sin \theta, r) : 0 \le r \le 1 \text{ and } 0 \le \theta \le 2\pi\}.$$

The normals are

$$\alpha_r \wedge \alpha_\theta = \begin{vmatrix} e_1 & e_2 & e_3 \\ \cos \theta & \sin \theta & 0 \\ -r \sin \theta & r \cos \theta & 0 \end{vmatrix} = (0, 0, r) \qquad \text{(interior)}$$

$$\beta_\theta \wedge \beta_z = \begin{vmatrix} e_1 & e_2 & e_3 \\ -\sin\theta & \cos\theta & 0 \\ 0 & 0 & 1 \end{vmatrix} = (\cos\theta, \sin\theta, 0) \qquad \text{(exterior)}$$

$$\gamma_r \wedge \gamma_\theta = \begin{vmatrix} e_1 & e_2 & e_3 \\ \cos\theta & \sin\theta & 1 \\ -r\sin\theta & r\cos\theta & 0 \end{vmatrix} = (-r\cos\theta, -r\sin\theta, r) \qquad \text{(exterior)}.$$

We therefore find

$$\iint_{\Sigma_1} (F \cdot \nu)\, ds = \int_0^1 \int_0^{2\pi} \left(r^2\cos^2\theta, r^2\sin^2\theta, 0\right) \cdot (0,0,-r)\, dr\, d\theta = 0$$

$$\iint_{\Sigma_2} (F \cdot \nu)\, ds = \int_0^1 \int_0^{2\pi} \left(\cos^2\theta, \sin^2\theta, z^2\right) \cdot (\cos\theta, \sin\theta, 0)\, d\theta\, dz = 0$$

$$\iint_{\Sigma_3} (F \cdot \nu)\, ds = \int_0^1 \int_0^{2\pi} \left(r^2\cos^2\theta, r^2\sin^2\theta, r^2\right) \cdot (-r\cos\theta, -r\sin\theta, r)\, dr\, d\theta$$

$$= 2\pi \int_0^1 r^3\, dr = \frac{\pi}{2}$$

and thus the result follows.

Exercise 6.9 (i) Evaluation of $\iiint_\Omega \operatorname{div} F\, dx\, dy\, dz$. We find that $\operatorname{div} F = 2z$. Using cylindrical coordinates, we deduce that (cf. Figure 6.4)

$$\Omega = \left\{ (r\cos\theta, r\sin\theta, z) : 0 < z < 2,\ -\frac{\pi}{2} < \theta < \frac{\pi}{2} \text{ and } 0 < 2r < 4 - z \right\}.$$

We, thus, obtain

$$\iiint_\Omega \operatorname{div} F\, dx\, dy\, dz = \int_0^2 \int_0^{2-z/2} \int_{-\frac{\pi}{2}}^{\frac{\pi}{2}} 2\, z\, r\, d\theta\, dz\, dr$$

$$= \pi \int_0^2 z\left(2 - \frac{z}{2}\right)^2 dz = \frac{11\pi}{3}.$$

(ii) Evaluation of $\iint_{\partial\Omega} (F \cdot \nu)\, ds$. We note that

$$\partial\Omega = \Sigma_1 \cup \Sigma_2 \cup \Sigma_3 \cup \Sigma_4$$

with

$$\Sigma_1 = \{\alpha(y,z) = (0,y,z) : z \in (0,2) \text{ and } 2|y| \le 4 - z\}$$

$$\Sigma_2 = \left\{ \beta\,(r,\theta) = (r\cos\theta, r\sin\theta, 0) : -\frac{\pi}{2} \le \theta \le \frac{\pi}{2} \text{ and } r \le 2 \right\}$$

$$\Sigma_3 = \left\{ \gamma\,(r,\theta) = (r\cos\theta, r\sin\theta, 2) : -\frac{\pi}{2} \le \theta \le \frac{\pi}{2} \text{ and } r \le 1 \right\}$$

$$\Sigma_4 = \left\{ \delta\,(r,\theta) = (r\cos\theta, r\sin\theta, 4-2r) : -\frac{\pi}{2} \le \theta \le \frac{\pi}{2} \text{ and } 1 \le r \le 2 \right\}.$$

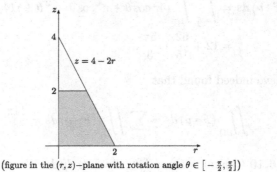

(figure in the (r,z)–plane with rotation angle $\theta \in \left[-\frac{\pi}{2},\frac{\pi}{2}\right]$)

Figure 6.4: Exercise 6.9

The normals are

$$\alpha_y \wedge \alpha_z = (1,0,0), \quad \beta_r \wedge \beta_\theta = \gamma_r \wedge \gamma_\theta = (0,0,r)$$

$$\delta_r \wedge \delta_\theta = (2r\cos\theta, 2r\sin\theta, r)$$

(the first two are interior normals and the others are exterior normals). We therefore obtain

$$\iint_{\Sigma_1} (F \cdot \nu)\,ds = \int_0^2 \int_{-(2-z/2)}^{(2-z/2)} (2,0,z^2) \cdot (-1,0,0)\,dy\,dz = -12$$

$$\iint_{\Sigma_2} (F \cdot \nu)\,ds = \int_{-\frac{\pi}{2}}^{\frac{\pi}{2}} \int_0^2 (2,0,r^3\cos\theta\sin^2\theta) \cdot (0,0,-r)\,dr\,d\theta = -\frac{64}{15}$$

$$\iint_{\Sigma_3} (F \cdot \nu)\,ds = \int_{-\frac{\pi}{2}}^{\frac{\pi}{2}} \int_0^1 (2,0,r^3\cos\theta\sin^2\theta + 4) \cdot (0,0,r)\,dr\,d\theta$$

$$= \int_{-\frac{\pi}{2}}^{\frac{\pi}{2}} \int_0^1 (r^4\cos\theta\sin^2\theta + 4r)\,dr\,d\theta = \frac{2}{15} + 2\pi$$

$$\iint_{\Sigma_4} (F \cdot \nu)\, ds$$

$$= \int_{-\frac{\pi}{2}}^{\frac{\pi}{2}} \int_1^2 \left(2, 0, r^3 \cos\theta \sin^2\theta + (4 - 2r)^2\right) \cdot (2r\cos\theta, 2r\sin\theta, r)\, dr\, d\theta$$

and thus

$$\iint_{\Sigma_4} (F \cdot \nu)\, ds = \int_{-\frac{\pi}{2}}^{\frac{\pi}{2}} \int_1^2 \left(4r\cos\theta + r^4 \cos\theta \sin^2\theta + r(4 - 2r)^2\right) dr\, d\theta$$

$$= 12 + \frac{62}{15} + \frac{5\pi}{3}.$$

We thus have indeed found that

$$\iint_{\partial\Omega} (F \cdot \nu)\, ds = \sum_{i=1}^{4} \iint_{\Sigma_i} (F \cdot \nu)\, ds = \frac{11\pi}{3}.$$

Exercise 6.10 (i) Evaluation of $\iiint_\Omega \operatorname{div} F\, dx\, dy\, dz$. We clearly have $\operatorname{div} F = 2x$. Using cylindrical coordinates, we get (cf. Figure 6.5)

$$\Omega = \left\{ (r\cos\theta, r\sin\theta, z) : \begin{array}{c} 0 < r < \sqrt{15}/4, \quad 0 < \theta < 2\pi, \\ \sqrt{4 - r^2} > z > 2 - \sqrt{1 - r^2} \end{array} \right\}.$$

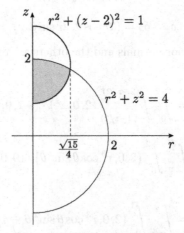

(figure in the (r, z)–plane)

Figure 6.5: Exercise 6.10

We, hence, obtain

$$\iiint_\Omega \operatorname{div} F \, dx \, dy \, dz = \int_0^{\sqrt{15}/4} \int_{2-\sqrt{1-r^2}}^{\sqrt{4-r^2}} \int_0^{2\pi} (2r\cos\theta)\, r \, d\theta \, dz \, dr = 0.$$

(ii) Evaluation of $\iint_{\partial\Omega} (F \cdot \nu)\, ds$. We have $\partial\Omega = \Sigma_1 \cup \Sigma_2$, where

$$\Sigma_1 = \left\{ \alpha(r,\theta) = \left(r\cos\theta, r\sin\theta, \sqrt{4-r^2} \right) : \begin{array}{c} 0 \le r \le \sqrt{15}/4 \\ 0 \le \theta \le 2\pi \end{array} \right\}$$

$$\Sigma_2 = \left\{ \beta(r,\theta) = \left(r\cos\theta, r\sin\theta, 2 - \sqrt{1-r^2} \right) : \begin{array}{c} 0 \le r \le \sqrt{15}/4 \\ 0 \le \theta \le 2\pi \end{array} \right\}.$$

The normals are then given by

$$\alpha_r \wedge \alpha_\theta = \begin{vmatrix} e_1 & e_2 & e_3 \\ \cos\theta & \sin\theta & \dfrac{-r}{\sqrt{4-r^2}} \\ -r\sin\theta & r\cos\theta & 0 \end{vmatrix}$$

$$= \left(\frac{r^2}{\sqrt{4-r^2}} \cos\theta, \frac{r^2}{\sqrt{4-r^2}} \sin\theta, r \right)$$

$$\beta_r \wedge \beta_\theta = \begin{vmatrix} e_1 & e_2 & e_3 \\ \cos\theta & \sin\theta & \dfrac{r}{\sqrt{1-r^2}} \\ -r\sin\theta & r\cos\theta & 0 \end{vmatrix}$$

$$= \left(\frac{-r^2}{\sqrt{1-r^2}} \cos\theta, \frac{-r^2}{\sqrt{1-r^2}} \sin\theta, r \right),$$

(the first one is an exterior normal and the second one is an interior normal). This leads to

$$\iint_{\Sigma_1} (F \cdot \nu)\, ds$$

$$= \int_0^{\sqrt{15}/4} \int_0^{2\pi} (r^2\cos^2\theta, 0, 0) \cdot \left(\frac{r^2}{\sqrt{4-r^2}} \cos\theta, \frac{r^2}{\sqrt{4-r^2}} \sin\theta, r \right) d\theta \, dr$$

$$= 0$$

$$\iint_{\Sigma_2} (F \cdot \nu)\, ds$$

$$= \int_0^{\sqrt{15}/4} \int_0^{2\pi} (r^2\cos^2\theta, 0, 0) \cdot \left(\frac{r^2}{\sqrt{1-r^2}} \cos\theta, \frac{r^2}{\sqrt{1-r^2}} \sin\theta, -r \right) d\theta \, dr$$

$$= 0$$

and thus
$$\iint_{\partial\Omega} (F \cdot \nu)\, ds = 0.$$

Exercise 6.11 Note that
$$\operatorname{div} F = 3 \quad \text{and} \quad \operatorname{div} G_i = 1 \text{ for } i = 1, 2, 3.$$

Using the divergence theorem, we get
$$\operatorname{Vol}(\Omega) = \iiint_{\Omega} dx\, dy\, dz = \frac{1}{3} \iiint_{\Omega} \operatorname{div} F\, dx\, dy\, dz = \frac{1}{3} \iint_{\partial\Omega} (F \cdot \nu)\, ds$$
$$= \iiint_{\Omega} \operatorname{div} G_i\, dx\, dy\, dz = \iint_{\partial\Omega} (G_i \cdot \nu)\, ds.$$

Exercise 6.12 (Green identities) (i) Theorem 1.2 gives
$$\operatorname{div}(v \operatorname{grad} u) = v\Delta u + (\operatorname{grad} u \cdot \operatorname{grad} v)$$
$$\operatorname{div}(v \operatorname{grad} u) - \operatorname{div}(u \operatorname{grad} v) = v\Delta u - u\Delta v.$$

(ii) The divergence theorem implies
$$\iiint_{\Omega} [v\Delta u + (\operatorname{grad} u \cdot \operatorname{grad} v)]\, dx\, dy\, dz$$
$$= \iiint_{\Omega} \operatorname{div}(v \operatorname{grad} u)\, dx\, dy\, dz$$
$$= \iint_{\partial\Omega} (v \operatorname{grad} u \cdot \nu)\, ds = \iint_{\partial\Omega} v\frac{\partial u}{\partial \nu}\, ds.$$

(iii) Using the divergence theorem, we find
$$\iiint_{\Omega} (v\Delta u - u\Delta v)\, dx\, dy\, dz = \iint_{\partial\Omega} (v \operatorname{grad} u - u \operatorname{grad} v) \cdot \nu\, ds$$
$$= \iint_{\partial\Omega} \left(v\frac{\partial u}{\partial \nu} - u\frac{\partial v}{\partial \nu} \right) ds.$$

Exercise* 6.13 (i) Since
$$\iiint_{\Omega} \Delta u\,(x, y, z)\, dx\, dy\, dz = \iint_{\partial\Omega} \frac{\partial u}{\partial \nu}\, ds$$
we have the necessary condition
$$\iiint_{\Omega} f\,(x, y, z)\, dx\, dy\, dz = \iint_{\partial\Omega} \varphi\, ds.$$

Note that if u is a solution to (N), then $u + c$ is also a solution, for every $c \in \mathbb{R}$.

(ii) Moreover, if u and v are two solutions, we get (as in Exercise 4.10) that $u - v = \text{constant}$. Therefore, the solution is unique up to a constant.

Chapter 7

Stokes theorem

Exercise 7.1 (i) Evaluation of $\iint_\Sigma \operatorname{curl} F \cdot ds$. We find

$$\operatorname{curl} F = \begin{vmatrix} e_1 & e_2 & e_3 \\ \dfrac{\partial}{\partial x} & \dfrac{\partial}{\partial y} & \dfrac{\partial}{\partial z} \\ x^2 y & z^2 & 0 \end{vmatrix} = \left(-2z, 0, -x^2\right)$$

and if $A = (0, 2\pi) \times (0, 1)$, then

$$\Sigma = \left\{ \sigma\left(\theta, z\right) = \left(z \cos \theta, z \sin \theta, z\right) : \left(\theta, z\right) \in \overline{A} \right\}.$$

The normal is

$$\sigma_\theta \wedge \sigma_z = \begin{vmatrix} e_1 & e_2 & e_3 \\ -z \sin \theta & z \cos \theta & 0 \\ \cos \theta & \sin \theta & 1 \end{vmatrix} = \left(z \cos \theta, z \sin \theta, -z\right).$$

We thus have

$$\iint_\Sigma \operatorname{curl} F \cdot ds$$

$$= \int_0^{2\pi} \int_0^1 \left(-2z, 0, -z^2 \cos^2 \theta\right) \cdot \left(z \cos \theta, z \sin \theta, -z\right) dz\, d\theta$$

$$= \int_0^{2\pi} \int_0^1 z^3 \cos^2 \theta\, dz\, d\theta = \pi \int_0^1 z^3 dz = \frac{\pi}{4}.$$

(ii) Evaluation of $\int_{\partial\Sigma} F \cdot dl$. We have

$$\sigma\left(\partial A\right) = \Gamma_1 \cup \Gamma_2 \cup \Gamma_3 \cup \Gamma_4$$

219

with

$$\Gamma_1 = \{\sigma\,(\theta,0) = (0,0,0)\}$$
$$\Gamma_2 = \{\sigma\,(2\pi,z) = (z,0,z) : z : 0 \to 1\}$$
$$\Gamma_3 = \{\sigma\,(\theta,1) = (\cos\theta,\sin\theta,1) : \theta : 2\pi \to 0\}$$
$$\Gamma_4 = \{\sigma\,(0,z) = (z,0,z) : z : 1 \to 0\} = -\Gamma_2\,.$$

We observe that $\partial\Sigma = \Gamma_3$, which is negatively oriented. We therefore get

$$\int_{\partial\Sigma} F \cdot dl = -\int_0^{2\pi} \left(\cos^2\theta\sin\theta,1,0\right)\cdot(-\sin\theta,\cos\theta,0)\,d\theta$$
$$= \int_0^{2\pi} \cos^2\theta\sin^2\theta\,d\theta = \frac{\pi}{4}\,.$$

Exercise 7.2 (i) Evaluation of $\iint_\Sigma \operatorname{curl} F \cdot ds$. We have

$$\operatorname{curl} F = \begin{vmatrix} e_1 & e_2 & e_3 \\ \dfrac{\partial}{\partial x} & \dfrac{\partial}{\partial y} & \dfrac{\partial}{\partial z} \\ x^2 y & z & x \end{vmatrix} = (-1,-1,-x^2)$$

and if $A = (0,1)\times(0,2\pi)$, then

$$\Sigma = \left\{\sigma\,(r,\theta) = \left(r^2\cos\theta, r^2\sin\theta, r\right) : (r,\theta) \in \overline{A}\right\}.$$

The normal is given by

$$\sigma_r \wedge \sigma_\theta = \begin{vmatrix} e_1 & e_2 & e_3 \\ 2r\cos\theta & 2r\sin\theta & 1 \\ -r^2\sin\theta & r^2\cos\theta & 0 \end{vmatrix} = \left(-r^2\cos\theta, -r^2\sin\theta, 2r^3\right)$$

and thus

$$\iint_\Sigma \operatorname{curl} F \cdot ds$$
$$= \int_0^1 \int_0^{2\pi} \left(-1,-1,-r^4\cos^2\theta\right)\cdot\left(-r^2\cos\theta, -r^2\sin\theta, 2r^3\right)\,dr\,d\theta$$
$$= -\int_0^1 \int_0^{2\pi} 2r^7\cos^2\theta\,dr\,d\theta = -\frac{\pi}{4}\,.$$

(ii) Evaluation of $\int_{\partial\Sigma} F \cdot dl$. We find that

$$\sigma(\partial A) = \Gamma_1 \cup \Gamma_2 \cup \Gamma_3 \cup \Gamma_4$$

where

$$\Gamma_1 = \{\sigma\,(r,0) = (r^2,0,r) : r : 0 \to 1\}$$
$$\Gamma_2 = \{\sigma\,(1,\theta) = (\cos\theta,\sin\theta,1) : \theta : 0 \to 2\pi\}$$
$$\Gamma_3 = \{\sigma\,(r,2\pi) = (r^2,0,r) : r : 1 \to 0\} = -\Gamma_1$$
$$\Gamma_4 = \{\sigma\,(0,\theta) = (0,0,0) : \theta : 2\pi \to 0\}.$$

We then deduce that $\partial\Sigma = \Gamma_2$, which is positively oriented. We finally have

$$\int_{\partial\Sigma} F \cdot dl = \int_0^{2\pi} (\cos^2\theta\sin\theta, 1, \cos\theta) \cdot (-\sin\theta, \cos\theta, 0)\, d\theta$$
$$= -\int_0^{2\pi} \cos^2\theta\sin^2\theta\, d\theta = -\frac{\pi}{4}.$$

Exercise 7.3 (i) Evaluation of $\iint_\Sigma \operatorname{curl} F \cdot ds$. An easy calculation gives

$$\operatorname{curl} F = \begin{vmatrix} e_1 & e_2 & e_3 \\ \dfrac{\partial}{\partial x} & \dfrac{\partial}{\partial y} & \dfrac{\partial}{\partial z} \\ x^2 y^3 & 1 & z \end{vmatrix} = (0,0,-3x^2y^2).$$

Moreover, if we let $A = (0,2\pi) \times (0,\pi/2)$, we then have

$$\Sigma = \{\sigma\,(\theta,\varphi) = R\,(\cos\theta\sin\varphi, \sin\theta\sin\varphi, \cos\varphi) : (\theta,\varphi) \in \overline{A}\}.$$

We find

$$\sigma_\theta \wedge \sigma_\varphi = R^2 \begin{vmatrix} e_1 & e_2 & e_3 \\ -\sin\theta\sin\varphi & \cos\theta\sin\varphi & 0 \\ \cos\theta\cos\varphi & \sin\theta\cos\varphi & -\sin\varphi \end{vmatrix}$$
$$= -R^2\sin\varphi(\cos\theta\sin\varphi, \sin\theta\sin\varphi, \cos\varphi)$$

and we infer that

$$\iint_\Sigma \operatorname{curl} F \cdot ds = 3R^6 \int_0^{2\pi} \int_0^{\pi/2} \cos^2\theta\sin^2\theta\sin^5\varphi\cos\varphi\, d\theta\, d\varphi$$
$$= \frac{R^6}{2} \int_0^{2\pi} \cos^2\theta\sin^2\theta\, d\theta = \frac{\pi R^6}{8}.$$

(ii) Evaluation of $\int_{\partial\Sigma} F \cdot dl$. We find

$$\sigma(\partial A) = \Gamma_1 \cup \Gamma_2 \cup \Gamma_3 \cup \Gamma_4$$

where

$$\Gamma_1 = \{\sigma(\theta, 0) = R(0,0,1) : \theta : 0 \to 2\pi\}$$

$$\Gamma_2 = \left\{\sigma(2\pi, \varphi) = R(\sin\varphi, 0, \cos\varphi) : \varphi : 0 \to \frac{\pi}{2}\right\}$$

$$\Gamma_3 = \left\{\sigma\left(\theta, \frac{\pi}{2}\right) = R(\cos\theta, \sin\theta, 0) : \theta : 2\pi \to 0\right\}$$

$$\Gamma_4 = \left\{\sigma(0, \varphi) = R(\sin\varphi, 0, \cos\varphi) : \varphi : \frac{\pi}{2} \to 0\right\} = -\Gamma_2.$$

We, hence, obtain that $\partial\Sigma = \Gamma_3$, which is negatively oriented. We eventually obtain

$$\int_{\partial\Sigma} F \cdot dl = -\int_0^{2\pi} (R^5 \cos^2\theta \sin^3\theta, 1, 0) \cdot (-R\sin\theta, R\cos\theta, 0)\, d\theta$$

$$= R^6 \int_0^{2\pi} \cos^2\theta \sin^4\theta\, d\theta = \frac{\pi R^6}{8}.$$

Exercise 7.4 (i) Evaluation of $\iint_\Sigma \operatorname{curl} F \cdot ds$. We first observe that

$$\operatorname{curl} F = \begin{vmatrix} e_1 & e_2 & e_3 \\ \dfrac{\partial}{\partial x} & \dfrac{\partial}{\partial y} & \dfrac{\partial}{\partial z} \\ -2y & xz & y \end{vmatrix} = (1 - x, 0, z + 2).$$

Setting $A = (2/3, 2) \times (-\pi/2, \pi/2)$, we get

$$\Sigma = \left\{\sigma(r, \theta) = \left(r\cos\theta, r\sin\theta, -\frac{3}{2}r + 3\right) : (r, \theta) \in \overline{A}\right\}.$$

We then find

$$\sigma_r \wedge \sigma_\theta = \begin{vmatrix} e_1 & e_2 & e_3 \\ \cos\theta & \sin\theta & -\dfrac{3}{2} \\ -r\sin\theta & r\cos\theta & 0 \end{vmatrix} = \left(\frac{3}{2}r\cos\theta, \frac{3}{2}r\sin\theta, r\right)$$

and thus

$$\iint_\Sigma \operatorname{curl} F \cdot ds$$

$$= \int_{2/3}^2 \int_{-\pi/2}^{\pi/2} \left(1 - r\cos\theta, 0, -\frac{3}{2}r + 5\right) \cdot \left(\frac{3}{2}r\cos\theta, \frac{3}{2}r\sin\theta, r\right)\, dr\, d\theta$$

$$= \int_{2/3}^2 \int_{-\pi/2}^{\pi/2} \left(\frac{3}{2}r\cos\theta - \frac{3}{2}r^2\cos^2\theta - \frac{3}{2}r^2 + 5r\right)\, d\theta\, dr = \frac{16}{3} + \frac{28\pi}{9}.$$

(ii) Evaluation of $\int_{\partial\Sigma} F \cdot dl$. Note that

$$\partial\Sigma = \sigma\left(\partial A\right) = \Gamma_1 \cup \Gamma_2 \cup \Gamma_3 \cup \Gamma_4$$

where

$$\Gamma_1 = \left\{ \sigma\left(r, -\frac{\pi}{2}\right) = \left(0, -r, -\frac{3}{2}r + 3\right) : r : \frac{2}{3} \to 2 \right\}$$

$$\Gamma_2 = \left\{ \sigma\left(2, \theta\right) = \left(2\cos\theta, 2\sin\theta, 0\right) : \theta : -\frac{\pi}{2} \to \frac{\pi}{2} \right\}$$

$$\Gamma_3 = \left\{ \sigma\left(r, \frac{\pi}{2}\right) = \left(0, r, -\frac{3}{2}r + 3\right) : r : 2 \to \frac{2}{3} \right\}$$

$$\Gamma_4 = \left\{ \sigma\left(\frac{2}{3}, \theta\right) = \left(\frac{2}{3}\cos\theta, \frac{2}{3}\sin\theta, 2\right) : \theta : \frac{\pi}{2} \to -\frac{\pi}{2} \right\}.$$

We get

$$\int_{\Gamma_1} F \cdot dl = \int_{2/3}^{2} (2r, 0, -r) \cdot \left(0, -1, -\frac{3}{2}\right) dr = \frac{8}{3}$$

$$\int_{\Gamma_2} F \cdot dl = \int_{-\pi/2}^{\pi/2} (-4\sin\theta, 0, 2\sin\theta) \cdot (-2\sin\theta, 2\cos\theta, 0) \, d\theta = 4\pi$$

$$\int_{\Gamma_3} F \cdot dl = -\int_{2/3}^{2} (-2r, 0, r) \cdot \left(0, 1, -\frac{3}{2}\right) dr = \frac{8}{3}$$

$$\int_{\Gamma_4} F \cdot dl = -\int_{-\pi/2}^{\pi/2} \left(-\frac{4}{3}\sin\theta, \frac{4}{3}\cos\theta, \frac{2}{3}\sin\theta\right) \cdot \left(-\frac{2}{3}\sin\theta, \frac{2}{3}\cos\theta, 0\right) d\theta$$

$$= -\int_{-\pi/2}^{\pi/2} \left(\frac{8}{9}\sin^2\theta + \frac{8}{9}\cos^2\theta\right) d\theta = -\frac{8}{9}\pi.$$

We thus obtain

$$\int_{\partial\Sigma} F \cdot dl = \frac{16}{3} + \frac{28}{9}\pi.$$

Exercise 7.5 (i) Evaluation of $\iint_{\Sigma} \operatorname{curl} F \cdot ds$. We have

$$\operatorname{curl} F = \begin{vmatrix} e_1 & e_2 & e_3 \\ \dfrac{\partial}{\partial x} & \dfrac{\partial}{\partial y} & \dfrac{\partial}{\partial z} \\ 0 & z^2 & 0 \end{vmatrix} = (-2z, 0, 0).$$

Using spherical coordinates, we let $A = (0, \pi/2) \times (\pi/6, \pi/3)$ and we obtain (cf. Figure 7.1)

$$\Sigma = \left\{ \sigma\left(\theta, \varphi\right) = (2\cos\theta\sin\varphi, 2\sin\theta\sin\varphi, 2\cos\varphi) : (\theta, \varphi) \in \overline{A} \right\}.$$

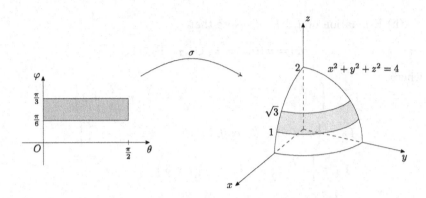

Figure 7.1: Exercise 7.5

The normal is then given by

$$\sigma_\theta \wedge \sigma_\varphi = \begin{vmatrix} e_1 & e_2 & e_3 \\ -2\sin\theta\sin\varphi & 2\cos\theta\sin\varphi & 0 \\ 2\cos\theta\cos\varphi & 2\sin\theta\cos\varphi & -2\sin\varphi \end{vmatrix}$$

$$= -4\sin\varphi\left(\cos\theta\sin\varphi, \sin\theta\sin\varphi, \cos\varphi\right).$$

We therefore get

$$\iint_\Sigma \text{curl}\, F \cdot ds$$

$$= \int_0^{\frac{\pi}{2}} \int_{\frac{\pi}{6}}^{\frac{\pi}{3}} -4\sin\varphi\left(-4\cos\varphi, 0, 0\right) \cdot \left(\cos\theta\sin\varphi, \sin\theta\sin\varphi, \cos\varphi\right) d\theta\, d\varphi$$

$$= 16 \int_{\frac{\pi}{6}}^{\frac{\pi}{3}} \int_0^{\frac{\pi}{2}} \cos\theta\cos\varphi\sin^2\varphi\, d\theta\, d\varphi = 2\sqrt{3} - \frac{2}{3}.$$

(ii) Evaluation of $\int_{\partial\Sigma} F \cdot dl$. We have $\partial\Sigma = \sigma\left(\partial A\right) = \bigcup_{i=1}^{4} \Gamma_i$, where

$$\Gamma_1 = \left\{ \sigma\left(\theta, \frac{\pi}{6}\right) = (\cos\theta, \sin\theta, \sqrt{3}) : \theta : 0 \to \frac{\pi}{2} \right\}$$

$$\Gamma_2 = \left\{ \sigma\left(\frac{\pi}{2}, \varphi\right) = (0, 2\sin\varphi, 2\cos\varphi) : \varphi : \frac{\pi}{6} \to \frac{\pi}{3} \right\}$$

$$\Gamma_3 = \left\{ \sigma\left(\theta, \frac{\pi}{3}\right) = (\sqrt{3}\cos\theta, \sqrt{3}\sin\theta, 1) : \theta : \frac{\pi}{2} \to 0 \right\}$$

$$\Gamma_4 = \left\{ \sigma\left(0, \varphi\right) = (2\sin\varphi, 0, 2\cos\varphi) : \varphi : \frac{\pi}{3} \to \frac{\pi}{6} \right\}.$$

An immediate calculation gives

$$\int_{\Gamma_1} F \cdot dl = \int_0^{\frac{\pi}{2}} (0,3,0) \cdot (-\sin\theta, \cos\theta, 0) \, d\theta = \int_0^{\frac{\pi}{2}} 3\cos\theta \, d\theta = 3$$

$$\int_{\Gamma_2} F \cdot dl = \int_{\frac{\pi}{6}}^{\frac{\pi}{3}} (0, 4\cos^2\varphi, 0) \cdot (0, 2\cos\varphi, -2\sin\varphi) \, d\varphi$$

$$= \int_{\frac{\pi}{6}}^{\frac{\pi}{3}} 8\cos^3\varphi \, d\varphi = 3\sqrt{3} - \frac{11}{3}$$

$$\int_{\Gamma_3} F \cdot dl = -\int_0^{\frac{\pi}{2}} (0,1,0) \cdot \left(-\sqrt{3}\sin\theta, \sqrt{3}\cos\theta, 0\right) d\theta = -\sqrt{3}$$

$$\int_{\Gamma_4} F \cdot dl = -\int_{\frac{\pi}{6}}^{\frac{\pi}{3}} (0, 4\cos^2\varphi, 0) \cdot (2\cos\varphi, 0, -2\sin\varphi) \, d\varphi = 0.$$

We therefore have the claim, namely

$$\int_{\partial\Sigma} F \cdot dl = \sum_{i=1}^4 \int_{\Gamma_i} F \cdot dl = 2\sqrt{3} - \frac{2}{3}.$$

Exercise 7.6 (i) Evaluation of $\iint_\Sigma \operatorname{curl} F \cdot ds$. We immediately find that

$$\operatorname{curl} F = \begin{vmatrix} e_1 & e_2 & e_3 \\ \dfrac{\partial}{\partial x} & \dfrac{\partial}{\partial y} & \dfrac{\partial}{\partial z} \\ 0 & 0 & y+z^2 \end{vmatrix} = (1,0,0).$$

We can write (cf. Figure 7.2)

$$\Sigma = \left\{ \sigma(\theta, \varphi) = (2\cos\theta\sin\varphi, 2\sin\theta\sin\varphi, 2\cos\varphi) : (\theta, \varphi) \in \overline{A} \right\}$$

where

$$A = \left\{ (\theta, \varphi) : 0 < \varphi \le \theta < \frac{\pi}{2} \right\}$$

since

$$\arccos\frac{z}{2} = \varphi \quad \text{and} \quad \arctan\frac{y}{x} = \theta.$$

We get

$$\sigma_\theta \wedge \sigma_\varphi = \begin{vmatrix} e_1 & e_2 & e_3 \\ -2\sin\theta\sin\varphi & 2\cos\theta\sin\varphi & 0 \\ 2\cos\theta\cos\varphi & 2\sin\theta\cos\varphi & -2\sin\varphi \end{vmatrix}$$

$$= -4\sin\varphi(\cos\theta\sin\varphi, \sin\theta\sin\varphi, \cos\varphi).$$

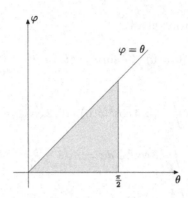

Figure 7.2: Exercise 7.6

We therefore obtain

$$\iint_\Sigma \operatorname{curl} F \cdot ds$$

$$= 4 \int_0^{\pi/2} \int_0^\theta (1,0,0) \cdot \left(-\cos\theta \sin^2\varphi, -\sin\theta \sin^2\varphi, \cos\varphi \sin\varphi\right) d\theta\, d\varphi$$

$$= -4 \int_0^{\pi/2} \cos\theta\, d\theta \int_0^\theta \sin^2\varphi\, d\varphi$$

$$= -4 \int_0^{\pi/2} \cos\theta \left[\frac{\theta}{2} - \frac{1}{4}\sin(2\theta)\right] d\theta = \frac{8}{3} - \pi.$$

(ii) Evaluation of $\displaystyle\int_{\partial\Sigma} F \cdot dl$. We have

$$\sigma\left(\partial A\right) = \Gamma_1 \cup \Gamma_2 \cup \Gamma_3$$

where

$$\Gamma_1 = \left\{\sigma\left(\theta, 0\right) = (0,0,2) : \theta : 0 \to \frac{\pi}{2}\right\}$$

$$\Gamma_2 = \left\{\sigma\left(\frac{\pi}{2}, \varphi\right) = (0, 2\sin\varphi, 2\cos\varphi) : \varphi : 0 \to \frac{\pi}{2}\right\}$$

$$\Gamma_3 = \left\{\sigma\left(\theta, \theta\right) = \left(2\cos\theta\sin\theta, 2\sin^2\theta, 2\cos\theta\right) : \theta : \frac{\pi}{2} \to 0\right\}.$$

We, hence, find that $\partial\Sigma = \Gamma_2 \cup \Gamma_3$ and therefore

$$\int_{\Gamma_2} F \cdot dl = \int_0^{\pi/2} \left(0, 0, 2\sin\varphi + 4\cos^2\varphi\right) \cdot \left(0, 2\cos\varphi, -2\sin\varphi\right) d\varphi$$

$$= -\int_0^{\pi/2} \left(4\sin^2\varphi + 8\cos^2\varphi\sin\varphi\right) d\varphi = -\left(\frac{8}{3} + \pi\right).$$

$$\int_{\Gamma_3} F \cdot dl$$

$$= -\int_0^{\pi/2} \left(0, 0, 2\sin^2\theta + 4\cos^2\theta\right) \cdot \left(2\cos(2\theta), 4\sin\theta\cos\theta, -2\sin\theta\right) d\theta$$

$$= \int_0^{\pi/2} \left(4\sin^3\theta + 8\cos^2\theta\sin\theta\right) d\theta = \frac{16}{3}.$$

We have thus proved that

$$\int_{\partial\Sigma} F \cdot dl = \frac{8}{3} - \pi.$$

Exercise 7.7 (i) Evaluation of $\iint_\Sigma \operatorname{curl} F \cdot ds$. We have

$$\operatorname{curl} F = \begin{vmatrix} e_1 & e_2 & e_3 \\ \dfrac{\partial}{\partial x} & \dfrac{\partial}{\partial y} & \dfrac{\partial}{\partial z} \\ 0 & x^2 & 0 \end{vmatrix} = (0, 0, 2x).$$

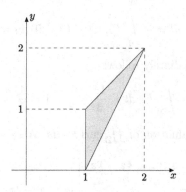

Figure 7.3: Exercise 7.7

We find (cf. Figure 7.3)

$$\Sigma = \left\{\sigma\left(x,y\right) = \left(x,y,0\right) \in \mathbb{R}^3 : \left(x,y\right) \in \overline{A}\right\}$$

$$A = \left\{\left(x,y\right) \in \mathbb{R}^2 : x \in (1,2) \text{ and } 2x-2 < y < x\right\}.$$

We, hence, have

$$\sigma_x \wedge \sigma_y = \begin{vmatrix} e_1 & e_2 & e_3 \\ 1 & 0 & 0 \\ 0 & 1 & 0 \end{vmatrix} = (0,0,1)$$

$$\iint_\Sigma \operatorname{curl} F \cdot ds = \int_1^2 \int_{2x-2}^x 2x \, dy \, dx = \frac{4}{3}.$$

(ii) Evaluation of $\int_{\partial\Sigma} F \cdot dl$. We find

$$\partial\Sigma = \sigma\left(\partial A\right) = \Gamma_1 \cup \Gamma_2 \cup \Gamma_3$$

where

$$\Gamma_1 = \left\{\sigma\left(x,2x-2\right) = \left(x,2x-2,0\right) : x : 1 \to 2\right\}$$
$$\Gamma_2 = \left\{\sigma\left(x,x\right) = \left(x,x,0\right) : x : 2 \to 1\right\}$$
$$\Gamma_3 = \left\{\sigma\left(1,y\right) = \left(1,y,0\right) : y : 1 \to 0\right\}.$$

A simple calculation leads to

$$\int_{\Gamma_1} F \cdot dl = \int_1^2 \left(0,x^2,0\right) \cdot (1,2,0) \, dx = \frac{14}{3}$$

$$\int_{\Gamma_2} F \cdot dl = -\int_1^2 \left(0,x^2,0\right) \cdot (1,1,0) \, dx = -\frac{7}{3}$$

$$\int_{\Gamma_3} F \cdot dl = -\int_0^1 \left(0,1,0\right) \cdot (0,1,0) \, dy = -1.$$

The result then immediately follows

$$\int_{\partial\Sigma} F \cdot dl = \frac{14}{3} - \frac{7}{3} - 1 = \frac{4}{3}.$$

Exercise 7.8 (i) Evaluation of $\iint_\Sigma \operatorname{curl} F \cdot ds$. We find

$$\operatorname{curl} F = \begin{vmatrix} e_1 & e_2 & e_3 \\ \dfrac{\partial}{\partial x} & \dfrac{\partial}{\partial y} & \dfrac{\partial}{\partial z} \\ xy & xz & x^2 \end{vmatrix} = (-x, -2x, z-x).$$

Setting $A = (0,1) \times (0, 2\pi)$, we obtain

$$\Sigma = \left\{ \sigma\left(r, \theta\right) = \left(r \cos \theta, r \sin \theta, 1 - r \cos \theta\right) : \left(r, \theta\right) \in \overline{A} \right\}$$

and, hence,

$$\sigma_r \wedge \sigma_\theta = \begin{vmatrix} e_1 & e_2 & e_3 \\ \cos \theta & \sin \theta & -\cos \theta \\ -r \sin \theta & r \cos \theta & r \sin \theta \end{vmatrix} = (r, 0, r).$$

We, hence, get

$$\iint_\Sigma \operatorname{curl} F \cdot ds = \int_0^1 \int_0^{2\pi} \left(-r \cos \theta, -2r \cos \theta, 1 - 2r \cos \theta\right) \cdot (r, 0, r) \, dr \, d\theta$$

$$= \int_0^1 \int_0^{2\pi} \left(-3r^2 \cos \theta + r\right) d\theta \, dr = \pi.$$

(ii) Evaluation of $\int_{\partial \Sigma} F \cdot dl$. We have

$$\sigma\left(\partial A\right) = \Gamma_1 \cup \Gamma_2 \cup \Gamma_3 \cup \Gamma_4$$

with

$$\Gamma_1 = \left\{\sigma\left(r, 0\right) = (r, 0, 1 - r) : r : 0 \to 1\right\}$$
$$\Gamma_2 = \left\{\sigma\left(1, \theta\right) = (\cos \theta, \sin \theta, 1 - \cos \theta) : \theta : 0 \to 2\pi\right\}$$
$$\Gamma_3 = \left\{\sigma\left(r, 2\pi\right) = (r, 0, 1 - r) : r : 1 \to 0\right\} = -\Gamma_1$$
$$\Gamma_4 = \left\{\sigma\left(0, \theta\right) = (0, 0, 1)\right\}.$$

We therefore have $\partial \Sigma = \Gamma_2$, which is positively oriented. We, hence, find

$$\int_{\partial \Sigma} F \cdot dl$$

$$= \int_0^{2\pi} \left(\cos \theta \sin \theta, \cos \theta (1 - \cos \theta), \cos^2 \theta\right) \cdot \left(-\sin \theta, \cos \theta, \sin \theta\right) d\theta$$

$$= \int_0^{2\pi} \left(-\cos \theta \sin^2 \theta + \cos^2 \theta - \cos^3 \theta + \cos^2 \theta \sin \theta\right) d\theta = \pi.$$

Exercise 7.9 (i) Evaluation of $\iint_\Sigma \operatorname{curl} F \cdot ds$. We have

$$\operatorname{curl} F = \begin{vmatrix} e_1 & e_2 & e_3 \\ \dfrac{\partial}{\partial x} & \dfrac{\partial}{\partial y} & \dfrac{\partial}{\partial z} \\ z & y & 0 \end{vmatrix} = (0, 1, 0).$$

We set (cf. Figure 7.4)

$$\Sigma = \left\{ \sigma\,(x,y) = (x,y,\sin(3x+2y)) \in \mathbb{R}^3 : (x,y) \in \overline{A} \right\}$$

$$A = \left\{ (x,y) \in \mathbb{R}^2 : x,y > 0,\ x+y < \frac{\pi}{2} \right\}.$$

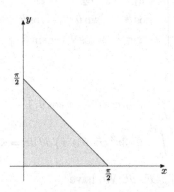

Figure 7.4: Exercise 7.9

We get

$$\sigma_x \wedge \sigma_y = \begin{vmatrix} e_1 & e_2 & e_3 \\ 1 & 0 & 3\cos(3x+2y) \\ 0 & 1 & 2\cos(3x+2y) \end{vmatrix} = (-3\cos(3x+2y), -2\cos(3x+2y), 1).$$

We then have

$$\iint_{\Sigma} \operatorname{curl} F \cdot ds$$

$$= \int_0^{\frac{\pi}{2}} \int_0^{\frac{\pi}{2}-x} (0,1,0) \cdot (-3\cos(3x+2y),\, -2\cos(3x+2y),\, 1)\, dx\, dy$$

$$= \int_0^{\frac{\pi}{2}} \left[-\sin(3x+2y) \right]_0^{-x+\pi/2} dx$$

and thus

$$\iint_{\Sigma} \operatorname{curl} F \cdot ds = \int_0^{\frac{\pi}{2}} (\sin(3x) - \sin(x+\pi))\, dx = \frac{4}{3}.$$

(ii) Evaluation of $\int_{\partial \Sigma} F \cdot dl$. We find

$$\partial \Sigma = \sigma\,(\partial A) = \Gamma_1 \cup \Gamma_2 \cup \Gamma_3$$

where

$$\Gamma_1 = \left\{ \sigma\left(x, 0\right) = \left(x, 0, \sin\left(3x\right)\right) : x : 0 \to \frac{\pi}{2} \right\}$$

$$\Gamma_2 = \left\{ \sigma\left(x, \frac{\pi}{2} - x\right) = \left(x, \frac{\pi}{2} - x, \sin\left(x + \pi\right)\right) : x : \frac{\pi}{2} \to 0 \right\}$$

$$\Gamma_3 = \left\{ \sigma\left(0, y\right) = \left(0, y, \sin\left(2y\right)\right) : y : \frac{\pi}{2} \to 0 \right\}.$$

We therefore obtain

$$\int_{\Gamma_1} F \cdot dl = \int_0^{\frac{\pi}{2}} \left(\sin\left(3x\right), 0, 0\right) \cdot \left(1, 0, 3\cos\left(3x\right)\right) dx = \frac{1}{3}$$

$$\int_{\Gamma_2} F \cdot dl = -\int_0^{\frac{\pi}{2}} \left(-\sin x, \frac{\pi}{2} - x, 0\right) \cdot \left(1, -1, -\cos x\right) dx$$

$$= \int_0^{\frac{\pi}{2}} \sin x \, dx + \int_0^{\frac{\pi}{2}} \left(\frac{\pi}{2} - x\right) dx = 1 + \frac{\pi^2}{8}$$

$$\int_{\Gamma_3} F \cdot dl = -\int_0^{\frac{\pi}{2}} \left(\sin\left(2y\right), y, 0\right) \cdot \left(0, 1, 2\cos\left(2y\right)\right) dy = -\frac{\pi^2}{8}$$

and we conclude that

$$\int_{\partial\Sigma} F \cdot dl = \frac{4}{3}.$$

Chapter 9

Holomorphic functions and Cauchy–Riemann equations

Exercise 9.1 (i) We have

$$\cos z = \frac{e^{iz} + e^{-iz}}{2} = \frac{e^{-y}e^{ix} + e^{y}e^{-ix}}{2}$$
$$= \frac{e^y + e^{-y}}{2}\cos x - i\frac{e^y - e^{-y}}{2}\sin x$$
$$= \cosh y \cos x - i \sinh y \sin x.$$

We note that the Cauchy–Riemann equations are satisfied since

$$u_x = -\cosh y \sin x = v_y$$
$$u_y = \sinh y \cos x = -v_x$$

and thus

$$(\cos z)' = u_x + iv_x = -(\cosh y \sin x + i \sinh y \cos x) = -\sin z.$$

(ii) We find that

$$\cosh z = \frac{e^z + e^{-z}}{2} = \frac{e^x e^{iy} + e^{-x}e^{-iy}}{2}$$
$$= \cosh x \cos y + i \sinh x \sin y.$$

The partial derivatives satisfy the Cauchy–Riemann equations

$$u_x = \sinh x \cos y = v_y$$
$$u_y = -\cosh x \sin y = -v_x$$

and thus (see (iii) below)

$$(\cosh z)' = u_x + iv_x = \sinh z.$$

(iii) We finally get

$$\sinh z = \frac{e^z - e^{-z}}{2} = \frac{e^x e^{iy} - e^{-x} e^{-iy}}{2}$$
$$= \sinh x \cos y + i \cosh x \sin y.$$

We have indeed obtained the Cauchy–Riemann equations

$$u_x = \cosh x \cos y = v_y$$
$$u_y = -\sinh x \sin y = -v_x$$

and consequently

$$(\sinh z)' = u_x + iv_x = \cosh z.$$

Exercise 9.2 The function $f(z) = (\operatorname{Re} z)^2$ is not holomorphic because the Cauchy–Riemann equations are not satisfied. We indeed have

$$u_x = 2x \quad u_y = 0$$
$$v_x = 0 \quad v_y = 0.$$

Exercise 9.3 The function $z \to 1 + z^2$ is holomorphic in $\mathbb{C}$. We know that the function $w \to \log w$ is holomorphic in the set

$$D = \mathbb{C} \backslash \{w \in \mathbb{C} : \operatorname{Im} w = 0, \ \operatorname{Re} w \le 0\}.$$

Setting $w = 1 + z^2$, we next find the set of $z \in \mathbb{C}$ such that

$$\operatorname{Im} w = 0 \quad \text{and} \quad \operatorname{Re} w \le 0.$$

If $z = x + iy$, then $z^2 = x^2 - y^2 + 2ixy$ and we have

$$\begin{cases} \operatorname{Im}(1 + z^2) = 2xy = 0 \\ \operatorname{Re}(1 + z^2) = 1 + x^2 - y^2 \le 0 \end{cases} \Leftrightarrow \begin{cases} x = \operatorname{Re} z = 0 \\ y^2 \ge 1 \end{cases}$$

(we have to exclude the case where $y = 0$ and $1 + x^2 \le 0$). The required set is

$$D = \mathbb{C} \backslash \{z \in \mathbb{C} : \operatorname{Re} z = 0 \text{ and } |\operatorname{Im} z| \ge 1\}.$$

Exercise 9.4 We are looking for a holomorphic function $f = u + iv$ such that
$$u(x, y) = e^{(x^2 - y^2)} \cos(2xy).$$

We write the Cauchy–Riemann equations in order to determine v

$$\begin{cases} u_x = v_y \\ u_y = -v_x \end{cases} \Leftrightarrow \begin{cases} v_y = 2e^{(x^2-y^2)}(x\cos(2xy) - y\sin(2xy)) \\ v_x = 2e^{(x^2-y^2)}(y\cos(2xy) + x\sin(2xy)). \end{cases}$$

Integrating the second equation, we find

$$v(x, y) = e^{(x^2 - y^2)} \sin(2xy) + \alpha(y).$$

Substituting this result in the first equation gives

$$\alpha'(y) = 0 \Rightarrow \alpha(y) = \alpha_0 = \text{constant}.$$

We therefore have

$$v(x, y) = e^{(x^2 - y^2)} \sin(2xy) + \alpha_0$$

and, for $\alpha_0 \in \mathbb{R}$,

$$f = u + iv = e^{x^2 - y^2 + i(2xy)} + i\alpha_0 = e^{z^2} + i\alpha_0.$$

Exercise 9.5 Using the Cauchy–Riemann equations, we find

$$\begin{cases} u_x = 2x - e^{-x}\cos y = v_y \\ u_y = -2y - e^{-x}\sin y = -v_x \end{cases} \Leftrightarrow \begin{cases} v_y = 2x - e^{-x}\cos y \\ v_x = 2y + e^{-x}\sin y. \end{cases}$$

We deduce from the second equation that

$$v(x, y) = 2xy - e^{-x}\sin y + \alpha(y).$$

Then, taking the derivative of v with respect to y and comparing with the first equation leads to

$$\alpha'(y) = 0 \Rightarrow \alpha(y) = \alpha_0 = \text{constant}$$

and thus

$$f = u + iv = (x^2 - y^2) + e^{-x}\cos y + i(2xy - e^{-x}\sin y) + i\alpha_0.$$

Since
$$z^2 = x^2 - y^2 + 2ixy \quad \text{and} \quad \cos y - i\sin y = e^{-iy},$$

we find, for $\alpha_0 \in \mathbb{R}$,
$$f(z) = z^2 + e^{-z} + i\alpha_0.$$

Exercise 9.6 (i) Appealing to Theorem 9.2, we deduce that $u, v \in C^2(\Omega)$ (in particular $v_{yx} = v_{xy}$). Using the Cauchy–Riemann equations

$$u_x = v_y \quad \text{and} \quad u_y = -v_x$$

we get, after differentiation with respect to x in the first equation and to y in the second equation,

$$u_{xx} = v_{yx} \quad \text{and} \quad u_{yy} = -v_{xy}.$$

We immediately deduce that

$$\Delta u = u_{xx} + u_{yy} = v_{yx} - v_{xy} = 0.$$

A similar reasoning gives

$$\Delta v = v_{xx} + v_{yy} = 0.$$

(ii) The following identity is a direct consequence of the Cauchy–Riemann equations

$$u_x v_x + u_y v_y = u_x(-u_y) + u_y u_x = 0.$$

(iii) We have

$$f'(z) = u_x + i v_x.$$

Then, using the Cauchy–Riemann equations, we get

$$|f'(z)|^2 = u_x^2 + v_x^2 = u_x v_y - u_y v_x.$$

Exercise 9.7 (i) $\Rightarrow$ (ii) f constant obviously implies that

$$\mathrm{Re}\,(f) = \text{constant}.$$

(ii) $\Rightarrow$ (iii) Since $\mathrm{Re}\,(f) = a = \text{constant}$ and if we let $f = a + i v(x, y)$, we get (using the Cauchy–Riemann equations) that

$$v_x = v_y = 0.$$

Hence $v = \mathrm{Im}\,(f)$ is constant.

(iii) $\Rightarrow$ (i) Assume now that $\mathrm{Im}\,(f) = b = \text{constant}$. Invoking again the Cauchy–Riemann equations, we get that $\mathrm{Re}\,(f) = a = \text{constant}$ and thus f is constant.

(iv) Since $f(z) = |z| \in \mathbb{R}$, we therefore deduce that f is not holomorphic.

Exercise 9.8 We let $x = r\cos\theta$ and $y = r\sin\theta$ and we find

$$u_r = \cos\theta\, u_x + \sin\theta\, u_y \quad\text{and}\quad u_\theta = -r\sin\theta\, u_x + r\cos\theta\, u_y\,.$$

Solving these equations with respect to u_x and u_y, we obtain

$$u_x = \cos\theta\, u_r - \frac{\sin\theta}{r}\, u_\theta \quad\text{and}\quad u_y = \sin\theta\, u_r + \frac{\cos\theta}{r}\, u_\theta\,.$$

In a similar way, we get

$$v_x = \cos\theta\, v_r - \frac{\sin\theta}{r}\, v_\theta \quad\text{and}\quad v_y = \sin\theta\, v_r + \frac{\cos\theta}{r}\, v_\theta\,.$$

The Cauchy–Riemann equations become

$$u_x = \cos\theta\, u_r - \frac{\sin\theta}{r}\, u_\theta = \sin\theta\, v_r + \frac{\cos\theta}{r}\, v_\theta = v_y$$

$$u_y = \sin\theta\, u_r + \frac{\cos\theta}{r}\, u_\theta = -\cos\theta\, v_r + \frac{\sin\theta}{r}\, v_\theta = -v_x\,.$$

We easily find that these imply

$$u_r = \frac{1}{r}\, v_\theta \quad\text{and}\quad v_r = -\frac{1}{r}\, u_\theta\,.$$

Exercise 9.9 The Cauchy–Riemann equations imply that $u_x = v_y$ and since $u_x + v_y = 0$, we deduce that $u_x = v_y = 0$. Therefore, u (and, hence, u_y) depends only on y, and v (and thus v_x) depends only on x. Since $u_y = -v_x$ for every x and y, we deduce that $u_y = -v_x = c$, where c is a constant. We thus have

$$u = cy + d_1 \quad\text{and}\quad v = -cx + d_2\,.$$

Letting $d_1 + id_2 = d$ we infer that

$$f = u + iv = -ic\,(x + iy) + (d_1 + id_2) = -icz + d.$$

Exercise 9.10 Since g is harmonic, we have

$$\Delta g = g_{uu} + g_{vv} = 0.$$

Since f is holomorphic, we obtain

$$u_x = v_y \quad\text{and}\quad u_y = -v_x\,.$$

Differentiating h, we find

$$h_x = g_u u_x + g_v v_x \quad\text{and}\quad h_y = g_u u_y + g_v v_y$$

and also

$$h_{xx} = g_{uu}u_x^2 + +2g_{uv}u_xv_x + g_{vv}v_x^2 + g_uu_{xx} + g_vv_{xx}$$
$$h_{yy} = g_{uu}u_y^2 + +2g_{uv}u_yv_y + g_{vv}v_y^2 + g_uu_{yy} + g_vv_{yy}.$$

Recalling (cf. Exercise 9.6) that the Cauchy–Riemann equations imply

$$\Delta u = \Delta v = 0, \quad u_x^2 + u_y^2 = v_x^2 + v_y^2, \quad u_xv_x + u_yv_y = 0,$$

we deduce

$$\Delta h = h_{xx} + h_{yy} = g_{uu}\left(u_x^2 + u_y^2\right) + 2g_{uv}\left(u_xv_x + u_yv_y\right)$$
$$+ g_{vv}\left(v_x^2 + v_y^2\right) + g_u\,\Delta u + g_v\,\Delta v$$
$$= \Delta g\left(u_x^2 + u_y^2\right) = 0.$$

Exercise 9.11 We let $f = U + iV$, where $U = u_x$ and $V = -u_y$. By hypothesis $u \in C^2$ and thus

$$U_y = u_{xy} = u_{yx} = -V_x.$$

Hence, one of the Cauchy–Riemann equations for f is satisfied. Since $u_{xx} + u_{yy} = 0$, we have

$$U_x = u_{xx} = -u_{yy} = V_y.$$

Thus, f is holomorphic.

Exercise* 9.12 Let $z_0 = x_0 + iy_0$. Using the suggestion, we consider first the case $z = z_0 + h$, with $h \in \mathbb{R}$. We write

$$f'(z_0) = \lim_{z \to z_0} \frac{f(z) - f(z_0)}{z - z_0} = \lim_{h \to 0} \frac{f(x_0 + h + iy_0) - f(x_0 + iy_0)}{h}$$
$$= \lim_{h \to 0} \frac{u(x_0 + h, y_0) - u(x_0, y_0) + i[v(x_0 + h, y_0) - v(x_0, y_0)]}{h}$$
$$= u_x(x_0, y_0) + iv_x(x_0, y_0) = u_x + iv_x.$$

Letting next $z = z_0 + ih$, with $h \in \mathbb{R}$, a similar computation gives

$$f'(z_0) = \frac{1}{i}u_y(x_0, y_0) + v_y(x_0, y_0) = v_y - iu_y.$$

The two identities give the Cauchy–Riemann equations, namely

$$u_x = v_y \quad \text{and} \quad u_y = -v_x.$$

Chapter 10

Complex integration

Exercise 10.1 We consider three cases.

Case 1: $0 \in \text{int}\,\gamma$. We apply Cauchy integral formula to $g(\xi) = e^{\xi^2}/2$ and we find

$$\int_\gamma \frac{e^{\xi^2}/2}{\xi - 0}\, d\xi = 2\pi i g(0) = \pi i.$$

Case 2: $0 \notin \overline{\text{int}\,\gamma}$. Cauchy theorem applied to $f(z) = e^{z^2}/2z$ gives immediately

$$\int_\gamma \frac{e^{z^2}}{2z}\, dz = 0.$$

Case 3: $0 \in \gamma$. In this case, the integral is not well defined.

Exercise 10.2 We apply Cauchy integral formula to $f(\xi) = \xi^2 \sin\xi$, $z = \pi/2$ and $n = 1$ (note that $f'(z) = 2z \sin z + z^2 \cos z$). We then get

$$\int_\gamma \frac{z^2 \sin z}{\left(z - \frac{\pi}{2}\right)^2}\, dz = \frac{2\pi i}{1!} f'\left(\frac{\pi}{2}\right) = 2\pi^2 i.$$

Exercise 10.3 (i) Let

$$\gamma = [1, 1+i] = \{\gamma(t) = 1 + it : t \in [0,1]\}.$$

We then find

$$\int_\gamma (z^2 + 1)\, dz = \int_0^1 \left((1+it)^2 + 1\right) i\, dt = i \int_0^1 (-t^2 + 2it + 2)\, dt$$

$$= \frac{5i}{3} - 1.$$

(ii) We use the following parametrization of the unit circle centered at the origin

$$\gamma : t \to e^{it} \quad \text{with } t \in [0, 2\pi].$$

We thus deduce that

$$\int_\gamma \mathrm{Re}\left(z^2\right) dz = \int_0^{2\pi} \mathrm{Re}\left(e^{2it}\right) ie^{it} \, dt = i \int_0^{2\pi} \cos\left(2t\right)\left(\cos t + i \sin t\right) dt = 0.$$

Exercise 10.4 *Case 1:* $0 \in \mathrm{int}\,\gamma$. We apply Cauchy integral formula to $f\left(\xi\right) \equiv 1$, $z = 0$ and $n = 1$. We find that $f' \equiv 0$ and therefore

$$\int_\gamma \frac{1}{z^2} \, dz = 0.$$

Case 2: $0 \notin \overline{\mathrm{int}\,\gamma}$. Cauchy theorem gives immediately

$$\int_\gamma \frac{1}{z^2} \, dz = 0.$$

Case 3: $0 \in \gamma$. In this case, the integral is not well defined.

Exercise 10.5 *Case 1:* $1 \in \mathrm{int}\,\gamma$. We apply Cauchy integral formula to $f\left(\xi\right) = 5\xi^2 - 3\xi + 2$, $z = 1$ and $n = 2$. We get $f''\left(1\right) = 10$ and therefore

$$\int_\gamma \frac{5z^2 - 3z + 2}{\left(z - 1\right)^3} \, dz = 10\pi i.$$

Case 2: $1 \notin \overline{\mathrm{int}\,\gamma}$. Cauchy theorem gives immediately

$$\int_\gamma \frac{5z^2 - 3z + 2}{\left(z - 1\right)^3} \, dz = 0.$$

Case 3: $1 \in \gamma$. In this case, the integral is not well defined.

Exercise 10.6 (i) We consider $f\left(\xi\right) = e^{2\xi}$, which is holomorphic in $\mathbb{C}$. Cauchy integral formula therefore implies

$$\int_\gamma \frac{f\left(\xi\right)}{\xi} \, d\xi = \int_\gamma \frac{e^{2\xi}}{\xi} \, d\xi = 2\pi i f\left(0\right) = 2\pi i.$$

Observe that a direct computation of the integral is difficult. Indeed, using the parametrization

$$\gamma\left(t\right) = 2e^{it} \quad 0 \leq t \leq 2\pi$$

we would have

$$\int_\gamma \frac{e^{2z}}{z}\, dz = \int_0^{2\pi} \frac{e^{4e^{it}}}{e^{it}} ie^{it}\, dt = i\int_0^{2\pi} e^{4\cos t} e^{4i\sin t}\, dt$$

which is indeed difficult to evaluate.

(ii) We let $f(\xi) = \xi^3 + 2\xi^2 + 2$ and $z = 2i$ in the Cauchy integral formula. We then immediately find

$$\int_\gamma \frac{f(z)}{z - 2i}\, dz = 2\pi i f(2i) = 16\pi - 12\pi i.$$

Without using Cauchy integral formula, we would take the parametrization

$$\gamma(t) = 2i + \frac{1}{4}e^{it} \quad 0 \le t \le 2\pi$$

and find again

$$\int_\gamma \frac{z^3 + 2z^2 + 2}{z - 2i}\, dz = \int_0^{2\pi} \frac{\left(2i + \frac{1}{4}\,e^{it}\right)^3 + 2\left(2i + \frac{1}{4}\,e^{it}\right)^2 + 2}{\frac{1}{4}e^{it}}\, \frac{i}{4}e^{it}\, dt$$

$$= 16\pi - 12\pi i.$$

(iii) Let $f(\xi) = \sin\left(2\xi^2 + 3\xi + 1\right)$ and $z = \pi$ in the Cauchy integral formula. We then have

$$\int_\gamma \frac{\sin\left(2z^2 + 3z + 1\right)}{z - \pi}\, dz = 2\pi i f(\pi) = 2\pi i \sin\left(2\pi^2 + 3\pi + 1\right).$$

Exercise 10.7 (i) Let $z = 2$, $n = 1$ and

$$f(\xi) = 3\xi^2 + 2\xi + \sin(\xi + 1).$$

We find that $f'(\xi) = 6\xi + 2 + \cos(\xi + 1)$. Cauchy integral formula gives

$$\int_\gamma \frac{3z^2 + 2z + \sin(z + 1)}{(z - 2)^2}\, dz = 2\pi i f'(2) = 2\pi i (14 + \cos 3).$$

(ii) We consider the function

$$f(\xi) = \frac{e^\xi}{\xi + 2}$$

which is holomorphic in $\overline{\operatorname{int}\gamma}$. Since $0 \in \operatorname{int}\gamma$, we have, by the Cauchy integral formula,

$$\int_\gamma \frac{e^z}{z(z + 2)}\, dz = 2\pi i f(0) = 2\pi i \frac{1}{2} = \pi i.$$

Exercise 10.8 (i) Let

$$f(\xi) = \frac{e^{\xi^2}}{\xi^2 + 4}.$$

This function is holomorphic in $\overline{\mathrm{int}\,\gamma}$ and furthermore $1 \in \mathrm{int}\,\gamma$. We then find

$$f'(\xi) = \frac{2\xi e^{\xi^2}\left(\xi^2 + 4\right) - 2\xi e^{\xi^2}}{\left(\xi^2 + 4\right)^2}$$

and thus $f'(1) = 8e/25$. Using Cauchy integral formula, we infer that

$$\int_\gamma \frac{e^{z^2}}{(z-1)^2 (z^2 + 4)}\, dz = 2\pi i f'(1) = \frac{16e\pi}{25} i.$$

(ii) We let

$$f(\xi) = \frac{e^{\xi^2}}{(\xi - 1)^2 (\xi + 2i)}$$

which is holomorphic in $\overline{\mathrm{int}\,\gamma}$ while $2i \in \mathrm{int}\,\gamma$. Cauchy integral formula gives

$$\int_\gamma \frac{e^{z^2}}{(z-1)^2 (z^2 + 4)}\, dz = \int_\gamma \frac{f(z)}{(z - 2i)}\, dz = 2\pi i f(2i) = \frac{-\pi e^{-4}}{2(3 + 4i)}.$$

(iii) In this case, it is easy to see that the integrand is holomorphic in $\overline{\mathrm{int}\,\gamma}$. Cauchy theorem therefore implies

$$\int_\gamma \frac{e^{z^2}}{(z-1)^2 (z^2 + 4)}\, dz = 0.$$

Exercise 10.9 Note that $f(z) = e^{-z^2}$ is holomorphic in $\mathbb{C}$ (as a composition of holomorphic functions). Since γ is a closed curve and f is holomorphic, we have

$$\int_\gamma f(z)\, dz = 0.$$

We parametrize the γ_i (cf. Figure 10.1) by

$$\gamma_1 = \{\gamma_1(t) = t : t : -a \to a\}, \quad \gamma_2 = \{\gamma_2(t) = a + it : t : 0 \to b\}$$

$$\gamma_3 = \{\gamma_3(t) = t + ib : t : a \to -a\}, \quad \gamma_4 = \{\gamma_4(t) = -a + it : t : b \to 0\}.$$

We obtain

$$\int_{\gamma_1} f(z)\, dz = \int_{-a}^{a} e^{-t^2}\, dt$$

and thus

$$\lim_{a \to +\infty} \int_{\gamma_1} f(z)\, dz = \int_{-\infty}^{+\infty} e^{-t^2}\, dt = \sqrt{\pi}.$$

We similarly get

$$\int_{\gamma_2} f(z)\, dz = i \int_0^b e^{-(a+it)^2}\, dt = i e^{-a^2} \int_0^b e^{-2ait+t^2}\, dt$$

which implies

$$\lim_{a \to +\infty} \int_{\gamma_2} f(z)\, dz = 0.$$

We furthermore have

$$\int_{\gamma_3} f(z)\, dz = -\int_{-a}^a e^{-(t+ib)^2}\, dt = -\int_{-a}^a e^{-(t^2+2ibt-b^2)}\, dt$$

$$= -e^{b^2} \int_{-a}^a e^{-t^2} \left(\cos(2bt) - i \sin(2bt) \right) dt$$

and thus

$$\lim_{a \to +\infty} \int_{\gamma_3} f(z)\, dz = -e^{b^2} \int_{-\infty}^{+\infty} e^{-t^2} \cos(2bt)\, dt + i e^{b^2} \int_{-\infty}^{+\infty} e^{-t^2} \sin(2bt)\, dt.$$

We finally get

$$\int_{\gamma_4} f(z)\, dz = -\int_0^b e^{-(-a+it)^2} i\, dt = -e^{-a^2} \int_0^b e^{2ait+t^2} i\, dt$$

which implies

$$\lim_{a \to +\infty} \int_{\gamma_4} f(z)\, dz = 0.$$

Since $\int_\gamma f(z)\, dz = 0$, letting a tend to infinity we obtain

$$e^{b^2} \int_{-\infty}^{+\infty} e^{-x^2} \cos(2bx)\, dx = \sqrt{\pi}$$

$$e^{b^2} \int_{-\infty}^{+\infty} e^{-x^2} \sin(2bx)\, dx = 0.$$

Exercise 10.10 Let $f = u + iv$ and $F = U + iV$. By hypothesis, we have

$$F'(z) = U_x + iV_x = V_y - iU_y = u + iv.$$

Let $\gamma : [a, b] \to \mathbb{C}$ be such that $\gamma(t) = \alpha(t) + i\beta(t)$. We then get

$$\int_\gamma f(z) \, dz = \int_a^b (u + iv) \cdot (\alpha' + i\beta') \, dt$$

$$= \int_a^b \left[u(\alpha(t), \beta(t)) \alpha'(t) - v(\alpha(t), \beta(t)) \beta'(t) \right] dt$$

$$+ i \int_a^b \left[u(\alpha(t), \beta(t)) \beta'(t) + v(\alpha(t), \beta(t)) \alpha'(t) \right] dt$$

and thus

$$\int_\gamma f(z) \, dz = \int_a^b \left[U_x \alpha' + U_y \beta' \right] dt + i \int_a^b \left[V_y \beta' + V_x \alpha' \right] dt.$$

Since

$$\frac{d}{dt} \left[U(\alpha(t), \beta(t)) \right] = U_x(\alpha(t), \beta(t)) \alpha'(t) + U_y(\alpha(t), \beta(t)) \beta'(t)$$

$$\frac{d}{dt} \left[V(\alpha(t), \beta(t)) \right] = V_x(\alpha(t), \beta(t)) \alpha'(t) + V_y(\alpha(t), \beta(t)) \beta'(t),$$

we have the claim, namely

$$\int_\gamma f(z) \, dz = \int_a^b \left\{ \frac{d}{dt} \left[U(\alpha(t), \beta(t)) \right] + i \frac{d}{dt} \left[V(\alpha(t), \beta(t)) \right] \right\} dt$$

$$= U(\alpha(b), \beta(b)) + iV(\alpha(b), \beta(b))$$

$$- \left[U(\alpha(a), \beta(a)) + iV(\alpha(a), \beta(a)) \right].$$

We, hence, infer that

$$\int_\gamma f(z) \, dz = F(\gamma(b)) - F(\gamma(a)).$$

If γ is a closed curve, then $\gamma(b) = \gamma(a)$ and thus

$$\int_\gamma f(z) \, dz = 0.$$

Exercise 10.11 (i) We know that $\log z$ is holomorphic in $\operatorname{Re} z > 0$ and that $(\log z)' = 1/z$. Using the previous exercise and considering $\gamma(t) = 2 + e^{it}$, with $t \in (0, 2\pi)$, we find

$$\int_\gamma f(z) \, dz = \int_0^{2\pi} \frac{i e^{it}}{2 + e^{it}} \, dt = \log(\gamma(2\pi)) - \log(\gamma(0)) = 0.$$

(ii) Since the function $F(z) = \log z$ is not holomorphic in the given disk, we cannot use the previous exercise. We therefore choose a parametrization, for example $\gamma(t) = e^{it}$, with $t \in (0, 2\pi)$, and we obtain

$$\int_\gamma f(z)\, dz = \int_0^{2\pi} f\left(e^{it}\right)\left(ie^{it}\right) dt = 2\pi i \neq 0.$$

Exercise* 10.12 (i) We prove the theorem under the assumption that γ is a regular curve of parametrization $\gamma : [a, b] \to \mathbb{C}$ such that $\gamma(t) = \alpha(t) + i\beta(t)$ (if γ is just piecewise regular, we proceed in a similar way). We write $f = u + iv$ and we find

$$\int_\gamma f(z)\, dz = \int_a^b (u + iv) \cdot (\alpha' + i\beta')\, dt$$

$$= \int_a^b [u(\alpha(t), \beta(t))\, \alpha'(t) - v(\alpha(t), \beta(t))\, \beta'(t)]dt$$

$$+ i \int_a^b [u(\alpha(t), \beta(t))\, \beta'(t) + v(\alpha(t), \beta(t))\, \alpha'(t)]dt.$$

(ii) We start by recalling Green theorem. Let $\Omega \subset \mathbb{R}^2$ be a regular domain with positively oriented boundary $\partial\Omega$ and $F \in C^1\left(\overline{\Omega}; \mathbb{R}^2\right)$ with $F = (f_1, f_2)$, then

$$\int_{\partial\Omega} F \cdot dl = \iint_\Omega \left(\frac{\partial f_2}{\partial x} - \frac{\partial f_1}{\partial y}\right) dx dy.$$

We let $\Omega = \operatorname{int}\gamma$ and we apply Green theorem first to $(f_1, f_2) = (u, -v)$ and then to $(f_1, f_2) = (v, u)$. Observe that since f is holomorphic, then the first-order partial derivatives of u and v exist and are continuous.

(iii) Using the Cauchy–Riemann equations ($u_x = v_y$ and $u_y = -v_x$), we find

$$\int_\gamma f(z)\, dz = \int_a^b [(u, -v) \cdot (\alpha', \beta') + i(v, u)(\alpha', \beta')]dt$$

$$= \iint_{\operatorname{int}\gamma} (-v_x - u_y)\, dx\, dy + i \iint_{\operatorname{int}\gamma} (u_x - v_y)\, dx\, dy$$

$$= 0.$$

Chapter 11

Laurent series

Exercise 11.1 (i) Consider $f(z) = \sin z$ at $z_0 = \pi/4$. We have

$$f\left(\frac{\pi}{4}\right) = f'\left(\frac{\pi}{4}\right) = \frac{\sqrt{2}}{2} \quad \text{and} \quad f''\left(\frac{\pi}{4}\right) = f'''\left(\frac{\pi}{4}\right) = -\frac{\sqrt{2}}{2}.$$

We then get

$$f(z) = \frac{\sqrt{2}}{2} \sum_{n=0}^{\infty} \frac{\alpha_n}{n!} \left(z - \frac{\pi}{4}\right)^n$$

with $\alpha_{4n} = \alpha_{4n+1} = -\alpha_{4n+2} = -\alpha_{4n+3} = 1$. We thus find that $z_0 = \pi/4$ is a regular point and that $\text{Res}_{\pi/4}(f) = 0$. Furthermore, the series converges to f for every $z \in \mathbb{C}$.

(ii) Consider $f(z) = \dfrac{\sin z}{z^3}$ at $z_0 = 0$. We know that

$$\sin z = \sum_{n=0}^{\infty} \frac{(-1)^n}{(2n+1)!} z^{2n+1},$$

which implies that

$$\frac{\sin z}{z^3} = \sum_{n=0}^{\infty} \frac{(-1)^n}{(2n+1)!} z^{2n-2} = \frac{1}{z^2} + \sum_{n=1}^{\infty} \frac{(-1)^n}{(2n+1)!} z^{2(n-1)}$$

$$= \frac{1}{z^2} + \sum_{n=0}^{\infty} \frac{(-1)^{n+1}}{(2n+3)!} z^{2n}.$$

Therefore, $z_0 = 0$ is a pole of order 2, $\text{Res}_0(f) = 0$ and the series is convergent for every $z \neq 0$.

(iii) Consider $f(z) = \dfrac{z}{1+z^2}$ at $z_0 = 1$. Since the function $h(z) = z$ is holomorphic in $\mathbb{C}$ and the function $g(z) = \dfrac{1}{1+z^2}$ is holomorphic except at $z_{1,2} = \pm i$, we immediately deduce that $z_0 = 1$ is a regular point. The Laurent series is thus a Taylor series, which implies that $\text{Res}_1(f) = 0$. Moreover, the theorem implies that the series is convergent provided

$$|z - 1| < \min\{|i - 1|, |i + 1|\} = \sqrt{2}.$$

We could find the Taylor series by computing $f^{(n)}(1)$ for arbitrary n but this is a tedious calculation. We use here a more efficient method which uses the geometric series of g. We first write

$$g(z) = \frac{1}{1 + z^2} = \frac{-1}{2i}\left(\frac{1}{i - z} + \frac{1}{i + z}\right)$$

$$= \frac{-1}{2i}\left(\frac{1}{(i - 1) - (z - 1)} + \frac{1}{(i + 1) + (z - 1)}\right)$$

$$= \frac{-1}{2i}\left[\frac{1}{i - 1} \cdot \frac{1}{1 - \frac{z-1}{i-1}} + \frac{1}{i + 1} \cdot \frac{1}{1 + \frac{z-1}{i+1}}\right].$$

We substitute first $q = (z - 1)/(i - 1)$ and then $q = -(z - 1)/(i + 1)$ in the formula of the geometric series and we get

$$g(z) = \frac{-1}{2i}\left[\frac{1}{i - 1}\sum_{n=0}^{\infty}\frac{1}{(i - 1)^n}(z - 1)^n + \frac{1}{i + 1}\sum_{n=0}^{\infty}\frac{(-1)^n}{(i + 1)^n}(z - 1)^n\right]$$

$$= \frac{-1}{2i}\sum_{n=0}^{\infty}\left(\frac{1}{(i - 1)^{n+1}} + \frac{(-1)^n}{(i + 1)^{n+1}}\right)(z - 1)^n$$

$$= \sum_{n=0}^{\infty}(-1)^n\frac{1}{2i}\left[\frac{1}{(1 - i)^{n+1}} - \frac{1}{(1 + i)^{n+1}}\right](z - 1)^n.$$

Let

$$c_n = \frac{(-1)^n}{2i}\left[\frac{1}{(1 - i)^{n+1}} - \frac{1}{(1 + i)^{n+1}}\right].$$

Since $1 - i = \sqrt{2}e^{-i\frac{\pi}{4}}$ and $1 + i = \sqrt{2}e^{i\frac{\pi}{4}}$, we get

$$c_n = (-1)^n\frac{\sin\left[(n + 1)\frac{\pi}{4}\right]}{2^{\frac{n+1}{2}}}.$$

Getting back to the study of f, we find

$$
\begin{aligned}
f(z) &= (z-1+1) \sum_{n=0}^{\infty} c_n (z-1)^n = \sum_{n=0}^{\infty} c_n (z-1)^{n+1} + \sum_{n=0}^{\infty} c_n (z-1)^n \\
&= \sum_{m=1}^{\infty} c_{m-1} (z-1)^m + c_0 + \sum_{n=1}^{\infty} c_n (z-1)^n \\
&= c_0 + \sum_{n=1}^{\infty} (c_{n-1} + c_n)(z-1)^n.
\end{aligned}
$$

(iv) Consider $f(z) = \sin(1/z)$ at $z_0 = 0$. Substituting $y = 1/z$ in the Taylor series of $\sin y$ we obtain

$$
\sin y = \sum_{n=0}^{\infty} \frac{(-1)^n}{(2n+1)!} y^{2n+1} \Rightarrow \sin\left(\frac{1}{z}\right) = \sum_{n=0}^{\infty} \frac{(-1)^n}{(2n+1)!} \frac{1}{z^{2n+1}}.
$$

We then find that $z_0 = 0$ is an essential singularity and that $\mathrm{Res}_0(f) = 1$. Furthermore, the series is convergent for every $z \neq 0$.

(v) Consider

$$
f(z) = \frac{z^2 + 2z + 1}{z+1} = \frac{(z+1)^2}{z+1} = z + 1.
$$

Contrary to appearance, $z_0 = -1$ is regular point (that is, it is a removable singularity). We therefore have $\mathrm{Res}_{-1}(f) = 0$ and the Taylor expansion at $z_0 = -1$ is then trivially $z + 1$. Moreover, the series is convergent for every $z \in \mathbb{C}$.

(vi) Writing

$$
f(z) = \frac{1}{(1-z)^3} = \frac{-1}{(z-1)^3}
$$

gives immediately the Laurent series at $z_0 = 1$. Thus, $z_0 = 1$ is a pole of order 3 and $\mathrm{Res}_1(f) = 0$. Hence, the series is convergent if $|z-1| > 0$.

(vii) Consider

$$
f(z) = \frac{z^2 + z + 1}{z^2 - 1} = 1 - \frac{1}{2(z+1)} + \frac{3}{2(z-1)}
$$

at $z_0 = 1$. The only other singularity of f is at $z = -1$ and the series is therefore convergent provided that $0 < |z-1| < 2$. To find the Laurent series, we use again the geometric series (where we let $q = -(z-1)/2$ and thus $|q| < 1 \Leftrightarrow |z - 1| < 2$) to write

$$\frac{1}{z+1} = \frac{1}{2 + (z-1)} = \frac{1}{2} \frac{1}{1 + \frac{z-1}{2}} = \frac{1}{2} \sum_{n=0}^{\infty} (-1)^n \left(\frac{z-1}{2}\right)^n$$

$$= \sum_{n=0}^{\infty} \frac{(-1)^n}{2^{n+1}} (z-1)^n .$$

We, hence, infer that

$$f(z) = \frac{3}{2} \frac{1}{z-1} + 1 - \frac{1}{2} \sum_{n=0}^{\infty} \frac{(-1)^n}{2^{n+1}} (z-1)^n$$

$$= \frac{3}{2} \frac{1}{z-1} + \frac{3}{4} + \sum_{n=1}^{\infty} \frac{(-1)^{n+1}}{2^{n+2}} (z-1)^n .$$

Therefore, $z_0 = 1$ is a pole of order 1 and $\mathrm{Res}_1(f) = 3/2$.

(viii) Consider $f(z) = \dfrac{1}{\sin z}$ at $z_0 = 0$. We write

$$f(z) = \frac{1}{z} \frac{z}{\sin z} .$$

We note that $g(z) = z/\sin z$ is holomorphic at $z_0 = 0$ (which is thus a removable singularity for g). We can then compute its Taylor expansion. Note first that g is even and therefore

$$g^{(2k+1)}(0) = 0 \quad \text{for every integer } k.$$

An elementary computation leads to $g(0) = 1$ and a slightly more difficult one gives $g''(0) = 1/3$. Therefore, the Taylor series of g at 0 is given by

$$g(z) = 1 + \frac{z^2}{6} + \sum_{k=2}^{\infty} \frac{g^{(2k)}(0)}{(2k)!} z^{2k}$$

and thus

$$f(z) = \frac{1}{z} + \frac{z}{6} + \sum_{k=2}^{\infty} \frac{g^{(2k)}(0)}{(2k)!} z^{2k-1} .$$

Therefore, $z_0 = 0$ is a pole of order 1 and $\mathrm{Res}_1(f) = 1$. Since the other singularities of f are at $k\pi$, $k \in \mathbb{Z}$, the series is convergent if $0 < |z| < \pi$.

Exercise 11.2 (i) *Case 1:* $z_1 = 0$. We start by finding the Taylor series of the function $(1 + z/2)^{-3}$, namely

$$\frac{1}{(1 + z/2)^3}$$
$$= 1 + (-3)\left(\frac{z}{2}\right) + \frac{(-3)(-4)}{2!}\left(\frac{z}{2}\right)^2 + \frac{(-3)(-4)(-5)}{3!}\left(\frac{z}{2}\right)^3 + \cdots.$$

We therefore find

$$\frac{1}{z(z+2)^3} = \frac{1}{8z(1 + z/2)^3}$$
$$= \frac{1}{8z} - \frac{3}{16} + \frac{3}{16}z - \frac{5}{32}z^2 + \cdots.$$

Hence, $z_1 = 0$ is a pole of order 1, $\text{Res}_0(f) = 1/8$ and the series is convergent for $0 < |z| < 2$.

(ii) *Case 2:* $z_2 = -2$. We let $z + 2 = u$ and we use the formula for the geometric series to get

$$\frac{1}{z(z+2)^3} = \frac{1}{(u-2)u^3} = \frac{1}{-2u^3(1 - u/2)}$$
$$= -\frac{1}{2u^3}\left\{1 + \frac{u}{2} + \left(\frac{u}{2}\right)^2 + \left(\frac{u}{2}\right)^3 + \cdots\right\}$$
$$= -\frac{1}{2u^3} - \frac{1}{4u^2} - \frac{1}{8u} - \sum_{n=0}^{\infty}\frac{u^n}{2^{n+4}}$$

and thus

$$\frac{1}{z(z+2)^3} = -\frac{1}{2(z+2)^3} - \frac{1}{4(z+2)^2} - \frac{1}{8(z+2)} - \sum_{n=0}^{\infty}\frac{(z+2)^n}{2^{n+4}}.$$

Therefore, $z_2 = -2$ is a pole of order 3, $\text{Res}_{-2}(f) = -1/8$ and the series is convergent for $0 < |z + 2| < 2$.

Exercise 11.3 (i) Consider

$$f(z) = \frac{\cos z}{z - \pi} \quad \text{at } z_0 = \pi.$$

It is sufficient to observe that $\cos z = -\cos(z - \pi)$. The Taylor series of $\cos z$ is given by

$$\cos z = -\cos(z - \pi) = -\sum_{n=0}^{\infty}\frac{(-1)^n}{(2n)!}(z - \pi)^{2n}$$

which implies

$$f(z) = \frac{\cos z}{z - \pi} = -\sum_{n=0}^{\infty} \frac{(-1)^n}{(2n)!} (z - \pi)^{2n-1}$$

$$= \frac{-1}{z - \pi} + \sum_{m=0}^{\infty} \frac{(-1)^m}{(2m+2)!} (z - \pi)^{2m+1}.$$

Thus, $z_0 = \pi$ is a pole of order 1, the radius of convergence is infinite (which means that series is convergent provided that $z \neq \pi$) and $\text{Res}_\pi(f) = -1$.

(ii) Consider $z^2 e^{1/z}$ at $z_0 = 0$. We immediately find that

$$z^2 e^{\frac{1}{z}} = z^2 \left(1 + \frac{1}{z} + \frac{1}{2!z^2} + \cdots \right) = z^2 + z + \frac{1}{2} + \frac{1}{3!z} + \cdots$$

$$= z^2 + z + \frac{1}{2} + \sum_{n=1}^{\infty} \frac{1}{(n+2)!} \frac{1}{z^n}$$

that is $z_0 = 0$ is an essential singularity, $\text{Res}_0(f) = 1/6$ and the series is convergent for every $z \neq 0$.

(iii) Let

$$f(z) = \frac{z^2}{(z-1)^2(z+3)} \quad \text{at } z_0 = 1.$$

Set

$$g(z) = \frac{z^2}{z+3}$$

and note that g is a holomorphic function around $z_0 = 1$. Since $g(1) = 1/4$, $g'(1) = 7/16$, $g''(1) = 9/32$ and $g'''(1) = -27/128$, the first terms of its Taylor series are

$$g(z) = \frac{1}{4} + \frac{7}{16}(z-1) + \frac{9}{64}(z-1)^2 - \frac{9}{256}(z-1)^3 + \cdots$$

and thus

$$f(z) = \frac{g(z)}{(z-1)^2} = \frac{1}{4(z-1)^2} + \frac{7}{16(z-1)} + \frac{9}{64} - \frac{9}{256}(z-1) + \cdots.$$

Therefore, $z_0 = 1$ is a pole of order 2, $\text{Res}_1(f) = 7/16$ and the series is convergent for $0 < |z - 1| < 4$.

Exercise 11.4 We use the Taylor expansion of e^y and we substitute $(z-1)^{-2}$ for y. We then get

$$f(z) = z^2 e^{(z-1)^{-2}} = z^2 \sum_{n=0}^{\infty} \frac{1}{n!(z-1)^{2n}} = ((z-1)+1)^2 \sum_{n=0}^{\infty} \frac{1}{n!(z-1)^{2n}}$$

$$= \sum_{n=0}^{\infty} \left(\frac{1}{n!(z-1)^{2n-2}} + \frac{2}{n!(z-1)^{2n-1}} + \frac{1}{n!(z-1)^{2n}} \right)$$

which implies

$$f(z) = \left[(z-1)^2 + 1 + \frac{1}{2!(z-1)^2} + \frac{1}{3!(z-1)^4} + \cdots \right]$$

$$+ \left[2(z-1) + \frac{2}{(z-1)} + \frac{2}{2!(z-1)^3} + \cdots \right]$$

$$+ \left[1 + \frac{1}{(z-1)^2} + \frac{1}{2!(z-1)^4} + \cdots \right]$$

and thus

$$f(z) = (z-1)^2 + 2(z-1) + 2 + \frac{2}{(z-1)} + \frac{3}{2(z-1)^2} + \frac{1}{(z-1)^3} + \cdots.$$

We finally find

$$f(z) = (z-1)^2 + 2(z-1) + 2 + \sum_{n=1}^{\infty} \left(\frac{n+2}{(n+1)!} \frac{1}{(z-1)^{2n}} + \frac{2}{n!} \frac{1}{(z-1)^{2n-1}} \right).$$

Therefore, we can deduce that 1 is an essential singularity, $\text{Res}_1(f) = 2$ and the series is convergent provided that $0 < |z-1|$ (i.e. $\forall z \neq 1$).

Exercise 11.5 Using the Taylor series of $\cos y$ and substituting $1/z$ for y gives

$$z \cos \left(\frac{1}{z} \right) = z \sum_{n=0}^{\infty} \frac{(-1)^n}{(2n)!} \frac{1}{z^{2n}} = z + \sum_{n=1}^{\infty} \frac{(-1)^n}{(2n)!} \frac{1}{z^{2n-1}}.$$

We conclude that 0 is an essential singularity, $\text{Res}_0(f) = -1/2$ and the series is convergent for every $z \neq 0$.

Exercise 11.6 Substituting $z - 1$ for y in the Taylor expansion of e^y gives

$$\frac{e^z}{(z-1)^2} = \frac{e^{z-1+1}}{(z-1)^2} = \frac{e}{(z-1)^2} \sum_{n=0}^{\infty} \frac{1}{n!} (z-1)^n$$

$$= e \sum_{n=0}^{\infty} \frac{1}{n!} (z-1)^{n-2} = \frac{e}{(z-1)^2} + \frac{e}{(z-1)} + e \sum_{m=0}^{\infty} \frac{(z-1)^m}{(m+2)!}.$$

Therefore, $z_0 = 1$ is a pole of order 2, $\text{Res}_1(f) = e$ and the series is convergent provided $z \neq 1$.

Exercise 11.7 The function $g(z) = \sqrt{z} = e^{(1/2)\log z}$ is holomorphic in $\mathbb{C}\backslash\mathbb{R}_-$, where

$$\mathbb{R}_- = \{z \in \mathbb{C} : \text{Re}\, z \leq 0 \text{ and } \text{Im}\, z = 0\}.$$

Furthermore, we have

$$g'(z) = \frac{1}{2}z^{-1/2}, \quad g''(z) = -\frac{1}{4}z^{-3/2},$$

$$g'''(z) = \frac{3}{8}z^{-5/2}, \quad g^{(iv)}(z) = -\frac{15}{16}z^{-7/2}$$

and, for arbitrary n,

$$g^{(n)}(z) = (-1)^{n+1}\frac{1 \times 3 \times 5 \times \cdots \times (2n-3)}{2^n}z^{-(2n-1)/2}, \quad n = 2, 3, 4, \cdots.$$

Thus, in the neighborhood of 1, we have

$$g(z) = 1 + \frac{1}{2}(z-1) + \sum_{n=2}^{\infty}(-1)^{n+1}\frac{1 \times 3 \times 5 \times \cdots \times (2n-3)}{2^n n!}(z-1)^n.$$

We then deduce that

$$\begin{aligned}
\frac{\sqrt{z}}{(z-1)^2} = {} & \frac{1}{(z-1)^2} + \frac{1}{2(z-1)} \\
& + \sum_{n=2}^{\infty}(-1)^{n+1}\frac{1 \times 3 \times 5 \times \cdots \times (2n-3)}{2^n n!}(z-1)^{n-2}
\end{aligned}$$

and thus

$$\begin{aligned}
\frac{\sqrt{z}}{(z-1)^2} = {} & \frac{1}{(z-1)^2} + \frac{1}{2(z-1)} \\
& + \sum_{k=0}^{\infty}(-1)^{k+1}\frac{1 \times 3 \times 5 \times \cdots \times (2k+1)}{2^{k+2}(k+2)!}(z-1)^k.
\end{aligned}$$

Therefore, $z_0 = 1$ is a pole of order 2, its residue is $1/2$ and the Laurent series converges provided $0 < |z-1| < 1$.

Exercise 11.8 We write

$$f(z) = \frac{1}{(z-1)^2} \frac{(z-1)^2}{\cos^2\left(\frac{\pi}{2}z\right)} \quad \text{and} \quad g(z) = \frac{(z-1)^2}{\cos^2\left(\frac{\pi}{2}z\right)}.$$

Using l'Hôpital rule we obtain

$$g(1) = \lim_{z \to 1} \frac{(z-1)^2}{\cos^2\left(\frac{\pi}{2}z\right)} = \left(\lim_{z \to 1} \frac{(z-1)}{\cos\left(\frac{\pi}{2}z\right)}\right)^2 = \left(\frac{1}{-\frac{\pi}{2}\sin\frac{\pi}{2}}\right)^2 = \frac{4}{\pi^2}. \quad (11.1)$$

Therefore, $z_0 = 1$ is a pole of order 2 and, moreover,

$$\begin{aligned}
\text{Res}_1(f) &= \lim_{z \to 1} \frac{d}{dz} \left[\frac{(z-1)^2}{\cos^2\left(\frac{\pi}{2}z\right)}\right] \\
&= \lim_{z \to 1} \left[\frac{2(z-1)\cos^2\left(\frac{\pi}{2}z\right) + \pi(z-1)^2\cos\left(\frac{\pi}{2}z\right)\sin\left(\frac{\pi}{2}z\right)}{\cos^4\left(\frac{\pi}{2}z\right)}\right] \\
&= \lim_{z \to 1} \frac{(z-1)}{\cos\left(\frac{\pi}{2}z\right)} \lim_{z \to 1} \frac{2\cos\left(\frac{\pi}{2}z\right) + \pi(z-1)\sin\left(\frac{\pi}{2}z\right)}{\cos^2\left(\frac{\pi}{2}z\right)}.
\end{aligned}$$

The equation (11.1) implies that the first limit is equal to $-2/\pi$. Using l'Hôpital rule again, we find that the second limit is 0. We therefore get

$$\text{Res}_1(f) = 0.$$

The Laurent series of f is therefore

$$\frac{4}{\pi^2(z-1)^2} + \sum_{n=0}^{\infty} \alpha_n(z-1)^n.$$

We also note that the other singularities are when $\cos(\pi z/2) = 0$, that is when $z = 2k + 1$ for $k \in \mathbb{Z}$. The series is therefore convergent when $0 < |z-1| < 2$.

Exercise 11.9 We write

$$f(z) = \frac{\log(1+z)}{\sin(z^2)} = \frac{p(z)}{q(z)}$$

with p and q holomorphic at $z_0 = 0$. Observe that $z_0 = 0$ is a pole of order 1 of f. Indeed, 0 is a zero of order 1 of p since

$$p(0) = 0 \quad \text{and} \quad p'(0) = 1 \neq 0$$

and 0 is a zero of order 2 of q since

$$q(0) = 0, \quad q'(0) = 0 \quad \text{and} \quad q''(0) = 2 \neq 0.$$

We, hence, deduce that the function $g(z) = zf(z)$ is holomorphic at $z_0 = 0$. Using l'Hôpital rule twice we get

$$
\begin{aligned}
g(0) &= \lim_{z \to 0} g(z) = \lim_{z \to 0} \frac{\log(1+z) + \frac{z}{1+z}}{2z \cos(z^2)} \\
&= \lim_{z \to 0} \frac{\log(1+z)}{2z \cos(z^2)} + \lim_{z \to 0} \frac{1}{2(1+z)\cos(z^2)} \\
&= \frac{1}{2} \lim_{z \to 0} \frac{\frac{1}{1+z}}{\cos(z^2) - 2z\sin(z^2)} + \frac{1}{2} = 1
\end{aligned}
$$

Since

$$f(z) = \frac{g(z)}{z} = \frac{1}{z} \sum_{n=0}^{\infty} \frac{g^{(n)}(0)}{n!} z^n = \frac{1}{z} + g'(0) + \frac{g''(0)}{2!} z + \cdots,$$

we have that the residue at 0 of $f(z)$ is 1. Since the nearest singularity of $\log(1+z)$ is at $z = -1$ and since the nearest zero of $\sin(z^2)$ is at $\sqrt{\pi}$, then the series is convergent for $0 < |z| < 1$.

Exercise 11.10 It is easy to see that

$$f(z) = e^{\frac{1}{z}} \sin \frac{1}{z} = \left(1 + \frac{1}{z} + \frac{1}{2!z^2} + \cdots\right)\left(\frac{1}{z} - \frac{1}{3!z^3} + \cdots\right) = \frac{1}{z} + \frac{1}{z^2} + \cdots.$$

Therefore, $z_0 = 0$ is an essential singularity of f, its residue is 1 and its radius of convergence is infinite.

Exercise 11.11 Let

$$g(z) = \frac{\sin z}{e^z - 1}.$$

Note that, according to l'Hôpital rule, we have

$$g(0) = \lim_{z \to 0} \frac{\cos z}{e^z} = 1.$$

We therefore have that g is holomorphic provided $|z| < 2\pi$ (indeed $e^z - 1 \neq 0$ $\Leftrightarrow z \neq 2n\pi i$). Since $f(z) = g(z)/z$, we deduce that

$$f(z) = \frac{\sin z}{z(e^z - 1)} = \frac{1}{z} + \cdots.$$

Thus, $z_0 = 0$ is a pole of order 1, its residue is 1 and the Laurent series converges if $0 < |z| < 2\pi$.

Exercise 11.12 Since $\sin z = -\sin(z - \pi)$, we write

$$\sin z = -\sin(z - \pi) = -\sum_{n=0}^{\infty} \frac{(-1)^n}{(2n+1)!} (z - \pi)^{2n+1}$$

$$= \sum_{n=0}^{\infty} \frac{(-1)^{n+1}}{(2n+1)!} (z - \pi)^{2n+1}.$$

We, hence, infer that

$$f(z) = \frac{\sin z}{(z - \pi)^2} = \sum_{n=0}^{\infty} \frac{(-1)^{n+1}}{(2n+1)!} (z - \pi)^{2n-1}$$

$$= \frac{-1}{(z - \pi)} + \sum_{n=1}^{\infty} \frac{(-1)^{n+1}}{(2n+1)!} (z - \pi)^{2n-1}.$$

We then deduce that $z_0 = \pi$ is a pole of order 1, the radius of convergence is infinite (which means that the series is convergent for every $z \neq \pi$) and $\operatorname{Res}_\pi(f) = -1$.

Exercise 11.13 Let

$$f(z) = \frac{\sin z}{\sin(z^2)} = \frac{p(z)}{q(z)}.$$

Observe that p and q are holomorphic at $z_0 = 0$. We find that $z_0 = 0$ is a pole of order 1 of f. Indeed, 0 is a zero of order 1 of p

$$p(0) = 0 \quad \text{and} \quad p'(0) = 1 \neq 0$$

while 0 is a zero of order 2 of q

$$q(0) = 0, \quad q'(0) = 0 \quad \text{and} \quad q''(0) = 2 \neq 0.$$

It is therefore sufficient to write

$$f(z) = \frac{1}{z} \frac{z \sin z}{\sin(z^2)} = \frac{1}{z} g(z),$$

where g is holomorphic at $z = 0$. Note that g is even and therefore

$$g^{(2k+1)}(0) = 0 \quad \text{for every integer } k.$$

We thus have

$$f(z) = \frac{g(0)}{z} + \sum_{k=1}^{\infty} \frac{g^{(2k)}(0)}{(2k)!} z^{2k-1}.$$

We next need to compute $g(0)$. We use l'Hôpital rule to obtain

$$g(0) = \lim_{z \to 0} \frac{z \sin z}{\sin(z^2)} = \lim_{z \to 0} \frac{z^2}{\sin(z^2)} \lim_{z \to 0} \frac{\sin z}{z} = 1.$$

We, hence, get

$$f(z) = \frac{1}{z} + \sum_{k=1}^{\infty} \frac{g^{(2k)}(0)}{(2k)!} z^{2k-1}$$

and $\mathrm{Res}_0(f) = 1$. We, moreover, have

$$\sin(z^2) = 0 \iff z^2 = k\pi \iff z = \begin{cases} \sqrt{k\pi} & \text{if } k \geq 0 \\ i\sqrt{|k|\,\pi} & \text{if } k < 0. \end{cases}$$

and thus the Laurent series is convergent when $0 < |z| < \sqrt{\pi}$.

Exercise 11.14 (i) The function $y \to \log y$ is holomorphic in

$$\mathbb{C} \setminus \{y \in \mathbb{C} : \mathrm{Im}\, y = 0 \text{ and } \mathrm{Re}\, y \leq 0\}.$$

If $y = 1 + z$, we find that f is holomorphic in

$$D = \mathbb{C} \setminus \{z \in \mathbb{C} : \mathrm{Im}\, z = 0 \text{ and } \mathrm{Re}\, z \leq -1\}.$$

(ii) If $f(z) = \log(1+z)$, then

$$f'(z) = \frac{1}{1+z}, \quad f''(z) = \frac{-1}{(1+z)^2} \quad \text{and} \quad f'''(z) = \frac{2}{(1+z)^3}.$$

More generally, we have

$$f^{(n)}(z) = (-1)^{n+1} \frac{(n-1)!}{(1+z)^n}.$$

We therefore find that

$$f(0) = 0 \quad \text{and} \quad f^{(n)}(0) = (-1)^{n+1}(n-1)! \text{ for every } n \geq 1,$$

$$f(i) = \log(1+i) \quad \text{and} \quad f^{(n)}(i) = (-1)^{n+1} \frac{(n-1)!}{(1+i)^n} \text{ for every } n \geq 1.$$

Thus, if $z_0 = 0$, we get

$$f(z) = \log(1+z) = \sum_{n=1}^{\infty} (-1)^{n+1} \frac{z^n}{n}$$

and the radius of convergence is $R = 1$ (that is, the series is convergent if $|z| < 1$). If $z_0 = i$, we deduce that

$$f(z) = \log(1+z) = \log(1+i) + \sum_{n=1}^{\infty} (-1)^{n+1} \frac{(z-i)^n}{n(1+i)^n}$$

and the radius of convergence is $R = \sqrt{2}$ (that is, the series is convergent if $|z - i| < \sqrt{2}$).

Exercise 11.15 (i) Clearly $z = \pm i$ are the only singularities of f and they are poles of order 1. Indeed

$$f(z) = \frac{\sin(z^2+1)}{(z^2+1)} \cdot \frac{1}{(z^2+1)} = g(z) \cdot \frac{1}{(z^2+1)}$$

and g is holomorphic (the singularities $\pm i$ of g are removable).

(ii) L'Hôpital rule leads to

$$\mathrm{Res}_i(f) = \lim_{z \to i} \left[(z-i) \frac{\sin(z^2+1)}{(z^2+1)} \cdot \frac{1}{(z+i)(z-i)} \right]$$

$$= \frac{1}{2i} \lim_{z \to i} \frac{\sin(z^2+1)}{(z^2+1)} = \frac{1}{2i} \lim_{z \to i} \frac{2z\cos(z^2+1)}{2z} = \frac{1}{2i}$$

$$\mathrm{Res}_{-i}(f) = \lim_{z \to -i} \frac{\sin(z^2+1)}{(z^2+1)} \cdot \frac{1}{(z-i)} = -\frac{1}{2i} \lim_{z \to -i} \frac{\sin(z^2+1)}{(z^2+1)} = -\frac{1}{2i}.$$

(iii) The radius of convergence is in both cases 2, that is, the series converges when $0 < |z - i| < 2$ and $0 < |z + i| < 2$ respectively.

Exercise 11.16 (i) Note that $z \to 1+z^2$ is holomorphic in $\mathbb{C}$. The function $w \to \log w$ is holomorphic in

$$\mathbb{C} \setminus \{w \in \mathbb{C} : \mathrm{Im}\, w = 0 \text{ and } \mathrm{Re}\, w \leq 0\}.$$

If $w = 1 + z^2$, we find

$$w = (1 + x^2 - y^2) + 2ixy$$

and

$$\begin{cases} \mathrm{Im}\, w = 0 & \Leftrightarrow & xy = 0 & \Leftrightarrow & x = 0 \text{ or } y = 0 \\ \mathrm{Re}\, w \leq 0 & \Leftrightarrow & 1 + x^2 \leq y^2. \end{cases}$$

It is easy to see that $y = 0$ is not allowed while if $x = 0$, then $y^2 \geq 1$. Summarizing, we have that f is holomorphic in

$$D = \mathbb{C} \setminus \{z \in \mathbb{C} : \mathrm{Re}\, z = 0 \text{ and } |\mathrm{Im}\, z| \geq 1\}.$$

(ii) We have $f(0) = 0$. Note that f is even and therefore

$$f^{(2k+1)}(0) = 0 \quad \text{for every integer } k.$$

We immediately obtain

$$f'(z) = \frac{2z}{1+z^2} \;\Rightarrow\; f''(z) = \frac{2-2z^2}{(1+z^2)^2} \;\Rightarrow\; f''(0) = 2.$$

We therefore find

$$f(z) = \log\left(1+z^2\right) = z^2 + \sum_{k=2}^{\infty} \frac{f^{(2k)}(0)}{(2k)!} z^{2k}.$$

The radius of convergence is 1, that is, the Taylor series converges if $|z| < 1$.

Exercise* 11.17 Since f is holomorphic in the neighborhood of z_0 and $f(z_0) = 0$, we have

$$f(z) = f(z_0) + f'(z_0)(z - z_0) + \frac{f''(z_0)}{2!}(z - z_0)^2 + \cdots$$

$$= (z - z_0)\left[f'(z_0) + \sum_{n=2}^{\infty} \frac{f^{(n)}(z_0)}{n!}(z - z_0)^{n-1}\right]$$

$$= (z - z_0)f'(z_0) + (z - z_0)F(z).$$

In a similar way, we get

$$g(z) = (z - z_0)\left[g'(z_0) + \sum_{n=2}^{\infty} \frac{g^{(n)}(z_0)}{n!}(z - z_0)^{n-1}\right]$$

$$= (z - z_0)g'(z_0) + (z - z_0)G(z).$$

Since $F(z_0) = G(z_0) = 0$ and $g'(z_0) \neq 0$, we deduce that

$$\lim_{z \to z_0} \frac{f(z)}{g(z)} = \lim_{z \to z_0} \frac{f'(z_0) + F(z)}{g'(z_0) + G(z)} = \frac{f'(z_0)}{g'(z_0)}.$$

Exercise* 11.18 (i) Since f is holomorphic in $\mathbb{C}$, we have

$$f(z) = \sum_{n=0}^{\infty} \frac{f^{(n)}(0)}{n!} z^n \quad \forall z \in \mathbb{C}.$$

Cauchy integral formula gives

$$f^{(n)}(0) = \frac{n!}{2\pi i} \int_{\gamma} \frac{f(z)}{z^{n+1}} \, dz.$$

If $\gamma(t) = Re^{it}$, we obtain

$$f^{(n)}(0) = \frac{n!}{2\pi i} \int_0^{2\pi} \frac{f(Re^{it})}{(Re^{it})^{n+1}} iRe^{it}dt = \frac{n!}{2\pi R^n} \int_0^{2\pi} e^{-int} f(Re^{it}) dt$$

and thus

$$\left|f^{(n)}(0)\right| \leq \frac{n!}{2\pi R^n} \int_0^{2\pi} \left|e^{-int}\right| \left|f(Re^{it})\right| dt = \frac{n!}{2\pi R^n} \int_0^{2\pi} \left|f(Re^{it})\right| dt.$$

Since f is bounded, say $|f(z)| \leq M$, we have

$$\left|f^{(n)}(0)\right| \leq \frac{Mn!}{2\pi} \cdot \frac{2\pi}{R^n}.$$

Letting $R \to \infty$ we get

$$f^{(n)}(0) = 0, \quad \forall n \geq 1.$$

We have therefore found that f is constant, since

$$f(z) = \sum_{n=0}^{\infty} \frac{f^{(n)}(0)}{n!} z^n = f(0), \quad \forall z.$$

(ii) The function

$$f(x) = \sin x = \sum_{n=0}^{\infty} (-1)^n \frac{x^{2n+1}}{(2n+1)!}$$

is bounded all over $\mathbb{R}$ but is not constant.

Chapter 12

Residue theorem and applications

Exercise 12.1 *Case 1:* $0 \in \text{int}\,\gamma$. We have

$$e^{\frac{1}{z^2}} = \sum_{n=0}^{\infty} \frac{1}{n!} \left(\frac{1}{z^2}\right)^n = \sum_{n=1}^{\infty} \frac{1}{n!} \frac{1}{z^{2n}} + 1.$$

We, hence, deduce that $\text{Res}_0\,(f) = 0$ and thus

$$\int_{\gamma} f(z)\,dz = 0.$$

(Note that the function is not holomorphic in $\text{int}\,\gamma$).

Case 2: $0 \notin \overline{\text{int}\,\gamma}$. Cauchy theorem implies $\int_{\gamma} f(z)\,dz = 0$.

Case 3: $0 \in \gamma$. In this case, the integral is not well defined.

Exercise 12.2 (i) Let

$$f(z) = \frac{1}{(z-i)(z+2)^2(z-4)}.$$

We first compute the residues at i, 4 and -2 which are poles of order 1 for the first two and of order 2 for the third one. Their residue are given by

$$\text{Res}_i\,(f) = \lim_{z \to i} [(z-i)\,f(z)] = \frac{1}{(i+2)^2(i-4)}$$

$$\text{Res}_4\,(f) = \lim_{z \to 4} [(z-4)\,f(z)] = \frac{1}{36(4-i)}$$

$$\text{Res}_{-2}\,(f) = \lim_{z \to -2} \frac{d}{dz} \left[(z+2)^2\,f(z)\right] = \lim_{z \to -2} \frac{d}{dz} \left(\frac{1}{(z-i)(z-4)}\right)$$

$$= \lim_{z \to -2} \left(\frac{-2z+4+i}{(z-i)^2(z-4)^2}\right) = \frac{8+i}{36(i+2)^2}.$$

263

We then consider five cases.

Case 1: $i, -2, 4 \notin \overline{\text{int}\,\gamma}$. Cauchy theorem implies

$$\int_\gamma f(z)\,dz = 0.$$

Case 2: Only one of the points $i, -2, 4$ belongs to $\text{int}\,\gamma$.

2.a) $i \in \text{int}\,\gamma$ but $-2, 4 \notin \overline{\text{int}\,\gamma}$. We find

$$\int_\gamma f(z)\,dz = 2\pi i \operatorname{Res}_i(f) = \frac{2\pi i}{(i+2)^2(i-4)}.$$

2.b) $-2 \in \text{int}\,\gamma$ but $i, 4 \notin \overline{\text{int}\,\gamma}$. We deduce that

$$\int_\gamma f(z)\,dz = 2\pi i \operatorname{Res}_{-2}(f) = \frac{\pi i(8+i)}{18(i+2)^2}.$$

2.c) $4 \in \text{int}\,\gamma$ but $i, -2 \notin \overline{\text{int}\,\gamma}$. We obtain

$$\int_\gamma f(z)\,dz = 2\pi i \operatorname{Res}_4(f) = \frac{\pi i}{18(4-i)}.$$

Case 3: Two of the points $i, -2, 4$ belong to $\text{int}\,\gamma$.

3.a) $i, -2 \in \text{int}\,\gamma$ but $4 \notin \overline{\text{int}\,\gamma}$. We obtain

$$\int_\gamma f(z)\,dz = 2\pi i(\operatorname{Res}_i(f) + \operatorname{Res}_{-2}(f)) = \frac{2\pi i}{(i+2)^2}\left(\frac{1}{i-4} + \frac{8+i}{36}\right).$$

3.b) $i, 4 \in \text{int}\,\gamma$ but $-2 \notin \overline{\text{int}\,\gamma}$. We get

$$\int_\gamma f(z)\,dz = 2\pi i\,(\operatorname{Res}_i(f) + \operatorname{Res}_4(f)) = \frac{2\pi i}{(i-4)}\left(\frac{1}{(i+2)^2} - \frac{1}{36}\right).$$

3.c) $-2, 4 \in \text{int}\,\gamma$ but $i \notin \overline{\text{int}\,\gamma}$. We find

$$\int_\gamma f(z)\,dz = 2\pi i\,(\operatorname{Res}_{-2}(f) + \operatorname{Res}_4(f)) = \frac{\pi i}{18}\left(\frac{8+i}{(i+2)^2} - \frac{1}{(i-4)}\right).$$

Case 4: $i, -2, 4 \in \text{int}\,\gamma$. In this case, we have

$$\int_\gamma f(z)\,dz = 2\pi i\,(\operatorname{Res}_i(f) + \operatorname{Res}_{-2}(f) + \operatorname{Res}_4(f))$$

$$= 2\pi i\left(\frac{1}{(i-4)(i+2)^2} + \frac{8+i}{36(i+2)^2} - \frac{1}{36(i-4)}\right) = 0.$$

Case 5: $i \in \gamma$ or $-2 \in \gamma$ or $4 \in \gamma$. The integral is not well defined.

(ii) Let

$$f(z) = \frac{z^2 + 2z + 1}{(z-3)^3}.$$

Observe that $z = 3$ is a pole of order 3 and

$$\text{Res}_3(f) = \frac{1}{2} \lim_{z \to 3} \frac{d^2}{dz^2} \left(z^2 + 2z + 1\right) = 1.$$

We consider three cases.

Case 1: $3 \notin \overline{\text{int}\,\gamma}$. Cauchy theorem gives

$$\int_\gamma f(z)\,dz = 0.$$

Case 2: $3 \in \text{int}\,\gamma$. We get

$$\int_\gamma f(z)\,dz = 2\pi i \,\text{Res}_3(f) = 2\pi i.$$

Case 3: $3 \in \gamma$. The integral is not well defined.

(iii) Let $f(z) = \dfrac{e^{(1/z)}}{z^2}$. We have

$$f(z) = \frac{e^{(1/z)}}{z^2} = \frac{1}{z^2} \sum_{n=0}^\infty \frac{1}{n!} \frac{1}{z^n} = \sum_{n=0}^\infty \frac{1}{n!} \frac{1}{z^{n+2}}.$$

Hence, 0 is an essential singularity and

$$\text{Res}_0(f) = 0.$$

Case 1: $0 \notin \overline{\text{int}\,\gamma}$. Cauchy theorem leads to

$$\int_\gamma f(z)\,dz = 0.$$

Case 2: $0 \in \text{int}\,\gamma$. We find that

$$\int_\gamma f(z)\,dz = 2\pi i \,\text{Res}_0(f) = 0.$$

Case 3: $0 \in \gamma$. The integral is not well defined.

Exercise 12.3 (i) The singularities of $f(z) = (\cosh z)^{-1}$ are when $\cosh z = 0$, that is when

$$e^z + e^{-z} = 0 \Leftrightarrow e^{2z} = -1 \Leftrightarrow z = i(2k+1)\pi/2.$$

The only singularity of f inside γ_R is then $i\pi/2$ and it is a pole of order 1. The residue is

$$\text{Res}_{i\frac{\pi}{2}}(f) = \lim_{z \to i\frac{\pi}{2}} \frac{z - i\frac{\pi}{2}}{\cosh z} = \frac{1}{\sinh\left(i\frac{\pi}{2}\right)} = \frac{1}{i}$$

and, therefore,

$$\int_{\gamma_R} f(z)\, dz = 2\pi i \, \text{Res}_{i\frac{\pi}{2}}(f) = 2\pi.$$

(ii) We next compute the integral along each side of the rectangle. We have

$$\int_{\gamma_R} f(z)\, dz = \sum_{i=1}^{4} \int_{\gamma_i} f(z)\, dz,$$

where

$$\gamma_1 = \{x : x : -R \to R\}, \quad \gamma_2 = \{R + iy : y : 0 \to \pi\},$$

$$\gamma_3 = \{x + i\pi : x : R \to -R\}, \quad \gamma_4 = \{-R + iy : y : \pi \to 0\}.$$

We, therefore, obtain

$$\int_{\gamma_1} \frac{dz}{\cosh z} = \int_{-R}^{R} \frac{dx}{\cosh x} \quad \text{and} \quad \int_{\gamma_3} \frac{dz}{\cosh z} = -\int_{-R}^{R} \frac{dx}{\cosh(x + i\pi)}.$$

Since

$$\cosh(x + i\pi) = \frac{1}{2}\left(e^{x+i\pi} + e^{-(x+i\pi)}\right) = -\frac{1}{2}\left(e^x + e^{-x}\right) = -\cosh x,$$

we deduce that

$$\int_{\gamma_1 \cup \gamma_3} \frac{dz}{\cosh z} = 2\int_{-R}^{R} \frac{dx}{\cosh x}.$$

Furthermore, as $R \to +\infty$, we have

$$\int_{\gamma_2} \frac{dz}{\cosh z} = \int_{0}^{\pi} \frac{i\, dy}{\cosh(R + iy)} \to 0$$

$$\int_{\gamma_4} \frac{dz}{\cosh z} = -\int_{0}^{\pi} \frac{i\, dy}{\cosh(-R + iy)} \to 0.$$

We have, hence, showed that

$$\lim_{R \to \infty} \int_{\gamma_R} f(z)\, dz = 2 \int_{-\infty}^{+\infty} \frac{dx}{\cosh x} = 2\pi$$

and, thus,

$$\int_{-\infty}^{+\infty} \frac{dx}{\cosh x} = \pi.$$

Exercise 12.4 Let

$$f(z) = \frac{1 - e^{i\pi z}}{z(z+i)(z-1)^2}.$$

The possible singularities are at $z = 0$, $z = -i$ and $z = 1$. Using l'Hôpital rule, we find

$$\lim_{z \to 0} f(z) = \frac{-i\pi}{i} = -\pi.$$

Hence, 0 is a regular point (i.e. a removable singularity). Moreover, $z = -i$ is clearly a pole of order 1 while $z = 1$ is a pole of order 2. We, therefore, find

$$\operatorname{Res}_{-i}(f) = \lim_{z \to -i} \frac{1 - e^{i\pi z}}{z(z-1)^2} = \frac{1 - e^{\pi}}{2}.$$

$$\operatorname{Res}_1(f) = \lim_{z \to 1} \frac{d}{dz}\left[\frac{1 - e^{i\pi z}}{z(z+i)} \right] = \lim_{z \to 1} \frac{-i\pi e^{i\pi z}(z(z+i)) - (1 - e^{i\pi z})(2z+i)}{z^2(z+i)^2}$$

$$= \frac{i\pi(1+i) - 2(2+i)}{2i} = \frac{\pi - 2}{2} + i\frac{\pi + 4}{2}.$$

We study five cases.

Case 1: $1, -i \notin \overline{\operatorname{int}\gamma}$. Cauchy theorem gives

$$\int_{\gamma} f(z)\, dz = 0.$$

Case 2: $1 \in \operatorname{int}\gamma$ but $-i \notin \overline{\operatorname{int}\gamma}$. We find

$$\int_{\gamma} f(z)\, dz = 2\pi i \operatorname{Res}_1(f) = -\pi(\pi + 4) + i\pi(\pi - 2).$$

Case 3: $-i \in \operatorname{int}\gamma$ but $1 \notin \overline{\operatorname{int}\gamma}$. We get

$$\int_{\gamma} f(z)\, dz = 2\pi i \operatorname{Res}_{-i}(f) = i\pi(1 - e^{\pi}).$$

Case 4: $1, -i \in \text{int}\,\gamma$. We deduce

$$\int_\gamma f(z)\,dz = 2\pi i(\text{Res}_1(f) + \text{Res}_{-i}(f)) = -\pi(\pi + 4) + i\pi(\pi - 1 - e^\pi).$$

Case 5: $1 \in \gamma$ or $-i \in \gamma$. The integral is not well defined.

Exercise 12.5 Let $f(z) = \tan z$. The only singularities of f inside the given disk are at $z = \pm\pi/2$. These are clearly poles of order 1. Proposition 11.5 then implies

$$\text{Res}_{\frac{\pi}{2}}(f) = \frac{\sin\left(\frac{\pi}{2}\right)}{-\sin\left(\frac{\pi}{2}\right)} = -1$$

$$\text{Res}_{-\frac{\pi}{2}}(f) = \frac{\sin\left(-\frac{\pi}{2}\right)}{-\sin\left(-\frac{\pi}{2}\right)} = -1.$$

Case 1: $\pm\dfrac{\pi}{2} \in \text{int}\,\gamma$. We have

$$\int_\gamma \tan z\,dz = 2\pi i\left(\text{Res}_{\frac{\pi}{2}}(f) + \text{Res}_{-\frac{\pi}{2}}(f)\right) = -4\pi i.$$

Case 2: $\dfrac{\pi}{2} \in \text{int}\,\gamma$ and $-\dfrac{\pi}{2} \notin \overline{\text{int}\,\gamma}$. We obtain

$$\int_\gamma \tan z\,dz = 2\pi i\,\text{Res}_{\frac{\pi}{2}}(f) = -2\pi i.$$

Case 3: $-\dfrac{\pi}{2} \in \text{int}\,\gamma$ and $\dfrac{\pi}{2} \notin \overline{\text{int}\,\gamma}$. We get

$$\int_\gamma \tan z\,dz = 2\pi i\,\text{Res}_{-\frac{\pi}{2}}(f) = -2\pi i.$$

Case 4: $\pm\dfrac{\pi}{2} \notin \overline{\text{int}\,\gamma}$. Cauchy theorem implies immediately

$$\int_\gamma \tan z\,dz = 0.$$

Case 5: $\dfrac{\pi}{2} \in \gamma$ or $-\dfrac{\pi}{2} \in \gamma$. The integral is not well defined.

Exercise 12.6 We have

$$f(\cos\theta, \sin\theta) = \frac{1}{\sqrt{5} - \sin\theta}.$$

Letting $z = e^{i\theta}$ we find

$$\widetilde{f}(z) = \frac{1}{iz\left(\sqrt{5} - \frac{z-1/z}{2i}\right)} = \frac{2}{-z^2 + 2iz\sqrt{5} + 1}$$

$$= \frac{-2}{\left(z - i\left(\sqrt{5} + 2\right)\right)\left(z - i\left(\sqrt{5} - 2\right)\right)}.$$

The singularities of $\widetilde{f}$ are at $z_1 = i(\sqrt{5} + 2)$ and $z_2 = i(\sqrt{5} - 2)$, but only z_2, which is a pole of order 1, belongs to the interior of the unit disk. We, therefore, get

$$\operatorname{Res}_{i(\sqrt{5}-2)}\left(\widetilde{f}\right) = \lim_{z \to i(\sqrt{5}-2)} \frac{-2}{z - i(\sqrt{5} + 2)} = \frac{1}{2i}$$

and, hence,

$$\int_0^{2\pi} \frac{d\theta}{\sqrt{5} - \sin\theta} = 2\pi i \operatorname{Res}_{i(\sqrt{5}-2)}\left(\widetilde{f}\right) = \pi.$$

Exercise 12.7 We let $z = e^{i\theta}$ and we find

$$\cos\theta = \frac{e^{i\theta} + e^{-i\theta}}{2} = \frac{z + \frac{1}{z}}{2} = \frac{z^2 + 1}{2z}$$

$$\cos(2\theta) = \frac{e^{2i\theta} + e^{-2i\theta}}{2} = \frac{z^4 + 1}{2z^2} \quad \text{and} \quad \sin(2\theta) = \frac{e^{2i\theta} - e^{-2i\theta}}{2i} = \frac{z^4 - 1}{2iz^2}.$$

For $f(\cos\theta, \sin\theta) = \cos\theta \sin(2\theta)(5 + 3\cos(2\theta))^{-1}$, we define

$$\widetilde{f}(z) = \frac{1}{iz} f\left(\frac{1}{2}\left(z + \frac{1}{z}\right), \frac{1}{2i}\left(z - \frac{1}{z}\right)\right) = \frac{-(z^2 + 1)(z^4 - 1)}{6z^2(z^2 + 3)(z^2 + \frac{1}{3})}.$$

We find that the only singularities inside the unit disk are 0, which is a pole of order 2, and $\pm i/\sqrt{3}$ which are poles of order 1. Their residues are

$$\operatorname{Res}_{i/\sqrt{3}}\left(\widetilde{f}\right) = \lim_{z \to \frac{i}{\sqrt{3}}} \frac{-(z^2 + 1)(z^4 - 1)}{6z^2(z^2 + 3)(z + \frac{i}{\sqrt{3}})} = \frac{i}{6\sqrt{3}}$$

$$\operatorname{Res}_{-i/\sqrt{3}}\left(\widetilde{f}\right) = \lim_{z \to \frac{-i}{\sqrt{3}}} \frac{-(z^2 + 1)(z^4 - 1)}{6z^2(z^2 + 3)(z - \frac{i}{\sqrt{3}})} = -\frac{i}{6\sqrt{3}}$$

$$\operatorname{Res}_0\left(\widetilde{f}\right) = -\frac{1}{2} \lim_{z \to 0} \frac{d}{dz}\left[\frac{z^6 + z^4 - z^2 - 1}{3z^4 + 10z^2 + 3}\right] = 0.$$

If γ denotes the unit circle, we have

$$\int_0^{2\pi} \frac{\cos\theta \sin(2\theta)}{5 + 3\cos(2\theta)}\, d\theta = \int_\gamma \tilde{f}(z)\, dz$$

$$= 2\pi i \left[\mathrm{Res}_{i/\sqrt{3}}\left(\tilde{f}\right) + \mathrm{Res}_{-i/\sqrt{3}}\left(\tilde{f}\right) + \mathrm{Res}_0\left(\tilde{f}\right) \right] = 0.$$

This result could have been obtained from symmetry considerations.

Exercise 12.8 Let $z = e^{i\theta}$. We deduce that

$$\cos\theta = \frac{e^{i\theta} + e^{-i\theta}}{2} = \frac{1}{2}\left(z + \frac{1}{z}\right)$$

$$\cos(2\theta) = \frac{e^{2i\theta} + e^{-2i\theta}}{2} = \frac{1}{2}\left(z^2 + \frac{1}{z^2}\right).$$

For $f(\cos\theta, \sin\theta) = (13 - 5\cos(2\theta))^{-1}\cos^2\theta$, we define

$$\tilde{f}(z) = \frac{1}{iz}\, f\left(\frac{1}{2}\left(z + \frac{1}{z}\right), \frac{1}{2i}\left(z - \frac{1}{z}\right) \right) = \frac{-\left(z^2 + 1\right)^2}{10iz\left(z^2 - 5\right)\left(z^2 - \dfrac{1}{5}\right)}.$$

The singularities of $\tilde{f}$ are at $z = 0$, $\pm 1/\sqrt{5}$ and $\pm\sqrt{5}$, and they are poles of order 1. We note that only the first three lie inside the unit circle γ. The corresponding residues are

$$\mathrm{Res}_0\left(\tilde{f}\right) = \lim_{z \to 0} z\tilde{f}(z) = \frac{-1}{10i}$$

$$\mathrm{Res}_{1/\sqrt{5}}\left(\tilde{f}\right) = \lim_{z \to 1/\sqrt{5}}\left(z - \frac{1}{\sqrt{5}}\right)\tilde{f}(z) = \frac{3}{40i}$$

$$\mathrm{Res}_{-1/\sqrt{5}}\left(\tilde{f}\right) = \lim_{z \to -1/\sqrt{5}}\left(z + \frac{1}{\sqrt{5}}\right)\tilde{f}(z) = \frac{3}{40i}.$$

Since

$$\int_\gamma \tilde{f}(z)\, dz = 2\pi i \left(\mathrm{Res}_0\left(\tilde{f}\right) + \mathrm{Res}_{1/\sqrt{5}}\left(\tilde{f}\right) + \mathrm{Res}_{-1/\sqrt{5}}\left(\tilde{f}\right) \right) = \frac{\pi}{10},$$

we find that

$$\int_0^{2\pi} \frac{\cos^2\theta}{13 - 5\cos(2\theta)}\, d\theta = \int_\gamma \tilde{f}(z)\, dz = \frac{\pi}{10}.$$

Exercise 12.9 We write $z = e^{i\theta}$ and we obtain

$$\sin(\theta/2) = \frac{e^{i\frac{\theta}{2}} - e^{-i\frac{\theta}{2}}}{2i} = \frac{z-1}{2iz^{1/2}}$$

$$\sin(5\theta/2) = \frac{e^{i\frac{5\theta}{2}} - e^{-i\frac{5\theta}{2}}}{2i} = \frac{z^5 - 1}{2iz^{5/2}}$$

and thus

$$\left(\frac{\sin(5\theta/2)}{\sin(\theta/2)}\right)^2 = \left(\frac{z^5 - 1}{(z-1)z^2}\right)^2 = \frac{(z^4 + z^3 + z^2 + z + 1)^2}{z^4}.$$

Setting

$$\tilde{f}(z) = \frac{(z^4 + z^3 + z^2 + z + 1)^2}{iz^5}$$

we find that $z = 0$ is the only singularity of $\tilde{f}$ and that this is a pole of order 5. Its residue is given by

$$\mathrm{Res}_0\left(\tilde{f}\right) = \frac{1}{4!} \lim_{z \to 0} \frac{d^4}{dz^4}\left[\frac{(z^4 + z^3 + z^2 + z + 1)^2}{i}\right] = \frac{5}{i}.$$

Finally, if γ denotes the unit circle, we have

$$\int_0^{2\pi} \left(\frac{\sin(5\theta/2)}{\sin(\theta/2)}\right)^2 d\theta = \int_\gamma \tilde{f}(z) \, dz = 2\pi i \, \mathrm{Res}_0\left(\tilde{f}\right) = 10\pi.$$

By a completely analogous reasoning one can prove that

$$\int_0^{2\pi} \left(\frac{\sin((n\theta)/2)}{\sin(\theta/2)}\right)^2 d\theta = 2n\pi,$$

since the corresponding $\tilde{f}$ is such that $\mathrm{Res}_0\left(\tilde{f}\right) = n/i$.

Exercise 12.10 (i) We first write

$$(z-1)^{2n} = z^{2n} - \binom{2n}{1}z^{2n-1} + \cdots - \binom{2n}{2n-1}z + 1$$

$$= \sum_{k=0}^{2n} (-1)^k \binom{2n}{k} z^k$$

and we find

$$(z-1)^{2n}\left(z^{2n} + 1\right) = \sum_{k=0}^{2n} (-1)^k \binom{2n}{k} z^{k+2n} + \sum_{k=0}^{2n} (-1)^k \binom{2n}{k} z^k$$

$$= \sum_{k=1}^{2n} (-1)^k \binom{2n}{k} z^{k+2n} + 2z^{2n} + \sum_{k=0}^{2n-1} (-1)^k \binom{2n}{k} z^k.$$

We thus have

$$h\left(z\right) = \frac{\left(z-1\right)^{2n}\left(z^{2n}+1\right)}{z^{2n+1}} = \sum_{k=1}^{2n}\left(-1\right)^{k}\binom{2n}{k}z^{k-1} + \frac{2}{z} + \sum_{k=0}^{2n-1}\binom{2n}{k}\frac{\left(-1\right)^{k}}{z^{2n+1-k}}.$$

We therefore see that $z = 0$ is a pole of order $2n + 1$ whose residue is 2.

(ii) As usual we let $z = e^{i\theta}$ and we find

$$\cos\theta = \frac{z + \dfrac{1}{z}}{2} = \frac{z^{2}+1}{2z} \quad\text{and}\quad \cos\left(n\theta\right) = \frac{z^{n} + \dfrac{1}{z^{n}}}{2} = \frac{z^{2n}+1}{2z^{n}}.$$

We therefore have

$$\left(1 - \cos\theta\right)^{n}\cos\left(n\theta\right) = \left(1 - \frac{z^{2}+1}{2z}\right)^{n}\frac{z^{2n}+1}{2z^{n}}$$

$$= \frac{\left(-1\right)^{n}\left(z-1\right)^{2n}\left(z^{2n}+1\right)}{2^{n+1}z^{2n}}.$$

Let

$$\widetilde{f}\left(z\right) = \frac{\left(-1\right)^{n}\left(z-1\right)^{2n}\left(z^{2n}+1\right)}{i2^{n+1}z^{2n+1}}.$$

Observe that

$$\widetilde{f}\left(z\right) = \frac{\left(-1\right)^{n}}{i2^{n+1}}h\left(z\right)$$

and thus

$$\text{Res}_{0}\left(\widetilde{f}\left(z\right)\right) = \frac{\left(-1\right)^{n}}{i2^{n}}.$$

If γ is the unit circle, we have

$$\int_{0}^{2\pi}\left(1 - \cos\theta\right)^{n}\cos\left(n\theta\right)d\theta = \int_{\gamma}\widetilde{f}\left(z\right)dz = 2\pi i\,\text{Res}_{0}\left(\widetilde{f}\right) = \frac{\left(-1\right)^{n}\pi}{2^{n-1}}.$$

Exercise 12.11 We let $z = e^{i\theta}$ and, hence, $\cos\theta = \dfrac{1}{2}\left(z + 1/z\right)$. We therefore find

$$f\left(z\right) = \frac{1}{1 - p\left(z + \dfrac{1}{z}\right) + p^{2}} = \frac{-z}{\left(pz - 1\right)\left(z - p\right)}.$$

Let

$$\widetilde{f}\left(z\right) = \frac{-1}{i(pz - 1)(z - p)}$$

and let γ be the unit circle. The function $\tilde{f}$ has singularities (in fact poles of order 1) at $z = p$ and $z = 1/p$. Since $p < 1$ only $z = p$ lies inside of γ. Its residue is

$$\text{Res}_p\left(\tilde{f}\right) = \lim_{z \to p} \frac{-(z-p)}{i(pz-1)(z-p)} = \frac{-1}{i(p^2-1)}.$$

We thus have

$$\int_0^{2\pi} \frac{d\theta}{1 - 2p\cos\theta + p^2} = \int_\gamma \tilde{f}(z)\ dz = 2\pi i\,\text{Res}_p\left(\tilde{f}\right) = \frac{2\pi}{1-p^2}.$$

Exercise 12.12 (i) If we let $z = r\,e^{i\theta}$, then the triangle inequality implies

$$\left|z^6 + 1\right| \geq \left||z^6| - 1\right| = \left|r^6 - 1\right|.$$

(ii) If $r > 1$, we write

$$\left|\int_{C_r} \frac{z^2}{1+z^6}\,dz\right| = \left|\int_0^\pi \frac{\left(r\,e^{i\theta}\right)^2}{1+\left(r\,e^{i\theta}\right)^6}\,ir\,e^{i\theta}\,d\theta\right|$$

$$\leq \int_0^\pi \frac{r^3}{r^6 - 1}\,d\theta = \frac{\pi r^3}{r^6 - 1} \to 0 \quad \text{if } r \to \infty.$$

(iii) Let $r > 1$ and $\gamma_r = C_r \cup L_r$ where C_r is as in (ii) and L_r is the line segment $[-r, r]$ on the real axis. The singularities of $f(z) = \dfrac{z^2}{1+z^6}$ are the zeroes of $1 + z^6$, which are given by

$$1 + z^6 = 0 \Leftrightarrow z^6 = -1 = e^{i(\pi + 2n\pi)}$$

and thus the zeroes are

$$z_n = e^{i\pi(2n+1)/6}, \quad n = 0, \cdots, 5.$$

Only z_0, z_1 and z_2 lie inside of γ_r and these are poles of order 1. Using Proposition 11.5 (with $p(z) = z^2$ and $q(z) = 1 + z^6$, which implies that $q'(z) = 6z^5$), we find that their residues are

$$\text{Res}_{z_n}\left(\frac{z^2}{1+z^6}\right) = \frac{1}{6z_n^3} = \frac{1}{6}e^{-i(2n+1)\pi/2} = \frac{(-1)^{n+1}}{6}i, \quad n = 0, 1, 2.$$

Using the residue theorem, we write

$$\int_{\gamma_r} f(z)\ dz = \int_{C_r} f(z)\ dz + \int_{L_r} f(z)\ dz = 2\pi i \sum_{n=0}^{2} \text{Res}_{z_n}(f) = \frac{\pi}{3}.$$

Since

$$\int_{L_r} f(z)\,dz = \int_{-r}^{r} \frac{x^2}{1+x^6}\,dx \to \int_{-\infty}^{+\infty} \frac{x^2}{1+x^6}\,dx \quad \text{as } r \to \infty$$

$$\int_{C_r} f(z)\,dz \to 0 \quad \text{as } r \to \infty$$

(cf. (ii) above) we deduce that

$$\int_{-\infty}^{+\infty} \frac{x^2}{1+x^6}\,dx = \lim_{r\to\infty} \int_{\gamma_r} f(z)\,dz = \frac{\pi}{3}.$$

Exercise 12.13 (i) Let $z = re^{i\theta} = r\cos\theta + ir\sin\theta$. We get

$$\left|e^{iz}\right| = \left|e^{ir\cos\theta - r\sin\theta}\right| = e^{-r\sin\theta} \le 1$$

since $r\sin\theta \ge 0$ ($r > 0$ and $\theta \in [0,\pi]$).

(ii) The triangle inequality implies

$$\left|16 + z^4\right| \ge \left|z^4\right| - 16 = r^4 - 16.$$

(iii) Using (i) and (ii), we find that if $r > 2$, then

$$\left|\frac{e^{iz}}{16 + z^4}\right| \le \frac{1}{r^4 - 16}.$$

(iv) Let $f(z) = \dfrac{e^{iz}}{16 + z^4}$. If $r \to \infty$, we then get

$$\left|\int_{C_r} f(z)\,dz\right| = \left|\int_0^{\pi} f(re^{i\theta})\, ire^{i\theta}\,d\theta\right| \le \int_0^{\pi} \left|f(re^{i\theta})\right| r\,d\theta$$

$$\le \frac{r}{r^4 - 16} \int_0^{\pi} d\theta \to 0.$$

(v) Let $r > 2$ and $\gamma_r = C_r \cup L_r$ where C_r is as in (iv) and L_r is the line segment $[-r, r]$ on the real axis. The singularities of f are given by

$$z^4 + 16 = 0 \Leftrightarrow z^4 = -16 = 16e^{i(\pi + 2n\pi)},$$

i.e.

$$z_n = 2e^{i\pi(2n+1)/4}, \quad n = 0, 1, 2, 3.$$

These are clearly poles of order 1 and only z_0 and z_1 lie inside of γ_r. Using Proposition 11.5 (let $p(z) = e^{iz}$ and $q(z) = 16 + z^4$, which implies $q'(z) = 4z^3$) we obtain

$$\text{Res}_{z_0}(f) = \frac{e^{iz_0}}{4z_0^3} = \frac{e^{i\sqrt{2}(1+i)}}{4\left(\sqrt{2}(1+i)\right)^3} = \frac{e^{\sqrt{2}(-1+i)}}{16\sqrt{2}(1+i)i}$$

$$\text{Res}_{z_1}(f) = \frac{e^{iz_1}}{4z_1^3} = \frac{e^{i\sqrt{2}(-1+i)}}{4\left(\sqrt{2}(-1+i)\right)^3} = \frac{e^{\sqrt{2}(-1-i)}}{16\sqrt{2}(1-i)i}$$

$$\text{Res}_{z_0}(f) + \text{Res}_{z_1}(f) = \frac{e^{-\sqrt{2}}}{16i\sqrt{2}}\left(\frac{\cos\sqrt{2} + i\sin\sqrt{2}}{1+i} + \frac{\cos\sqrt{2} - i\sin\sqrt{2}}{1-i}\right)$$

$$= \frac{e^{-\sqrt{2}}}{16i\sqrt{2}}\left(\cos\sqrt{2} + \sin\sqrt{2}\right).$$

The residue theorem then implies

$$\int_{\gamma_r} \frac{e^{iz}}{16+z^4}\, dz = \int_{C_r} \frac{e^{iz}}{16+z^4}\, dz + \int_{L_r} \frac{e^{iz}}{16+z^4}\, dz$$

$$= 2\pi i\left(\text{Res}_{z_0}(f) + \text{Res}_{z_1}(f)\right)$$

$$= \pi \frac{e^{-\sqrt{2}}}{8\sqrt{2}}\left(\cos\sqrt{2} + \sin\sqrt{2}\right).$$

Since

$$\lim_{r\to\infty} \int_{L_r} f(z)\, dz = \lim_{r\to\infty} \int_{-r}^{r} \frac{e^{ix}}{16+x^4}\, dx$$

$$= \int_{-\infty}^{+\infty} \frac{\cos x}{16+x^4}\, dx + i\int_{-\infty}^{+\infty} \frac{\sin x}{16+x^4}\, dx,$$

the previous identity combined with (iv) give

$$\int_{-\infty}^{+\infty} \frac{\cos x}{16+x^4}\, dx = \pi \frac{e^{-\sqrt{2}}}{8\sqrt{2}}\left(\cos\sqrt{2} + \sin\sqrt{2}\right)$$

$$\int_{-\infty}^{+\infty} \frac{\sin x}{16+x^4}\, dx = 0.$$

Chapter 13

Conformal mapping

Exercise 13.1 (i) Let $z = x + iy$. We write

$$f(z) = \frac{x - iy}{(x + iy)(x - iy)} = u + iv$$

and we find

$$u = \frac{x}{x^2 + y^2} \quad \text{and} \quad v = \frac{-y}{x^2 + y^2}.$$

(ii) Setting $w = 1/z$ we obtain

$$z = \frac{1}{w} = \frac{1}{u + iv} = \frac{u - iv}{(u + iv)(u - iv)}$$

and thus

$$x = \frac{u}{u^2 + v^2} \quad \text{and} \quad y = \frac{-v}{u^2 + v^2}.$$

(iii) We now find $f(A_i)$.

$- A_1$ is the line whose equation is $x = y$. Therefore, the above identities for u and v lead to

$$u = \frac{x}{2x^2} = \frac{1}{2x} \quad \text{and} \quad v = \frac{-x}{2x^2} = \frac{-1}{2x},$$

which implies

$$u + v = 0.$$

Therefore, $f(A_1)$ is a line passing through the origin (cf. Figure 13.1), namely

$$f(A_1) = \{w = u + iv : \operatorname{Im} w = -\operatorname{Re} w\}.$$

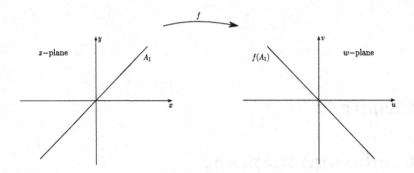

Figure 13.1: Exercise 13.1: A_1 and $f(A_1)$

– A_2 is the line given by $\operatorname{Re} z = x = 1$. We thus find

$$x = 1 = \frac{u}{u^2 + v^2} \implies u^2 + v^2 = u$$

which leads to

$$\left(u - \frac{1}{2}\right)^2 + v^2 = \frac{1}{4}.$$

Thus, $f(A_2)$ is a circle passing through the origin (cf. Figure 13.2)

$$f(A_2) = \left\{ w = u + iv : \left| w - \frac{1}{2} \right|^2 = \left(u - \frac{1}{2}\right)^2 + v^2 = \frac{1}{4} \right\}.$$

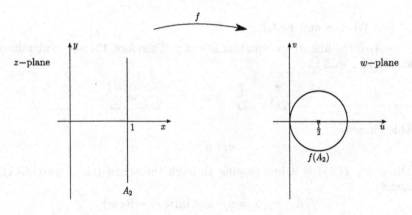

Figure 13.2: Exercise 13.1: A_2 and $f(A_2)$

– A_3 is a circle passing through the origin and whose equation is given by

$$(x-1)^2 + y^2 = 1$$

(that is $x^2 + y^2 = 2x$). We therefore have

$$\left(\frac{u}{u^2+v^2}\right)^2 + \left(\frac{-v}{u^2+v^2}\right)^2 = \frac{2u}{u^2+v^2}.$$

Thus $u = 1/2$ and $f(A_3)$ is a line which does not pass through the origin (cf. Figure 13.3)

$$f(A_3) = \left\{ w \in \mathbb{C} : \operatorname{Re} w = \frac{1}{2} \right\}.$$

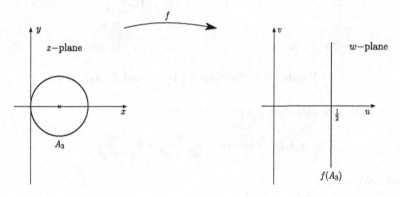

Figure 13.3: Exercise 13.1: A_3 and $f(A_3)$

– A_4 is a circle, which does not pass through the origin and which is defined by the equation

$$(x-1)^2 + y^2 = 4.$$

We, hence, deduce that

$$\left(\frac{u}{u^2+v^2} - 1\right)^2 + \frac{v^2}{(u^2+v^2)^2} = 4.$$

We therefore have

$$u^2 - 2u(u^2+v^2) + (u^2+v^2)^2 + v^2 = 4(u^2+v^2)^2$$

and thus

$$3(u^2+v^2) + 2u - 1 = 0 \Rightarrow \left(u + \frac{1}{3}\right)^2 + v^2 = \frac{4}{9}.$$

We finally find that $f(A_4)$ is a circle which does not pass through the origin (cf. Figure 13.4)

$$f(A_4) = \left\{ w \in \mathbb{C} : \left| w + \frac{1}{3} \right| = \frac{2}{3} \right\}.$$

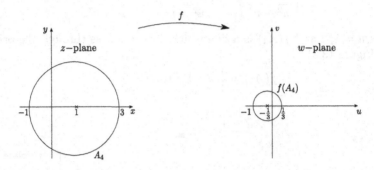

Figure 13.4: Exercise 13.1: A_4 and $f(A_4)$

(iv) We just saw that if $f(z) = 1/z$, then

$$x + iy \xrightarrow{f} u + iv = \frac{x}{x^2 + y^2} + i \frac{-y}{x^2 + y^2}$$

and conversely

$$x = \frac{u}{u^2 + v^2} \quad \text{and} \quad y = \frac{-v}{u^2 + v^2}.$$

Consider different cases.

Case 1. The image of a line passing through the origin is a line passing through the origin. We indeed have

$$\alpha x + \beta y = 0 \ \Rightarrow \ \alpha u - \beta v = 0.$$

Case 2. The image of a line which does not pass through the origin is a circle passing through the origin. More precisely, if $\gamma \neq 0$ and if

$$\alpha x + \beta y = \frac{\gamma}{2},$$

then

$$\frac{\alpha u}{u^2 + v^2} - \frac{\beta v}{u^2 + v^2} = \frac{\gamma}{2}.$$

We, hence, have

$$\gamma(u^2 + v^2) - 2\alpha u + 2\beta v = 0$$

and thus

$$\left(u - \frac{\alpha}{\gamma}\right)^2 + \left(v + \frac{\beta}{\gamma}\right)^2 = \frac{\alpha^2 + \beta^2}{\gamma^2}.$$

Case 3. The image of a circle passing through the origin

$$(x - \alpha)^2 + (y - \beta)^2 = \gamma^2$$

with $\gamma^2 = \alpha^2 + \beta^2$, is a line which does not pass through the origin. This follows from

$$\left(\frac{u}{u^2 + v^2} - \alpha\right)^2 + \left(\frac{-v}{u^2 + v^2} - \beta\right)^2 = \gamma^2$$

which implies

$$\gamma^2(u^2 + v^2)^2 = \left(\alpha^2 + \beta^2\right)\left(u^2 + v^2\right)^2 + u^2 + v^2 - 2\left(\alpha u - \beta v\right)\left(u^2 + v^2\right)$$

and therefore

$$2\beta v - 2\alpha u + 1 = 0.$$

Case 4. The image of a circle

$$(x - \alpha)^2 + (y - \beta)^2 = \gamma^2, \qquad \gamma^2 \neq \alpha^2 + \beta^2,$$

which does not pass through the origin, is a circle which does not pass through the origin. More precisely, we find, using the previous computation,

$$\left(\gamma^2 - \alpha^2 - \beta^2\right)\left(u^2 + v^2\right)^2 = \left(u^2 + v^2\right)\left(1 - 2\alpha u + 2\beta v\right)$$

which gives

$$u^2 + \frac{2\alpha u}{\gamma^2 - \alpha^2 - \beta^2} + v^2 - \frac{2\beta v}{\gamma^2 - \alpha^2 - \beta^2} = \frac{1}{\gamma^2 - \alpha^2 - \beta^2}$$

and thus

$$\left(u + \frac{\alpha}{\gamma^2 - \alpha^2 - \beta^2}\right)^2 + \left(v - \frac{\beta}{\gamma^2 - \alpha^2 - \beta^2}\right)^2 = \frac{\gamma^2}{\left(\gamma^2 - \alpha^2 - \beta^2\right)^2}.$$

Exercise 13.2 (i) We are looking for a map of the form

$$w(z) = \frac{az + b}{cz + d}.$$

We therefore have

$$w(0) = \frac{b}{d} = -1, \quad w(1 + i) = \frac{a(1 + i) + b}{c(1 + i) + d} = 1$$

and

$$w(1-i) = \frac{a(1-i)+b}{c(1-i)+d} = -1+2i$$

that is

$$b = -d, \quad b = \frac{c-a}{2}(1+i) \quad \text{and} \quad c = a\frac{1-i}{1+i} = -ai.$$

Letting $c = 1$, we get

$$a = i, \quad b = 1 \quad \text{and} \quad d = -1.$$

We thus find

$$w(z) = \frac{iz+1}{z-1}.$$

(ii) The required map is $z \to 2/z$. Indeed, reasoning as in Exercise 13.1, we note that if $w(z) = 2/z$, then $x+iy \to u+iv$, with

$$u^2 + v^2 = \frac{4}{x^2+y^2}.$$

We furthermore see that it maps the disk $\{z \in \mathbb{C} : |z| < 1\}$ into the exterior of $\{w \in \mathbb{C} : |w| > 2\}$.

Exercise 13.3 Let

$$f(z) = \frac{az+b}{cz+d}.$$

To the three points $\{-1, 0, 1\}$ of $\partial\Omega$ we associate the points $\{-1, i, 1\}$ of ∂D via f (cf. Figure 13.5).

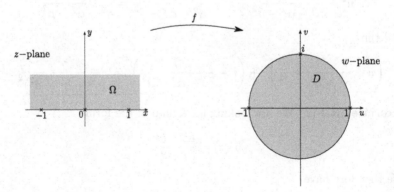

Figure 13.5: Exercise 13.3

We, hence, have

$$f(-1) = \frac{-a+b}{-c+d} = -1, \quad f(0) = \frac{b}{d} = i \quad \text{and} \quad f(1) = \frac{a+b}{c+d} = 1.$$

We easily find that

$$c = b = ai \quad \text{and} \quad d = a.$$

This leads to

$$f(z) = \frac{az + ai}{aiz + a} = \frac{z+i}{iz+1}.$$

It remains to check if the interior of Ω has been sent into the interior of D. For example, we have

$$2i \in \Omega \implies f(2i) = \frac{3i}{-1} = -3i \notin D.$$

Therefore, the required map (since we know that $f(z) = 1/z$ maps the exterior of the unit disk into its interior and conversely) is

$$g(z) = \frac{iz+1}{z+i}.$$

Exercise 13.4 We can proceed in two ways.

(I) (i) We write $f = f_2 \circ f_1$, where

$$f_1 : \Omega \to A = \{z \in \mathbb{C} : |z| > 1\}$$

$$f_2 : A \to D.$$

We have already seen that $f_2(z) = 1/z$ maps A onto D. In a similar way, we immediately see that

$$f_1(z) = \frac{z - 2(1+i)}{2}$$

maps Ω into A. We therefore obtain

$$f(z) = f_2(f_1(z)) = \frac{2}{z - 2(1+i)}.$$

(ii) Its inverse is given by

$$w = \frac{2}{z - 2(1+i)} \implies zw = 2(1+i)w + 2$$

which leads to

$$z = \frac{2(1+i)w + 2}{w}.$$

(II) (i) If one does not see the immediate decomposition, we fix any three points on $\partial\Omega$ and their images on ∂D. We then check if the map sends an interior point of Ω into an interior point of D (if this is not the case, it is sufficient to inverse the function). We are therefore looking for a function

$$g(z) = \frac{az+b}{cz+d}$$

such that

$$g(2i) = -1, \quad g(2) = -i \quad \text{and} \quad g(4+2i) = 1.$$

After an elementary computation, we find

$$g(z) = \frac{az - 2(1+i)a}{2a} = \frac{z - 2(1+i)}{2}.$$

We note that $0 \in \Omega$ and

$$g(0) = -(1+i) \notin D$$

since $|g(0)| = \sqrt{2} > 1$. The required transformation is thus

$$f(z) = \frac{1}{g(z)} = \frac{2}{z - 2(1+i)}.$$

We have therefore found the same map as above, but we could have found another one if we had chosen different points. For example, we could have chosen

$$f(2i) = -1, \quad f(2) = i \quad \text{and} \quad f(4+2i) = 1.$$

(ii) The inverse function depends on the transformation we have found in (i).

Exercise 13.5 (i) *Case 1:* $c \neq 0$ and $ad - bc \neq 0$. We first show that f is one-to-one and onto.

- We begin by showing the injectivity. We have

$$f(z_1) = f(z_2) \Leftrightarrow \frac{az_1+b}{cz_1+d} = \frac{az_2+b}{cz_2+d}$$

which leads to

$$acz_1z_2 + adz_1 + bcz_2 + bd = acz_1z_2 + bcz_1 + adz_2 + bd$$

and then

$$(ad-bc)z_2 = (ad-bc)z_1 \Leftrightarrow z_1 = z_2.$$

(Note that we have used the hypothesis $ad - bc \neq 0$.)

– We next prove that f is onto. We have

$$f(z) = w = \frac{az + b}{cz + d}$$

if and only if

$$(az + b) = w(cz + d) \quad \Leftrightarrow \quad (cw - a)z = b - dw$$

if and only if

$$z = f^{-1}(w) = \frac{-dw + b}{cw - a}.$$

Since $ad - bc \neq 0$, we obtain

$$\frac{-dw + b}{cw - a} \neq \frac{-d}{c}$$

and thus

$$\frac{-dw + b}{cw - a} \in \Omega = \mathbb{C} \backslash \{-d/c\}.$$

We, hence, have

$$f(\Omega) = \Omega^* = \mathbb{C} \backslash \{a/c\}.$$

– Furthermore, f is holomorphic as the quotient of two holomorphic functions (while the denominator does not vanish). The derivative is given by

$$f'(z) = \frac{ad - bc}{(cz + d)^2} \neq 0.$$

Case 2: $c = 0$ and $ad - bc \neq 0$. We then have $d \neq 0$ and

$$f(z) = \frac{a}{d} z + \frac{b}{d}.$$

Clearly, f is conformal from $\mathbb{C}$ onto $\mathbb{C}$.

(ii) When $ad - bc = 0$, we immediately observe that $f'(z) = 0$ and thus f is constant. We indeed have

$$f(z) = \frac{az + b}{cz + d} = \frac{b(cz + d)}{d(cz + d)} = \frac{b}{d}.$$

Exercise 13.6 (i) Let $f = u + iv$. We have to find u and v. We obtain

$$u + iv = x + iy + \frac{1}{x + iy} = x + iy + \frac{x - iy}{x^2 + y^2}$$

and thus

$$u = x\left(1 + \frac{1}{x^2 + y^2}\right) \quad \text{and} \quad v = y\left(1 - \frac{1}{x^2 + y^2}\right).$$

(ii) *Case 1: $a = 1$.* In this case, we have $u = 2x$ and $v = 0$. Since $|z| = a$ implies that $-a \leq x \leq a$, we deduce that the image of the given circle under f is the line segment (on the real axis) $[-2a, 2a] = [-2, 2]$.

Case 2: $a \neq 1$. We have

$$u = x\left(1 + \frac{1}{a^2}\right) \quad \text{and} \quad v = y\left(1 - \frac{1}{a^2}\right)$$

and thus

$$\frac{a^4 u^2}{(a^2 + 1)^2} + \frac{a^4 v^2}{(a^2 - 1)^2} = x^2 + y^2 = a^2$$

which is an ellipse of equation

$$\frac{u^2}{(a^2 + 1)^2} + \frac{v^2}{(a^2 - 1)^2} = \frac{1}{a^2}.$$

Exercise 13.7 (i) We have to check that $f : A \to \Omega$ has the following properties.

– f is one-to-one. Let $z_1, z_2 \in A$ such that $f(z_1) = f(z_2)$. We then have $e^{z_1} = e^{z_2}$ and thus $z_2 = z_1 + 2in\pi$. This equality implies

$$\text{Re}\, z_2 = \text{Re}\, z_1 \quad \text{and} \quad \text{Im}\, z_2 = \text{Im}\, z_1 + 2n\pi.$$

Since $0 < \text{Im}\, z_1, \text{Im}\, z_2 < \pi$, we deduce that $z_1 = z_2$ and thus f is one-to-one.

– f is onto. Let $w \in \Omega$. We want to find $z \in A$ such that $f(z) = w$. It is sufficient to take $z = \log w$. We indeed have $\text{Im}\, w > 0 \Rightarrow \text{Im}\, z \in (0, \pi)$.

– f is holomorphic. We know that $f(z) = e^z$ is holomorphic $\forall z \in \mathbb{C}$ and that

$$f'(z) \neq 0 \quad \forall z \in \mathbb{C}.$$

The function $f : A \to \Omega$ is therefore a conformal mapping.

(ii) Using Exercise 13.3 (with $g(z) = \frac{iz+1}{z+i}$) and the above question, we have

$$h = g \circ f : A \xrightarrow{f} \Omega \xrightarrow{g} D,$$

i.e. the function

$$h(z) = g(f(z)) = \frac{ie^z + 1}{e^z + i}$$

is a conformal mapping from A onto D.

Exercise 13.8 (i) We are looking for a holomorphic function $g = u + iv$. Such a g satisfies the Cauchy–Riemann equations, that is

$$v_x = -u_y \quad \text{and} \quad v_y = u_x.$$

We therefore should have

$$\operatorname{grad} v = (v_x, v_y) = (-u_y, u_x) = \Phi = (\varphi, \psi).$$

Since $\mathbb{R}^2$ is convex (and thus simply connected), a necessary and sufficient condition for such a v to exist is that $\operatorname{curl} \Phi = \psi_x - \varphi_y = 0$. Since u is harmonic, we have

$$\operatorname{curl} \Phi = (u_x)_x + (u_y)_y = u_{xx} + u_{yy} = \Delta u = 0.$$

We have therefore found v and, hence, a holomorphic function g such that $u = \operatorname{Re} g$.

(ii) Since u is harmonic, (i) implies the existence of a holomorphic g such that $u = \operatorname{Re} g$. We therefore have

$$u \circ f = \operatorname{Re}(g \circ f)$$

and $g \circ f$ is holomorphic. The function $u \circ f$ is thus harmonic (cf. Exercise 9.6).

Exercise 13.9 (i) We can use the fact that $h(O) = D$ (cf. below), Proposition 13.3 and note that

$$O = \{z : \operatorname{Im} z > 0\} \subset \mathbb{C} \setminus \{-i\}$$

$$D = \{z : |z| < 1\} \subset \mathbb{C} \setminus \{1\}$$

to get the claim, namely that h is conformal from O onto D. However, we show explicitly that h is a conformal mapping. We therefore have to show the next four properties.

– h is holomorphic on $\mathbb{C} \setminus \{-i\}$ and therefore in O.

– h is one-to-one since

$$h(z_1) = h(z_2) \;\Rightarrow\; \frac{z_1 - i}{z_1 + i} = \frac{z_2 - i}{z_2 + i} \;\Rightarrow\; z_1 = z_2.$$

– h is onto since

$$w = \frac{z - i}{z + i} \;\Rightarrow\; zw + iw = z - i \;\Rightarrow\; z(w - 1) = -i(w + 1)$$

and thus

$$z = -i\frac{w+1}{w-1} = h^{-1}(w).$$

In terms of components we have

$$h^{-1}(w) = h^{-1}(\alpha + i\beta) = -i\frac{(\alpha+1)+i\beta}{(\alpha-1)+i\beta} = \frac{\beta - i(\alpha+1)}{(\alpha-1)+i\beta}$$

$$= \frac{-2\beta}{(\alpha-1)^2 + \beta^2} + i\frac{(1-\alpha^2-\beta^2)}{(\alpha-1)^2 + \beta^2} = \operatorname{Re} h^{-1} + i\operatorname{Im} h^{-1}.$$

Note that since $\alpha^2 + \beta^2 < 1$ (because $w \in D$), we have $\operatorname{Im} h^{-1}(w) > 0$ and thus $h^{-1}(w) \in O$.

– We finally have

$$h'(z) = 2i/(z+i)^2 \neq 0.$$

(ii) To check that $g : \Omega \to O$ is a conformal mapping, we need to establish the following four properties.

– $g(z) = z^2$ is clearly holomorphic.

– g is one-to-one in Ω. We have to show that if $z_1, z_2 \in \Omega$ and if $g(z_1) = g(z_2)$, then $z_1 = z_2$. We first observe that

$$g(z_1) = g(z_2) \implies \begin{cases} x_1^2 - y_1^2 = x_2^2 - y_2^2 \\ x_1 y_1 = x_2 y_2. \end{cases}$$

Since $x_i, y_i > 0$ (and thus $x_i, y_i \neq 0$), we have

$$\begin{cases} y_2 = \frac{x_1 y_1}{x_2} \\ x_1^2 - y_1^2 = x_2^2 - (\frac{x_1 y_1}{x_2})^2 \end{cases} \implies \begin{cases} y_2 = \frac{x_1 y_1}{x_2} \\ x_2^4 - x_1^2 x_2^2 = x_1^2 y_1^2 - x_2^2 y_1^2. \end{cases}$$

This leads to

$$\left. \begin{array}{c} y_2 = \frac{x_1 y_1}{x_2} \\ x_2^2(x_2^2 - x_1^2) = -y_1^2(x_2^2 - x_1^2) \end{array} \right\} \underset{x_2 \neq 0,\ y_1 \neq 0}{\implies} x_1^2 = x_2^2 \underset{x_1,\ x_2 > 0}{\implies} x_1 = x_2$$

and thus

$$\begin{cases} x_1 = x_2 \\ y_1 = y_2 \end{cases} \implies z_1 = z_2.$$

– g is onto. Indeed, let $w = \alpha + i\beta \in O$, i.e. $\beta > 0$, and $z = x + iy$. We then deduce

$$w = z^2 \implies \alpha + i\beta = x^2 - y^2 + i2xy \implies \begin{cases} \alpha = x^2 - y^2 \\ \beta = 2xy \end{cases}$$

and therefore

$$x = \left(\frac{\sqrt{\alpha^2 + \beta^2} + \alpha}{2} \right)^{1/2} \quad \text{and} \quad y = \left(\frac{\sqrt{\alpha^2 + \beta^2} - \alpha}{2} \right)^{1/2}.$$

We could also let $z = |z| e^{i\theta}$ and $w = |w| e^{i\varphi}$ and write

$$w = z^2 \Rightarrow |w| e^{i\varphi} = |z|^2 e^{2i\theta} \Rightarrow \begin{cases} |w| = |z|^2 \\ \varphi = 2\theta + 2k\pi. \end{cases}$$

Then, after some computations, we would find, as before,

$$\begin{cases} \operatorname{Re} g^{-1}(\alpha, \beta) = \left(\frac{\sqrt{\alpha^2 + \beta^2} + \alpha}{2} \right)^{1/2} \\ \operatorname{Im} g^{-1}(\alpha, \beta) = \left(\frac{\sqrt{\alpha^2 + \beta^2} - \alpha}{2} \right)^{1/2}. \end{cases}$$

We have indeed found that $g^{-1}(w) \in \Omega$.

– We finally obtain

$$g'(z) = 2z \neq 0 \quad \text{if } z \in \Omega.$$

(iii) It is sufficient to take

$$f = h \circ g : \Omega \xrightarrow{g} O \xrightarrow{h} D$$

which means that

$$f(z) = \frac{z^2 - i}{z^2 + i}.$$

Exercise* 13.10 If such a map f exists, then f would be holomorphic and bounded (since $|f(z)| < 1$). Liouville theorem would thus imply that f is constant, which is absurd.

Chapter 14

Fourier series

Exercise 14.1 (i) Some elementary calculations lead to

$$a_n = \frac{1}{\pi} \int_0^{2\pi} e^{(x-\pi)} \cos(nx)\, dx = \frac{1}{\pi} \int_{-\pi}^{\pi} e^y \cos(ny + n\pi)\, dy$$

$$= \frac{(-1)^n}{\pi} \int_{-\pi}^{\pi} e^y \cos(ny)\, dy = \frac{2 \sinh \pi}{(1+n^2)\pi}.$$

$$b_n = \frac{1}{\pi} \int_0^{2\pi} e^{(x-\pi)} \sin(nx)\, dx = \frac{1}{\pi} \int_{-\pi}^{\pi} e^y \sin(ny + n\pi)\, dy$$

$$= \frac{(-1)^n}{\pi} \int_{-\pi}^{\pi} e^y \sin(ny)\, dy = \frac{-2n \sinh \pi}{(1+n^2)\pi}.$$

(ii) Using Dirichlet theorem, we find that if

$$Ff(x) = \frac{\sinh \pi}{\pi} + \frac{\sinh \pi}{\pi} \sum_{n=1}^{\infty} \left(\frac{2\cos(nx)}{1+n^2} - \frac{2n \sin(nx)}{1+n^2} \right),$$

then

$$Ff(x) = f(x) \quad \text{if } x \in (0, 2\pi).$$

(iii) In particular, if $x = \pi$, we have

$$1 = \frac{\sinh \pi}{\pi} \left(1 + \sum_{n=1}^{\infty} \frac{2(-1)^n}{1+n^2} \right)$$

and thus

$$\sum_{n=1}^{\infty} \frac{(-1)^n}{1+n^2} = \frac{\pi}{2 \sinh \pi} - \frac{1}{2},$$

291

that is,

$$\sum_{n=2}^{\infty} \frac{(-1)^n}{1+n^2} = \frac{\pi}{2\sinh\pi} = \frac{\pi}{e^\pi - e^{-\pi}}.$$

Exercise 14.2 (i) We have

$$a_0 = \frac{1}{\pi} \int_0^{2\pi} (x-\pi)^2 dx = \frac{1}{\pi} \int_{-\pi}^{\pi} y^2 dy = \frac{2}{3}\pi^2$$

$$a_n = \frac{1}{\pi} \int_0^{2\pi} (x-\pi)^2 \cos(nx)\, dx = \frac{(-1)^n}{\pi} \int_{-\pi}^{\pi} y^2 \cos(ny)\, dy = \frac{4}{n^2}$$

$$b_n = \frac{1}{\pi} \int_0^{2\pi} (x-\pi)^2 \sin(nx)\, dx = \frac{(-1)^n}{\pi} \int_{-\pi}^{\pi} y^2 \sin(ny)\, dy = 0.$$

The Fourier series is then

$$Ff(x) = \frac{\pi^2}{3} + \sum_{n=1}^{\infty} \frac{4\cos(nx)}{n^2}.$$

(ii) Since f is continuous at $x = 0$ and $x = 2\pi$, Dirichlet theorem ensures that

$$Ff(x) = f(x) \quad \text{if } x \in [0, 2\pi].$$

(iii) Therefore, taking respectively $x = 0$ ($f(0) = \pi^2$) and $x = \pi$ ($f(\pi) = 0$) gives

$$\sum_{n=1}^{\infty} \frac{4}{n^2} = \frac{2\pi^2}{3} \Rightarrow \sum_{n=1}^{\infty} \frac{1}{n^2} = \frac{\pi^2}{6}.$$

$$\sum_{n=1}^{\infty} 4\frac{(-1)^n}{n^2} = -\frac{\pi^2}{3} \Rightarrow \sum_{n=1}^{\infty} \frac{(-1)^n}{n^2} = -\frac{\pi^2}{12}.$$

Exercise 14.3 Note that f is odd, therefore implying that $a_n = 0$. On the other hand, since f is 2π–periodic we have

$$b_n = \frac{1}{\pi} \int_0^{2\pi} f(x)\sin(nx)\, dx = \frac{1}{\pi} \int_{-\pi/2}^{3\pi/2} f(x)\sin(nx)\, dx$$

$$= \frac{1}{\pi} \int_{-\pi/2}^{\pi/2} \sin x \sin(nx)\, dx = \begin{cases} 1/2 & \text{if } n = 1 \\ -\frac{2n}{\pi}\frac{\cos(\pi n/2)}{n^2-1} & \text{if } n \geq 2. \end{cases}$$

We therefore have, provided $n \geq 2$ ($b_1 = 1/2$),

$$b_n = \begin{cases} 0 & \text{if } n = 2k+1 \text{ (i.e. odd)} \\ \dfrac{-2n}{\pi(n^2-1)} & \text{if } n = 4k \text{ (i.e. a multiple of 4)} \\ \dfrac{2n}{\pi(n^2-1)} & \text{if } n = 4k+2 \text{ (i.e. even but not a multiple of 4).} \end{cases}$$

Exercise 14.4 By definition, we have

$$c_n = \frac{1}{2} \int_0^2 x e^{-i\pi n x}\, dx.$$

If $n = 0$, we find $c_0 = 1$. If $n \neq 0$, we obtain

$$c_n = \frac{1}{2} \left[\frac{x\, e^{-i\pi n x}}{-i\pi n} \right]_0^2 + \frac{1}{2i\pi n} \int_0^2 e^{-i\pi n x}\, dx = \frac{i}{\pi n}$$

and thus

$$F f(x) = 1 + \frac{i}{\pi} \sum_{n \neq 0} \frac{e^{i\pi n x}}{n}.$$

Note that since we allow positive as well as negative n, the right hand side of the above identity is a real-valued function.

Exercise 14.5 (i) Since f is odd, we have

$$f(x) = \begin{cases} x(\pi - x) & \text{if } x \in (0, \pi) \\ x(\pi + x) & \text{if } x \in (-\pi, 0). \end{cases}$$

Thus $a_n = 0$ and

$$b_n = \frac{2}{2\pi} \int_{-\pi}^{\pi} f(x) \sin(nx)\, dx = \frac{2}{\pi} \int_0^{\pi} x(\pi - x) \sin(nx)\, dx.$$

We, hence, get

$$\int_0^{\pi} x(\pi - x) \sin(nx)\, dx = \left[-x(\pi - x) \frac{\cos(nx)}{n} \right]_0^{\pi} + \int_0^{\pi} \frac{\cos(nx)}{n} (\pi - 2x)\, dx$$

$$= \left[(\pi - 2x) \frac{\sin(nx)}{n^2} \right]_0^{\pi} + 2 \int_0^{\pi} \frac{\sin(nx)}{n^2}\, dx$$

$$= \left[-\frac{2}{n^3} \cos(nx) \right]_0^{\pi} = -\frac{2}{n^3} (-1)^n + \frac{2}{n^3}.$$

We then find

$$b_n = \frac{4}{\pi n^3}\left[1 - (-1)^n\right] = \begin{cases} 0 & \text{if } n \text{ is even} \\ \dfrac{8}{\pi n^3} & \text{if } n \text{ is odd} \end{cases}$$

and we finally obtain that

$$Ff(x) = \sum_{n=1}^{\infty} \frac{8\sin\left((2n-1)x\right)}{\pi\left(2n-1\right)^3}.$$

(ii) Using Parseval identity, we find

$$\sum_{n=1}^{\infty} \frac{64}{\pi^2(2n-1)^6} = \frac{1}{\pi}\int_{-\pi}^{\pi}\left|f\left(x\right)\right|^2 dx = \frac{2}{\pi}\int_{0}^{\pi} x^2\left(\pi - x\right)^2 dx = \frac{\pi^4}{15}$$

and thus

$$\sum_{n=1}^{\infty} \frac{1}{(2n-1)^6} = \frac{\pi^6}{960}.$$

Exercise 14.6 We have

$$f\left(x\right) = \cos^2 x = \frac{1}{2} + \frac{1}{2}\cos\left(2x\right)$$

which implies that the Fourier coefficients are

$$a_0 = 1, \quad a_2 = \frac{1}{2}, \quad a_n = 0 \text{ if } n \neq 0, 2 \quad \text{and} \quad b_n = 0, \ \forall n.$$

Using the $2\pi-$periodicity of f and the Parseval identity, we get

$$\int_{-\pi}^{\pi}\left(f\left(x\right)\right)^2 dx = \int_{0}^{2\pi}\left(f\left(x\right)\right)^2 dx = \pi\left[\frac{1}{2} + \frac{1}{4}\right] = \frac{3\pi}{4}.$$

Exercise 14.7 (i) Observe that f is even, which implies that $b_n = 0$. We thus find

$$a_0 = \frac{2}{2\pi}\int_{0}^{2\pi} f\left(x\right)\,dx = \frac{1}{\pi}\int_{-\pi}^{\pi}\left|x\right|\,dx = \frac{2}{\pi}\int_{0}^{\pi} x\,dx = \pi,$$

$$a_n = \frac{2}{2\pi}\int_{0}^{2\pi} f\left(x\right)\cos\left(nx\right)\,dx = \frac{2}{\pi}\int_{0}^{\pi} x\cos\left(nx\right)\,dx$$

$$= \frac{2}{\pi}\left[\frac{x}{n}\sin\left(nx\right)\right]_{0}^{\pi} - \frac{2}{n\pi}\int_{0}^{\pi}\sin\left(nx\right)\,dx = \frac{2}{\pi n^2}[(-1)^n - 1].$$

We then obtain

$$a_n = \begin{cases} 0 & \text{if } n \text{ is even} \\ \dfrac{-4}{\pi n^2} & \text{if } n \text{ is odd} \end{cases}$$

and finally

$$Ff(x) = \frac{\pi}{2} - \sum_{n=0}^{\infty} \frac{4}{\pi (2n+1)^2} \cos\left((2n+1)x\right).$$

(ii) Using Parseval identity, we get

$$\frac{2}{2\pi} \int_0^{2\pi} (f(x))^2 \, dx = \frac{1}{\pi} \int_{-\pi}^{\pi} x^2 dx = \frac{2\pi^2}{3} = \frac{\pi^2}{2} + \sum_{n=0}^{\infty} \frac{16}{\pi^2 (2n+1)^4}$$

and we deduce that

$$\frac{16}{\pi^2} \sum_{n=0}^{\infty} \frac{1}{(2n+1)^4} = \frac{\pi^2}{6}$$

which leads to

$$\sum_{n=0}^{\infty} \frac{1}{(2n+1)^4} = \frac{\pi^4}{96}.$$

The above result gives

$$\sum_{n=1}^{\infty} \frac{1}{n^4} = \sum_{n=1}^{\infty} \frac{1}{(2n)^4} + \sum_{n=0}^{\infty} \frac{1}{(2n+1)^4} = \frac{1}{16} \sum_{n=1}^{\infty} \frac{1}{n^4} + \frac{\pi^4}{96}$$

and thus

$$\sum_{n=1}^{\infty} \frac{1}{n^4} = \frac{\pi^4}{90}.$$

Exercise 14.8 (i) Observe that the given function is even, which implies that $b_n = 0$. On the other hand, we have

$$\begin{aligned} a_n &= \frac{1}{\pi} \int_0^{2\pi} |\cos x| \cos(nx) \, dx \\ &= \frac{1}{\pi} \int_{-\pi/2}^{\pi/2} \cos x \cos(nx) \, dx - \frac{1}{\pi} \int_{\pi/2}^{3\pi/2} \cos x \cos(nx) \, dx. \end{aligned}$$

Substituting $x - \pi$ for y in the second integral gives (recall that $\cos(z + n\pi) = (-1)^n \cos z$)

$$\int_{\pi/2}^{3\pi/2} \cos x \cos(nx) \, dx = \int_{-\pi/2}^{\pi/2} \cos(y + \pi) \cos(n(y + \pi)) \, dy$$

$$= (-1)^{n+1} \int_{-\pi/2}^{\pi/2} \cos y \cos(ny) \, dy.$$

We therefore have

$$a_n = \frac{1 + (-1)^n}{\pi} \int_{-\pi/2}^{\pi/2} \cos y \cos (ny)\, dy$$

and we immediately deduce that if n is odd, then $a_n = 0$. Setting $n = 2k$ and using the fact that

$$2 \cos y \cos (2ky) = \cos \left[(2k+1)\, y \right] + \cos \left[(2k-1)\, y \right]$$

leads us to

$$a_{2k} = \frac{1}{\pi} \left[\frac{\sin \left((2k+1)\, y \right)}{2k+1} + \frac{\sin \left((2k-1)\, y \right)}{2k-1} \right]_{-\pi/2}^{\pi/2} = \frac{4}{\pi} \frac{(-1)^{k+1}}{4k^2 - 1}.$$

Finally, the Fourier series is

$$Ff(x) = \frac{2}{\pi} + \frac{4}{\pi} \sum_{k=1}^{\infty} \frac{(-1)^{k+1}}{4k^2 - 1} \cos (2kx).$$

(ii) Choosing $x = \pi/2$, we have $\cos (2kx) = \cos (k\pi) = (-1)^k$ and $f(\pi/2) = 0$. We finally find

$$\sum_{k=1}^{\infty} \frac{1}{4k^2 - 1} = \frac{1}{2}.$$

Exercise 14.9 Since f is 2π−periodic and even, we have $b_n = 0$ and

$$a_n = \frac{2}{2\pi} \int_0^{2\pi} f(t) \cos (nt)\, dt = \frac{2}{\pi} \int_0^{\pi} \sin (3t) \cos (nt)\, dt$$
$$= \frac{1}{\pi} \int_0^{\pi} \left[\sin \left((3-n)\, t \right) + \sin \left((3+n)\, t \right) \right] dt.$$

If $n \neq 3$ (if $n = 3$, we immediately get $a_3 = 0$), we then find

$$a_n = \frac{1}{\pi} \left[-\frac{\cos \left((3-n)\, t \right)}{3-n} - \frac{\cos \left((3+n)\, t \right)}{3+n} \right]_0^{\pi}$$
$$= \frac{1}{\pi} \left[-\frac{(-1)^{3-n}}{3-n} - \frac{(-1)^{3+n}}{3+n} + \frac{1}{3-n} + \frac{1}{3+n} \right]$$

and thus

$$a_n = \begin{cases} 0 & \text{if } n \text{ is odd} \\ \dfrac{1}{\pi} \left[\dfrac{2}{3-n} + \dfrac{2}{3+n} \right] = \dfrac{12}{\pi (9 - n^2)} & \text{if } n \text{ is even.} \end{cases}$$

The Fourier series is therefore given by

$$Ff(t) = \frac{2}{3\pi} + \frac{12}{\pi} \sum_{k=1}^{\infty} \frac{\cos(2kt)}{9 - 4k^2}.$$

Exercise 14.10 (i) Since the function is even, we have $b_n = 0$. We, moreover, get

$$a_n = \frac{2}{2\pi} \int_{-\pi}^{\pi} \cos(\alpha x) \cos(nx)\, dx = \frac{2}{\pi} \int_0^{\pi} \cos(\alpha x) \cos(nx)\, dx$$
$$= \frac{1}{\pi} \int_0^{\pi} [\cos((\alpha + n)x) + \cos((\alpha - n)x)]\, dx$$

and thus

$$a_n = \frac{1}{\pi} \left[\frac{\sin((\alpha + n)\pi)}{\alpha + n} + \frac{\sin((\alpha - n)\pi)}{\alpha - n} \right]$$
$$= \frac{(-1)^n \sin(\alpha\pi)}{\pi} \left[\frac{1}{\alpha + n} + \frac{1}{\alpha - n} \right] = \frac{(-1)^n \sin(\alpha\pi)}{\pi} \frac{2\alpha}{\alpha^2 - n^2}.$$

We therefore have, for every $x \in [-\pi, \pi)$,

$$Ff(x) = \frac{\sin(\alpha\pi)}{\alpha\pi} + \frac{2\alpha \sin(\alpha\pi)}{\pi} \sum_{n=1}^{\infty} \frac{(-1)^{n-1}}{n^2 - \alpha^2} \cos(nx).$$

(ii) Since f is continuous, letting $x = \pi$, we obtain

$$\cos(\alpha\pi) = \frac{2\alpha \sin(\alpha\pi)}{\pi} \left[\frac{1}{2\alpha^2} + \sum_{n=1}^{\infty} \frac{(-1)^{n-1}(-1)^n}{n^2 - \alpha^2} \right].$$

For every $\alpha \notin \mathbb{Z}$, we therefore have

$$\sum_{n=1}^{\infty} \frac{1}{n^2 - \alpha^2} = \frac{1}{2\alpha^2} - \frac{\pi}{2\alpha \tan(\alpha\pi)}.$$

Exercise 14.11 (i) Using complex notations, we have

$$c_n = \frac{1}{2\pi} \int_{-\pi}^{\pi} f(x) e^{-inx}\, dx$$
$$= \frac{1}{2\pi} \int_{-\pi}^{-\pi/2} (-\pi - x)e^{-inx}\, dx + \frac{1}{2\pi} \int_{-\pi/2}^{\pi/2} x e^{-inx}\, dx$$
$$+ \frac{1}{2\pi} \int_{\pi/2}^{\pi} (\pi - x) e^{-inx}\, dx.$$

We make the substitutions $u = x + \pi$ and $v = x - \pi$ and we get

$$\int_{-\pi}^{-\pi/2} (-\pi - x)e^{-inx}dx = \int_{0}^{\pi/2} -ue^{in\pi}e^{-inu}du$$

$$= (-1)^{n+1} \int_{0}^{\pi/2} ue^{-inu}du$$

$$\int_{\pi/2}^{\pi} (\pi - x)e^{-inx}dx = \int_{-\pi/2}^{0} -ve^{-in\pi}e^{-inv}dv$$

$$= (-1)^{n+1} \int_{-\pi/2}^{0} ve^{-inv}dv$$

and thus

$$c_n = \frac{1 + (-1)^{n+1}}{2\pi} \int_{-\pi/2}^{\pi/2} xe^{-inx}dx$$

$$= \frac{1 + (-1)^{n+1}}{2\pi} \left[\left(\frac{ix}{n} + \frac{1}{n^2} \right) e^{-inx} \right]_{-\pi/2}^{\pi/2}.$$

We immediately find that if n is even, then $c_n = 0$ and if n is odd (i.e. $n = 2k - 1$), then

$$e^{-i(2k-1)\pi/2} = -e^{i(2k-1)\pi/2} = (-1)^k i.$$

We thus get

$$c_{2k-1} = \frac{(-1)^k \, 2i}{\pi(2k-1)^2}.$$

Since f is continuous we have, for every $x \in [-\pi/2, \pi/2]$,

$$x = \frac{2i}{\pi} \sum_{k=-\infty}^{+\infty} \frac{(-1)^k}{(2k-1)^2} e^{i(2k-1)x}.$$

(ii) If we choose $x \in [-a, a]$, we have $(\pi/2a)x \in [-\pi/2, \pi/2]$ and thus

$$x = \frac{4}{\pi^2} ia \sum_{k=-\infty}^{+\infty} \frac{(-1)^k}{(2k-1)^2} e^{(i\pi(2k-1)x)/2a}.$$

(iii) In particular, for $x = a$, we obtain

$$\sum_{k=-\infty}^{+\infty} \frac{1}{(2k-1)^2} = \frac{\pi^2}{4}.$$

Exercise 14.12 (i) A direct computation gives

$$F_3 f(x) = \pi + 2 \sin x - \sin(2x) + \frac{2}{3} \sin(3x)$$

and thus

$$F_3 f(-\pi) = \pi \qquad f(-\pi) = 0 \qquad \Rightarrow \quad f(-\pi) - F_3 f(-\pi) = -\pi$$

$$F_3 f\left(-\frac{\pi}{2}\right) = \pi - \frac{4}{3} \quad f\left(-\frac{\pi}{2}\right) = \frac{\pi}{2} \quad \Rightarrow \quad f\left(-\frac{\pi}{2}\right) - F_3 f\left(-\frac{\pi}{2}\right) = \frac{4}{3} - \frac{\pi}{2}$$

$$F_3 f(0) = \pi \qquad f(0) = \pi \qquad \Rightarrow \quad f(0) - F_3 f(0) = 0$$

$$F_3 f\left(\frac{\pi}{2}\right) = \pi + \frac{4}{3} \quad f\left(\frac{\pi}{2}\right) = \frac{3\pi}{2} \quad \Rightarrow \quad f\left(\frac{\pi}{2}\right) - F_3 f\left(\frac{\pi}{2}\right) = \frac{\pi}{2} - \frac{4}{3}$$

$$F_3 f(\pi) = \pi \qquad f(\pi) = 0 \qquad \Rightarrow \quad f(\pi) - F_3 f(\pi) = -\pi.$$

Note that the 2π−periodicity implies $f(\pi) = f(-\pi)$ and thus $f(\pi) = 0$.
On the other hand, we have $f(\pi - 0) = 2\pi$ while $f(-\pi + 0) = f(-\pi) = 0$.

(ii) The 2π−periodicity and the next exercise imply

$$
\begin{aligned}
\int_0^{2\pi} |f(x) - F_3 f(x)|^2 \, dx &= \int_{-\pi}^{\pi} (f(x))^2 \, dx - \pi \left[\frac{a_0^2}{2} + \sum_{n=1}^{3} (a_n^2 + b_n^2) \right] \\
&= \int_{-\pi}^{\pi} (x + \pi)^2 dx - \pi \left[\frac{(2\pi)^2}{2} + 4 + 1 + \frac{4}{9} \right] \\
&= \frac{2}{3} \pi^3 - \frac{49}{9} \pi \approx 3.567.
\end{aligned}
$$

Exercise* 14.13 (i) We first compute

$$
\begin{aligned}
\int_0^{2\pi} F_N f(x) \cos(kx) \, dx &= \frac{a_0}{2} \int_0^{2\pi} \cos(kx) \, dx \\
&\quad + \sum_{n=1}^{N} a_n \int_0^{2\pi} \cos(nx) \cos(kx) \, dx \\
&\quad + \sum_{n=1}^{N} b_n \int_0^{2\pi} \sin(nx) \cos(kx) \, dx.
\end{aligned}
$$

We, hence, have, for every $k = 0, 1, \cdots, N$,

$$\int_0^{2\pi} F_N f(x) \cos(kx) \, dx = \pi a_k \quad \text{and} \quad \int_0^{2\pi} F_N f(x) \sin(kx) \, dx = \pi b_k \,.$$

Using again the definition of $F_N f$ and the previous result, we get

$$\int_0^{2\pi} (F_N f(x))^2 \, dx = \frac{a_0}{2} \int_0^{2\pi} F_N f(x) \, dx$$

$$+ \sum_{k=1}^{N} a_k \int_0^{2\pi} F_N f(x) \cos(kx) \, dx$$

$$+ \sum_{k=1}^{N} b_k \int_0^{2\pi} F_N f(x) \sin(kx) \, dx$$

and thus

$$\int_0^{2\pi} (F_N f(x))^2 \, dx = \pi \left\{ \frac{a_0^2}{2} + \sum_{k=1}^{N} (a_k^2 + b_k^2) \right\}.$$

In a similar way, we find

$$\int_0^{2\pi} f(x) F_N f(x) \, dx = \frac{a_0}{2} \int_0^{2\pi} f(x) \, dx + \sum_{n=1}^{N} a_n \int_0^{2\pi} f(x) \cos(nx) \, dx$$

$$+ \sum_{n=1}^{N} b_n \int_0^{2\pi} f(x) \sin(nx) \, dx.$$

Using the definition of a_n and b_n, we deduce that

$$\int_0^{2\pi} f(x) F_N f(x) \, dx = \frac{a_0^2}{2} \pi + \pi \sum_{n=1}^{N} (a_n^2 + b_n^2).$$

Gathering all these results leads to

$$\int_0^{2\pi} |f(x) - F_N f(x)|^2 \, dx = \int_0^{2\pi} (f(x))^2 \, dx - 2 \int_0^{2\pi} f(x) F_N f(x) \, dx$$

$$+ \int_0^{2\pi} (F_N f(x))^2 \, dx$$

and thus

$$\int_0^{2\pi} |f(x) - F_N f(x)|^2 \, dx = \int_0^{2\pi} (f(x))^2 \, dx - \pi \left[\frac{a_0^2}{2} + \sum_{n=1}^{N} (a_n^2 + b_n^2) \right].$$

(ii) It is sufficient to observe that

$$\int_0^{2\pi} |f(x) - F_N f(x)|^2 \, dx \geq 0.$$

The previous equality therefore implies

$$\frac{a_0^2}{2} + \sum_{n=1}^{N} \left(a_n^2 + b_n^2\right) \leq \frac{1}{\pi} \int_0^{2\pi} \left(f(x)\right)^2 dx.$$

We get the claimed inequality by letting N tend to infinity, i.e.

$$\frac{a_0^2}{2} + \sum_{n=1}^{\infty} \left(a_n^2 + b_n^2\right) \leq \frac{1}{\pi} \int_0^{2\pi} \left(f(x)\right)^2 dx.$$

Exercise* 14.14 (i) Let $n \geq 1$. Using the definition of a_n and integration by parts, we obtain

$$
\begin{aligned}
a_n &= \frac{2}{T} \int_0^T f(t) \cos\left(\frac{2n\pi}{T}t\right) dt \\
&= \frac{2}{T} \left[f(t) \frac{\sin\left(\frac{2n\pi}{T}t\right)}{\frac{2n\pi}{T}} \right]_0^T - \frac{1}{n\pi} \int_0^T f'(t) \sin\left(\frac{2n\pi}{T}t\right) dt \\
&= -\frac{1}{n\pi} \int_0^T f'(t) \sin\left(\frac{2n\pi}{T}t\right) dt.
\end{aligned}
$$

We then get

$$|a_n| \leq \frac{1}{n\pi} \int_0^T |f'(t)| \left|\sin\left(\frac{2n\pi}{T}t\right)\right| dt \leq \frac{1}{n\pi} \int_0^T |f'(t)| \, dt = \frac{c}{n}$$

with (recall that f is C^1 and thus f' is bounded)

$$c = \frac{1}{\pi} \int_0^T |f'(t)| \, dt.$$

We proceed in a similar way for b_n.

(ii) The proof of this part is very similar to the preceding one. To get the claim, one needs to make k integration by parts.

Chapter 15

Fourier transform

Exercise 15.1 Since $f = 0$ if $x < 0$, we have

$$\mathfrak{F}(f)(\alpha) = \frac{1}{\sqrt{2\pi}} \int_0^{+\infty} e^{-x-i\alpha x} dx = \frac{1}{\sqrt{2\pi}} \frac{1}{1+i\alpha} = \frac{1}{\sqrt{2\pi}} \frac{1-i\alpha}{1+\alpha^2}.$$

Exercise 15.2 (i) We first compute

$$\mathfrak{F}_c(f)(\alpha) = \sqrt{\frac{2}{\pi}} \int_0^{+\infty} f(x) \cos(\alpha x) dx.$$

We find

$$\begin{aligned}
\mathfrak{F}_c(f)(\alpha) &= \sqrt{\frac{2}{\pi}} \int_0^{+\infty} e^{-x} \cos x \cos(\alpha x) dx \\
&= \sqrt{\frac{2}{\pi}} \int_0^{+\infty} e^{-x} \left[\frac{\cos((\alpha+1)x) + \cos((\alpha-1)x)}{2} \right] dx
\end{aligned}$$

and thus

$$\begin{aligned}
\mathfrak{F}_c(f)(\alpha) &= \sqrt{\frac{2}{\pi}} \left[-\frac{e^{-x}[\cos((\alpha-1)x) + (1-\alpha)\sin((\alpha-1)x)]}{2(1+(\alpha-1)^2)} \right. \\
&\left. + \frac{e^{-x}[(1+\alpha)\sin((\alpha+1)x) - \cos((\alpha+1)x)]}{2(1+(\alpha+1)^2)} \right]_0^{+\infty} \\
&= \sqrt{\frac{2}{\pi}} \frac{2+\alpha^2}{(2-2\alpha+\alpha^2)(2+2\alpha+\alpha^2)}.
\end{aligned}$$

(ii) We next compute

$$\mathfrak{F}_s(f)(\alpha) = -i\sqrt{\frac{2}{\pi}} \int_0^{+\infty} f(x) \sin(\alpha x) dx.$$

We get

$$\mathfrak{F}_s(f)(\alpha) = -i\sqrt{\frac{2}{\pi}} \int_0^{+\infty} e^{-x} \cos x \sin(\alpha x)\, dx$$

$$= -i\sqrt{\frac{2}{\pi}} \int_0^{+\infty} e^{-x} \frac{\sin((\alpha+1)x) + \sin((\alpha-1)x)}{2}\, dx$$

and thus

$$\mathfrak{F}_s(f)(\alpha) = -i\sqrt{\frac{2}{\pi}} \left[\frac{e^{-x}[(1-\alpha)\cos((\alpha-1)x) - \sin((\alpha-1)x)]}{2(1+(\alpha-1)^2)} \right.$$

$$\left. - \frac{e^{-x}[(1+\alpha)\cos((\alpha+1)x) + \sin((\alpha+1)x)]}{2(1+(\alpha+1)^2)} \right]_0^{+\infty}$$

$$= -i\sqrt{\frac{2}{\pi}} \frac{\alpha^3}{(2-2\alpha+\alpha^2)(2+2\alpha+\alpha^2)}.$$

Exercise 15.3 We write

$$\mathfrak{F}(f')(\alpha) = \frac{1}{\sqrt{2\pi}} \int_{-\infty}^{+\infty} f'(y) e^{-i\alpha y}\, dy$$

$$= \frac{1}{\sqrt{2\pi}} \left\{ f(y) e^{-i\alpha y} \Big|_{-\infty}^{+\infty} + i\alpha \int_{-\infty}^{+\infty} f(y) e^{-i\alpha y}\, dy \right\}.$$

By hypothesis, we have $\lim_{|y|\to\infty} |f(y)| = 0$. Moreover, since $|e^{-i\alpha y}| = 1$, we get

$$f(y) e^{-i\alpha y} \Big|_{-\infty}^{+\infty} = 0$$

and thus

$$\mathfrak{F}(f')(\alpha) = i\alpha \frac{1}{\sqrt{2\pi}} \int_{-\infty}^{+\infty} f(y) e^{-i\alpha y}\, dy = i\alpha \mathfrak{F}(f)(\alpha).$$

Iterating the process, we have indeed shown that

$$\mathfrak{F}(f^{(n)})(\alpha) = (i\alpha)^n \mathfrak{F}(f)(\alpha) \quad \forall \alpha \in \mathbb{R} \text{ and } n \in \mathbb{N}.$$

Exercise 15.4 Let $a > 0$ (a completely similar argument holds if $a < 0$). Let

$$h(x) = e^{-ibx} f(ax).$$

If we let $z = ay$, we obtain

$$\mathfrak{F}(h)(\alpha) = \frac{1}{\sqrt{2\pi}} \int_{-\infty}^{+\infty} e^{-iby} f(ay) e^{-i\alpha y}\, dy$$

$$= \frac{1}{\sqrt{2\pi}} \int_{-\infty}^{+\infty} \frac{1}{a} e^{-i\frac{\alpha+b}{a}z} f(z)\, dz = \frac{1}{a} \mathfrak{F}(f)\left(\frac{\alpha+b}{a}\right).$$

Exercise 15.5 (i) We write

$$\sqrt{2\pi}\,\mathfrak{F}\left(f\right)(\alpha) = \int_{-\infty}^{+\infty} f\left(y\right)\cos\left(\alpha y\right) dy - i\int_{-\infty}^{+\infty} f\left(y\right)\sin\left(\alpha y\right) dy.$$

Since f is even, we find

$$y \to f\left(y\right)\cos\left(\alpha y\right) \text{ is even } \quad \text{and} \quad y \to f\left(y\right)\sin\left(\alpha y\right) \text{ is odd.}$$

We thus deduce

$$\int_{-\infty}^{+\infty} f\left(y\right)\cos\left(\alpha y\right) dy = 2\int_{0}^{+\infty} f\left(y\right)\cos\left(\alpha y\right) dy$$

$$\int_{-\infty}^{+\infty} f\left(y\right)\sin\left(\alpha y\right) dy = 0.$$

We finally get

$$\widehat{f}\left(\alpha\right) = \mathfrak{F}\left(f\right)(\alpha) = \sqrt{\frac{2}{\pi}}\int_{0}^{+\infty} f\left(y\right)\cos\left(\alpha y\right) dy$$

(in particular, $\widehat{f}$ is even). Using a similar reasoning, we find that if f is odd, then

$$\widehat{f}\left(\alpha\right) = \mathfrak{F}\left(f\right)(\alpha) = -i\sqrt{\frac{2}{\pi}}\int_{0}^{+\infty} f\left(y\right)\sin\left(\alpha y\right) dy$$

(in particular, $\widehat{f}$ is odd).

(ii) The inversion formula gives

$$\begin{aligned} \sqrt{2\pi} f\left(x\right) &= \int_{-\infty}^{+\infty} \widehat{f}\left(\alpha\right) e^{i\alpha x}\, d\alpha \\ &= \int_{-\infty}^{+\infty} \widehat{f}\left(\alpha\right)\cos\left(\alpha y\right) dy + i\int_{-\infty}^{+\infty} \widehat{f}\left(\alpha\right)\sin\left(\alpha y\right) dy. \end{aligned}$$

We just saw that $\widehat{f}$ is even when f is even. Therefore, the same reasoning as before leads us to

$$f\left(x\right) = \sqrt{\frac{2}{\pi}}\int_{0}^{+\infty} \widehat{f}\left(\alpha\right)\cos\left(\alpha x\right) d\alpha.$$

In a similar way we deduce that $\widehat{f}$ is odd if f is odd and thus

$$f\left(x\right) = i\sqrt{\frac{2}{\pi}}\int_{0}^{+\infty} \widehat{f}\left(\alpha\right)\sin\left(\alpha x\right) d\alpha.$$

Exercise 15.6 This exercise is essentially part of the previous one. We use the definition

$$\mathfrak{F}(f)(\alpha) = \widehat{f}(\alpha) = \frac{1}{\sqrt{2\pi}} \int_{-\infty}^{+\infty} f(y) e^{-i\alpha y} dy$$

to deduce

$$\mathfrak{F}(\mathfrak{F}(f))(t) = \mathfrak{F}(\widehat{f})(t) = \widehat{\widehat{f}}(t) = \frac{1}{\sqrt{2\pi}} \int_{-\infty}^{+\infty} \widehat{f}(\alpha) e^{-i\alpha t} d\alpha.$$

By hypothesis f is even. Hence, using the suggestion (or the previous exercise), we deduce that $\widehat{f}$ is also even, i.e. $\widehat{f}(\alpha) = \widehat{f}(-\alpha)$. Using then the inversion formula, we get

$$\begin{aligned} \mathfrak{F}(\mathfrak{F}(f))(t) &= \widehat{\widehat{f}}(t) = \frac{1}{\sqrt{2\pi}} \int_{-\infty}^{+\infty} \widehat{f}(\alpha) e^{-i\alpha t} d\alpha = \frac{1}{\sqrt{2\pi}} \int_{-\infty}^{+\infty} \widehat{f}(-\alpha) e^{-i\alpha t} d\alpha \\ &= \frac{1}{\sqrt{2\pi}} \int_{-\infty}^{+\infty} \widehat{f}(\alpha) e^{i\alpha t} d\alpha = f(t). \end{aligned}$$

Exercise 15.7 We only discuss the case $n = 1$ (the same reasoning leads to the general case $n \geq 1$). We also suppose that all the next formal computations are valid (and we let $g(x) = xf(x)$)

$$\begin{aligned} \mathfrak{F}'(f)(\alpha) &= \frac{d}{d\alpha} \left[\frac{1}{\sqrt{2\pi}} \int_{-\infty}^{+\infty} f(x) e^{-i\alpha x} dx \right] = \frac{1}{\sqrt{2\pi}} \int_{-\infty}^{+\infty} f(x) \frac{d}{d\alpha} \left[e^{-i\alpha x} \right] dx \\ &= \frac{1}{\sqrt{2\pi}} \int_{-\infty}^{+\infty} (-ix) f(x) e^{-i\alpha x} dx = \frac{-i}{\sqrt{2\pi}} \int_{-\infty}^{+\infty} xf(x) e^{-i\alpha x} dx \\ &= -i\mathfrak{F}(g)(\alpha), \end{aligned}$$

i.e.

$$\mathfrak{F}(g)(\alpha) = i\,\mathfrak{F}'(f)(\alpha).$$

Exercise* 15.8 Since the hypotheses allow us to permute the integrals, we find

$$\begin{aligned} \mathfrak{F}(f * g)(\alpha) &= \frac{1}{\sqrt{2\pi}} \int_{-\infty}^{+\infty} \left[\int_{-\infty}^{+\infty} f(y-t) g(t)\, dt \right] e^{-i\alpha y} dy \\ &= \frac{1}{\sqrt{2\pi}} \int_{-\infty}^{+\infty} \left[\int_{-\infty}^{+\infty} f(y-t) e^{-i\alpha y} dy \right] g(t)\, dt. \end{aligned}$$

We let $y - t = z$ and we have

$$\mathfrak{F}(f * g)(\alpha) = \frac{1}{\sqrt{2\pi}} \int_{-\infty}^{+\infty} \left[\int_{-\infty}^{+\infty} f(z) e^{-i\alpha(t+z)} dz \right] g(t) dt$$

$$= \frac{1}{\sqrt{2\pi}} \int_{-\infty}^{+\infty} f(z) e^{-i\alpha z} dz \int_{-\infty}^{+\infty} g(t) e^{-i\alpha t} dt$$

$$= \sqrt{2\pi} \, \mathfrak{F}(f)(\alpha) \, \mathfrak{F}(g)(\alpha).$$

Chapter 16

Laplace transform

Exercise 16.1 (i) Using the definition, we find, for $\operatorname{Re} z > 0$,

$$\mathfrak{L}(f)(z) = \int_0^{+\infty} \cos(kt)e^{-zt}dt$$

$$= \left[\frac{\cos(kt)e^{-zt}}{-z}\right]_0^{+\infty} - \frac{k}{z}\int_0^{+\infty} \sin(kt)e^{-zt}dt$$

$$= \frac{1}{z} - \frac{k}{z}\left\{\left[\frac{\sin(kt)e^{-zt}}{-z}\right]_0^{+\infty} + \frac{k}{z}\int_0^{+\infty} \cos(kt)\,e^{-zt}dt\right\}$$

$$= \frac{1}{z} - \frac{k^2}{z^2}\mathfrak{L}(f)(z)$$

which implies

$$\mathfrak{L}(f)(z) = \int_0^{+\infty} \cos(kt)e^{-zt}dt = \frac{z}{z^2 + k^2}\,.$$

(ii) In a similar way we get

$$\mathfrak{L}(f)(z) = \int_0^{+\infty} te^{\alpha t}e^{-zt}dt = \int_0^{+\infty} te^{(\alpha-z)t}dt$$

$$= \left[\frac{t\,e^{(\alpha-z)t}}{\alpha-z}\right]_0^{+\infty} - \frac{1}{\alpha-z}\int_0^{+\infty} e^{(\alpha-z)t}dt$$

$$= -\frac{1}{\alpha-z}\int_0^{+\infty} e^{(\alpha-z)t}dt = -\frac{1}{(\alpha-z)^2}\left[e^{(\alpha-z)t}\right]_0^{+\infty}$$

and thus (the argument is valid for $\operatorname{Re}(\alpha - z) < 0 \Rightarrow \operatorname{Re}(z) > \alpha$)

$$\mathfrak{L}(f)(z) = \frac{1}{(z-\alpha)^2}\,.$$

(iii) We finally obtain

$$\mathfrak{L}(f_\alpha)(z) = \int_0^\alpha \frac{1}{\alpha} e^{-zt} dt = \frac{1}{\alpha} \left[\frac{e^{-zt}}{-z} \right]_0^\alpha = \frac{1}{\alpha} \left[\frac{1 - e^{-\alpha z}}{z} \right].$$

From l'Hôpital rule we find

$$\lim_{\alpha \to 0} \mathfrak{L}(f_\alpha)(z) = 1.$$

Exercise 16.2 We first decompose F as

$$F(z) = \frac{1}{b - a} \left[\frac{1}{z - b} - \frac{1}{z - a} \right].$$

Since

$$\mathfrak{L}(e^{at})(z) = \frac{1}{z - a} \quad \text{and} \quad \mathfrak{L}(e^{bt})(z) = \frac{1}{z - b},$$

we get

$$\frac{1}{b - a} \left[\mathfrak{L}(e^{bt}) - \mathfrak{L}(e^{at}) \right] = F(z)$$

which implies

$$f(t) = \frac{1}{b - a} [e^{bt} - e^{at}].$$

Exercise 16.3 We use the linearity and the explicit computations of the Laplace transform in order to write

$$\mathfrak{L}(3\cos(6t) - 5\sin(6t)) = 3\mathfrak{L}(\cos(6t)) - 5\mathfrak{L}(\sin(6t)) = \frac{3z}{z^2 + 36} - \frac{30}{z^2 + 36}.$$

Since

$$\mathfrak{L}(e^{-2t})(z) = \frac{1}{z + 2}$$

we have from the shift formula (with $a = 1$ and $b = 2$)

$$\mathfrak{L}(e^{-2t}\{3\cos(6t) - 5\sin(6t)\}) = \frac{3(z + 2) - 30}{(z + 2)^2 + 36} = \frac{3z - 24}{z^2 + 4z + 40}.$$

Exercise 16.4 (i) Appealing to Exercise 16.1 we have

$$\mathfrak{L}^{-1}(F)(t) = 4\mathfrak{L}^{-1}\left(\frac{z}{z^2 + 64} \right)(t) = 4\cos(8t).$$

(ii) We write

$$\frac{z}{(z + 1)(z + 2)} = \frac{2}{z + 2} - \frac{1}{z + 1}$$

and thus

$$\mathfrak{L}^{-1}\left(\frac{z}{(z+2)(z+1)}\right)(t) = 2\mathfrak{L}^{-1}\left(\frac{1}{z+2}\right)(t) - \mathfrak{L}^{-1}\left(\frac{1}{z+1}\right)(t)$$
$$= 2e^{-2t} - e^{-t}.$$

Exercise 16.5 We proceed as in Example 16.7.

Step 1. We define

$$\widetilde{F}(z) = \frac{z^2}{(z^2+1)^2}e^{zt}.$$

The singularities of $\widetilde{F}$ are at $z = \pm i$ and they are poles of order 2. After some easy calculation, we find

$$\mathrm{Res}_i(\widetilde{F}) = \lim_{z\to i}\frac{d}{dz}\left[\frac{z^2 e^{zt}}{(z+i)^2}\right] = \frac{te^{it} - ie^{it}}{4}$$

$$\mathrm{Res}_{-i}(\widetilde{F}) = \lim_{z\to -i}\frac{d}{dz}\left[\frac{z^2 e^{zt}}{(z-i)^2}\right] = \frac{te^{-it} + ie^{-it}}{4}.$$

Step 2. We proceed exactly as in Example 16.7 (here, we have to take $\gamma > 0$, for example $\gamma = 1$) with $\Gamma_r = C_r \cup L_r$ and

$$C_r = \{z \in \mathbb{C} : |z - 1| = r \text{ and } \mathrm{Re}\, z < 1\}$$

$$L_r = \{z \in \mathbb{C} : \mathrm{Re}\, z = 1 \text{ and } -r < \mathrm{Im}\, z < r\}.$$

The residue theorem gives

$$\frac{1}{2\pi i}\int_{\Gamma_r}\widetilde{F}(z)\,dz = \frac{te^{it} - ie^{it}}{4} + \frac{te^{-it} + ie^{-it}}{4}$$

$$= \frac{t}{2}\frac{e^{it} + e^{-it}}{2} + \frac{1}{2}\frac{e^{it} - e^{-it}}{2i} = \frac{t}{2}\cos t + \frac{1}{2}\sin t.$$

Step 3. As in Example 16.7, we have

$$\lim_{r\to\infty}\int_{C_r}\widetilde{F}(z)\,dz = 0$$

and therefore

$$\mathfrak{L}^{-1}(F)(t) = \frac{1}{2\pi}\int_{-\infty}^{+\infty}F(1+is)e^{(1+is)t}\,ds = \frac{1}{2\pi i}\lim_{r\to\infty}\int_{L_r}F(z)e^{zt}\,dz$$

$$= \frac{t}{2}\cos t + \frac{1}{2}\sin t.$$

Finally, using Theorem 16.3, we find that

$$f(t) = \begin{cases} \frac{t}{2}\cos t + \frac{1}{2}\sin t & \text{if } t \geq 0 \\ 0 & \text{if } t < 0 \end{cases}$$

is the function we are looking for.

Exercise 16.6 We first note that the function $h(t) = tf(t)$ is a piecewise continuous function ($h(t) \equiv 0$ if $t < 0$). If $\operatorname{Re} z > \gamma_0$ (i.e. $z \in O$), we then get

$$\int_0^{+\infty} |f(t)| e^{-\gamma_0 t}\, dt < \infty \;\Rightarrow\; \int_0^{+\infty} |h(t) e^{-tz}|\, dt < \infty$$

(observe that this implication is not necessarily true if $\operatorname{Re} z = \gamma_0$). The following computation can easily be made rigorous

$$F'(z) = \frac{d}{dz} \int_0^{+\infty} f(t) e^{-tz}\, dt = \int_0^{+\infty} f(t)\, \frac{d}{dz}\left(e^{-tz}\right) dt$$

$$= -\int_0^{+\infty} tf(t) e^{-tz}\, dt = -\mathfrak{L}(h)(z) \quad \forall z \in O.$$

Exercise 16.7 (i) We have

$$\mathfrak{L}(f')(z) = \int_0^{+\infty} f'(t) e^{-tz}\, dt = \left[f(t) e^{-tz} \right]_0^{+\infty} - \int_0^{+\infty} f(t)\left(-z e^{-tz}\right) dt.$$

Since $\operatorname{Re} z > \gamma_0$, we deduce that

$$\left| f(t) e^{-tz} \right| = |f(t)| e^{-t\operatorname{Re} z} = |f(t)| e^{-\gamma_0 t} e^{-(\operatorname{Re} z - \gamma_0)t}$$

and thus

$$\left| f(t) e^{-tz} \right| \leq c\, e^{-(\operatorname{Re} z - \gamma_0)t} \to 0 \quad \text{if } t \to \infty.$$

We, hence, find

$$\mathfrak{L}(f')(z) = z \int_0^{+\infty} f(t) e^{-tz}\, dt - f(0) = z\, \mathfrak{L}(f)(z) - f(0).$$

(Note that we use the hypothesis $\int_0^{+\infty} |f'(t)| e^{-\gamma_0 t}\, dt < \infty$ to ensure that $\mathfrak{L}(f')$ is well defined whenever $\operatorname{Re} z > \gamma_0$.)

(ii) For the general case, we proceed by induction (we already established the case $n = 1$ in (i)). Assume that the result holds for $n - 1$, and let us show the case n. We find (as before $\left| f^{(k)}(t) e^{-tz} \right| \to 0$, if $t \to \infty$ and $\operatorname{Re} z > \gamma_0$)

$$
\begin{aligned}
\mathcal{L}(f^{(n)})(z) &= \int_0^{+\infty} f^{(n)}(t) e^{-tz}\, dt \\
&= \left[f^{(n-1)}(t) e^{-tz} \right]_0^{+\infty} - \int_0^{+\infty} f^{(n-1)}(t)(-ze^{-tz})dt \\
&= z \left\{ z^{n-1} \mathcal{L}(f)(z) - \sum_{k=0}^{n-2} z^k f^{(n-k-2)}(0) \right\} - f^{(n-1)}(0)
\end{aligned}
$$

and thus

$$
\begin{aligned}
\mathcal{L}(f^{(n)})(z) &= z^n \mathcal{L}(f)(z) - \sum_{k=1}^{n-1} z^k f^{(n-k-1)}(0) - f^{(n-1)}(0) \\
&= z^n \mathcal{L}(f)(z) - \sum_{k=0}^{n-1} z^k f^{(n-k-1)}(0).
\end{aligned}
$$

Exercise 16.8 Since

$$
\varphi(t) = \int_0^t f(s)\, ds,
$$

we have $\varphi \in C^1(\mathbb{R}_+)$, $\varphi(0) = 0$, $\varphi'(t) = f(t)$. Furthermore, we have by hypothesis

$$
\int_0^{+\infty} |\varphi'(t)| e^{-\gamma_0 t}\, dt < \infty \quad \text{and} \quad \int_0^{+\infty} |\varphi(t)| e^{-\gamma_0 t}\, dt < \infty.
$$

Since $z \in O$ and $\gamma_0 \geq 0$, we deduce that $z \neq 0$. We can therefore apply Theorem 16.2 (iii)

$$
\mathcal{L}(f)(z) = \mathcal{L}(\varphi')(z) = z\mathcal{L}(\varphi)(z) - \varphi(0) = z\mathcal{L}(\varphi)(z)
$$

and thus

$$
\mathcal{L}(\varphi)(z) = \frac{\mathcal{L}(f)(z)}{z} \quad \forall z \in O.
$$

Exercise 16.9 We immediately have that if $\varphi(t) = e^{-bt} f(at)$, then

$$
\begin{aligned}
\mathcal{L}(\varphi)(z) &= \int_0^{+\infty} e^{-bt} f(at) e^{-tz}\, dt = \int_0^{+\infty} f(at) e^{-t(b+z)}\, dt \\
&= \int_0^{+\infty} \frac{1}{a} f(s) e^{-s\frac{(b+z)}{a}}\, ds = \frac{1}{a} \mathcal{L}(f)\left(\frac{b+z}{a} \right)
\end{aligned}
$$

which is well defined if $\operatorname{Re}\left(\frac{b+z}{a} \right) \geq \gamma_0$.

Exercise 16.10 Since

$$|f(t)|e^{-\gamma_0 t} \le c \quad \forall t \ge 0,$$

we have, if $\operatorname{Re} z > \gamma_0$, that

$$|F(z)| = \left| \int_0^{+\infty} f(t) e^{-tz} dt \right| = \left| \int_0^{+\infty} f(t) e^{-t(\operatorname{Re} z + i \operatorname{Im} z)} dt \right|$$

$$\le \int_0^{+\infty} |f(t)| e^{-t \operatorname{Re} z} dt \le c \int_0^{+\infty} e^{-t(\operatorname{Re} z - \gamma_0)} dt = \frac{c}{\operatorname{Re} z - \gamma_0} .$$

Letting $\operatorname{Re} z$ tend to ∞ we have the claim.

Exercise 16.11 We have already seen (cf. Exercise 16.7) that

$$zF(z) - f(0) = \mathcal{L}(f')(z) = \int_0^{+\infty} f'(t) e^{-tz} dt.$$

(i) Applying Exercise 16.10, to f' we obtain

$$\lim_{\operatorname{Re} z \to \infty} \mathcal{L}(f')(z) = \lim_{\operatorname{Re} z \to \infty} \int_0^{+\infty} f'(t) e^{-tz} dt = 0.$$

The result now immediately follows, i.e.

$$\lim_{\operatorname{Re} z \to \infty} zF(z) = f(0) .$$

(ii) Note that the hypotheses imply that

$$\lim_{z \to 0, \, \operatorname{Re} z \ge 0} \int_0^{+\infty} f'(t) e^{-tz} dt = \int_0^{+\infty} f'(t) dt = \lim_{t \to \infty} f(t) - f(0)$$

which implies

$$\lim_{z \to 0, \, \operatorname{Re} z \ge 0} zF(z) = \lim_{t \to \infty} f(t).$$

Exercise* 16.12 Observe that since $f \equiv 0$ and $g \equiv 0$ if $t < 0$, we have

$$(f * g)(t) = \int_{-\infty}^{+\infty} f(t-s) g(s) \, ds = \int_0^t f(t-s) g(s) \, ds = \int_0^t f(s) g(t-s) ds.$$

Under our hypotheses, we can permute the integrals. We therefore get

$$\mathcal{L}(f * g)(z) = \int_0^{+\infty} \left[\int_0^t f(s) g(t-s) ds \right] e^{-zt} dt$$

$$= \int_0^{+\infty} ds \left[\int_s^{+\infty} f(s) g(t-s) e^{-zt} dt \right].$$

Letting $t - s = x$, we obtain

$$\mathcal{L}(f * g)(z) = \int_0^{+\infty} f(s) \left[\int_0^{+\infty} g(x) e^{-z(x+s)} dx \right] ds$$

$$= \int_0^{+\infty} f(s) e^{-zs} ds \int_0^{+\infty} g(x) e^{-zx} dx$$

$$= \mathcal{L}(f)(z) \, \mathcal{L}(g)(z).$$

Exercise* 16.13 (i) We have

$$\mathcal{L}(f)(z) = \int_0^{+\infty} f(t) e^{-zt} dt = \sum_{n=0}^{\infty} \int_{nT}^{(n+1)T} f(t) e^{-zt} dt$$

$$= \int_0^T f(t) e^{-zt} dt + \int_T^{2T} f(t) e^{-zt} dt + \int_{2T}^{3T} f(t) e^{-zt} dt + \cdots.$$

Letting $t = u + nT$, we find (using the periodicity of f, that is $f(u + nT) = f(u)$)

$$\int_{nT}^{(n+1)T} f(t) e^{-zt} dt = \int_0^T f(u + nT) e^{-z(u+nT)} du = e^{-nTz} \int_0^T f(u) e^{-zu} du.$$

We therefore obtain

$$\mathcal{L}(f)(z) = \left(\sum_{n=0}^{\infty} e^{-nTz} \right) \int_0^T f(u) e^{-zu} du.$$

Since $\mathrm{Re}\, z > 0$, we have $\left| e^{-Tz} \right| = e^{-T\,\mathrm{Re}\,z} < 1$ and we use the suggestion to get

$$\mathcal{L}(f)(z) = \frac{1}{1 - e^{-zT}} \int_0^T f(t) e^{-zt} dt.$$

(ii) Using the result of (i) with $T = 2\pi$, we find

$$\mathcal{L}(f)(z) = \frac{1}{1 - e^{-2\pi z}} \int_0^{2\pi} f(t) e^{-zt} dt = \frac{1}{1 - e^{-2\pi z}} \int_0^{\pi} e^{-zt} \sin t \, dt$$

$$= \frac{1}{1 - e^{-2\pi z}} \left[\frac{e^{-zt}(-z \sin t - \cos t)}{z^2 + 1} \right]_0^{\pi} = \frac{1}{(1 - e^{-\pi z})(z^2 + 1)}.$$

Exercise* 16.14 Recall that f is continuous, $f(0) = 0$ (and $f \equiv 0$ if $t < 0$). Let $g(t) = f(t) e^{-\gamma t}$. For $s \in \mathbb{R}$, we have

$$\mathfrak{F}(g)(s) = \widehat{g}(s) = \frac{1}{\sqrt{2\pi}} \int_{-\infty}^{+\infty} g(t) e^{-its} dt$$

$$= \frac{1}{\sqrt{2\pi}} \int_{-\infty}^{+\infty} f(t) e^{-(\gamma+is)t} dt = \frac{1}{\sqrt{2\pi}} F(\gamma + is).$$

Since all the hypotheses of Theorem 15.3 of the inversion formula for Fourier transform are satisfied, we get, for every $t \in \mathbb{R}$,

$$\frac{1}{\sqrt{2\pi}} \int_{-\infty}^{+\infty} \widehat{g}(s) e^{its} ds = g(t)$$

which leads to

$$\frac{1}{\sqrt{2\pi}} \int_{-\infty}^{+\infty} \frac{1}{\sqrt{2\pi}} F(\gamma + is) e^{its} ds = e^{-\gamma t} f(t).$$

We therefore obtain

$$\frac{1}{2\pi} \int_{-\infty}^{+\infty} F(\gamma + is) e^{(\gamma+is)t} ds = f(t).$$

Chapter 17

Applications to ordinary differential equations

Exercise 17.1 We proceed formally (to make the following calculations rigorous, we need to make appropriate assumptions on the coefficients and on f). We let

$$F(z) = \mathfrak{L}(f)(z) \quad \text{and} \quad Y(z) = \mathfrak{L}(y)(z)$$

be the Laplace transforms of f and y respectively. We then use the properties to get

$$
\begin{aligned}
\mathfrak{L}(y^{(n)})(z) &= z^n\,\mathfrak{L}(y)(z) - \sum_{k=0}^{n-1} z^k\, y^{(n-k-1)}(0) \\
&= z^n\,Y(z) - \sum_{k=0}^{n-1} z^k\, y_{n-k-1}
\end{aligned}
$$

and similarly for the other derivatives. For example, for the first derivative we have

$$\mathfrak{L}(y')(z) = z\mathfrak{L}(y)(z) - y(0) = zY(z) - y_0\,.$$

We, moreover, find

$$\mathfrak{L}\left(\sum_{l=0}^{n} a_l y^{(l)}\right)(z) = \sum_{l=0}^{n} a_l\, \mathfrak{L}\left(y^{(l)}\right)(z) = \mathfrak{L}(f)(z) = F(z),$$

317

that is

$$\sum_{l=1}^{n} a_l \left[z^l Y(z) - \sum_{k=0}^{l-1} z^k y_{l-k-1} \right] + a_0 Y(z)$$

$$= \sum_{l=0}^{n} a_l z^l Y(z) - \sum_{l=1}^{n} \sum_{k=0}^{l-1} a_l z^k y_{l-k-1} = F(z).$$

We therefore obtain

$$Y(z) = \frac{F(z) + \sum_{l=1}^{n} \sum_{k=0}^{l-1} a_l z^k y_{l-k-1}}{\sum_{l=0}^{n} a_l z^l}.$$

We get the solution to the system using the inverse of the Laplace transform, i.e.

$$y(t) = \mathcal{L}^{-1}(Y)(t).$$

Exercise 17.2 *First method.* We use Laplace transform to solve

$$\begin{cases} y''(t) + \lambda y(t) = 0 \\ y(0) = y_0 \quad \text{and} \quad y'(0) = y_1. \end{cases}$$

Let $Y(z) = \mathcal{L}(y)(z)$. We know that

$$\mathcal{L}(y'')(z) = z^2 Y(z) - z y_0 - y_1.$$

We therefore have

$$z^2 Y - z y_0 - y_1 + \lambda Y = 0 \Rightarrow Y = \frac{z y_0 + y_1}{z^2 + \lambda}.$$

We study separately the cases $\lambda = 0$, $\lambda < 0$ and $\lambda > 0$. We use here the result established in Example 17.2.

Case 1: $\lambda = 0$. The solution is trivially given by

$$y(t) = y_0 + y_1 t.$$

Since we also want $y(0) = y(2\pi)$ and $y'(0) = y'(2\pi)$, we find $y_1 = 0$. Therefore, if $\lambda = 0$, we have the following non-trivial solution

$$y(t) = \alpha_0.$$

Case 2: $\lambda < 0$, that is $\lambda = -\mu^2$. We have seen, in Example 17.2, that

$$y(t) = y_0 \cosh(\mu t) + \frac{y_1}{\mu} \sinh(\mu t).$$

We, moreover, want $y(0) = y(2\pi)$ and $y'(0) = y'(2\pi)$, that is

$$\begin{cases} y_0 \cosh(2\pi\mu) + \dfrac{y_1}{\mu} \sinh(2\pi\mu) = y_0 \\ y_0\mu \sinh(2\pi\mu) + y_1 \cosh(2\pi\mu) = y_1 \end{cases}$$

which is impossible unless $y_0 = y_1 = 0$ (that is, y is the trivial solution). In conclusion, if $\lambda < 0$, there is no non-trivial solution to the problem.

Case 3: $\lambda > 0$, that is, $\lambda = \mu^2$. As in Example 17.2, we find that

$$y(t) = y_0 \cos(\mu t) + \frac{y_1}{\mu} \sin(\mu t).$$

On the other hand, we want

$$y(0) = y(2\pi) = y_0 \quad \text{and} \quad y'(0) = y'(2\pi) = y_1$$

which leads to the system

$$\begin{cases} y_0 \cos(2\pi\mu) + \dfrac{y_1}{\mu} \sin(2\pi\mu) = y_0 \\ -y_0\mu \sin(2\pi\mu) + y_1 \cos(2\pi\mu) = y_1. \end{cases}$$

The system has a non-trivial solution only if $\mu = n \in \mathbb{N}$ and thus $\lambda = n^2$. Therefore, if $\lambda = n^2$ ($n = 1, 2, 3, \cdots$), the problem has the following solution

$$y(t) = \alpha_n \cos(nt) + \beta_n \sin(nt).$$

Second method. We proceed with formal computations. We are looking for solutions of the form

$$y(x) = \frac{a_0}{2} + \sum_{n=1}^{\infty} (a_n \cos(\alpha_n x) + b_n \sin(\alpha_n x))$$

$\alpha_n > 0$. Differentiating the function y, we obtain

$$y''(x) + \lambda y(x) = \lambda \frac{a_0}{2} + \sum_{n=1}^{\infty} (\lambda - \alpha_n^2)(a_n \cos(\alpha_n x) + b_n \sin(\alpha_n x)) = 0.$$

We then consider three cases.

Case 1: $\lambda = 0$. We deduce that $a_n = b_n = 0$, $\forall n \geq 1$ and thus $y(x) = a_0/2$. Since

$$y(0) = y(2\pi),$$

there exists a non-trivial solution which is given by

$$y(x) \equiv a = \text{constant}.$$

Case 2: $\lambda \neq 0$ and $\lambda \neq \alpha_n^2$, $\forall n \in \mathbb{N}$. We here have $a_n = b_n = 0$. There is therefore no non-trivial solution, i.e.

$$y(x) \equiv 0.$$

Case 3: $\lambda \neq 0$ and $\exists \bar{n} : \lambda = \alpha_{\bar{n}}^2$. We then have $a_n = b_n = 0$, $\forall n \neq \bar{n}$, while for $n = \bar{n}$, we have

$$y(x) = a_{\bar{n}} \cos(\alpha_{\bar{n}} x) + b_{\bar{n}} \sin(\alpha_{\bar{n}} x).$$

If we further want

$$y(0) = y(2\pi) \quad \text{and} \quad y'(0) = y'(2\pi),$$

we find

$$\begin{cases} a_{\bar{n}} = a_{\bar{n}} \cos(\alpha_{\bar{n}} 2\pi) + b_{\bar{n}} \sin(\alpha_{\bar{n}} 2\pi) \\ b_{\bar{n}} \alpha_{\bar{n}} = -a_{\bar{n}} \alpha_{\bar{n}} \sin(\alpha_{\bar{n}} 2\pi) + b_{\bar{n}} \alpha_{\bar{n}} \cos(\alpha_{\bar{n}} 2\pi). \end{cases}$$

This is possible only if $\alpha_{\bar{n}} = n \in \mathbb{N}$ and thus $\lambda = n^2$. The given problem therefore has a solution of the form

$$y(x) = a_n \cos(nx) + b_n \sin(nx).$$

Exercise 17.3 We first solve the problem

$$\begin{cases} v''(x) + 2v'(x) + (1+\mu)v(x) = 0 \\ v(0) = 0 \quad \text{and} \quad v'(0) = \alpha. \end{cases} \tag{17.1}$$

As in Example 17.1, we find that, if V denotes the Laplace transform of v, then

$$V(z) = \frac{\alpha}{z^2 + 2z + 1 + \mu} = \frac{\alpha}{\sqrt{|\mu|}} \frac{\sqrt{|\mu|}}{(z+1)^2 + \mu}.$$

We, hence, obtain that the solutions to (17.1) are given by

$$v(t) = \begin{cases} \alpha t e^{-t} & \text{if } \mu = 0 \\ \dfrac{\alpha}{\sqrt{|\mu|}} e^{-t} \sin\left(t\sqrt{|\mu|}\right) & \text{if } \mu > 0 \\ \dfrac{\alpha}{\sqrt{|\mu|}} e^{-t} \sinh\left(t\sqrt{|\mu|}\right) & \text{if } \mu < 0. \end{cases}$$

To solve our problem, we must have $v(\pi) = 0$. We thus infer that $\mu > 0$ and we get

$$\sin\left(\pi\sqrt{|\mu|}\right) = 0 \Leftrightarrow \mu = n^2.$$

The non-trivial solutions to our problem are finally given by $\mu = n^2$ and (α_n are constants)

$$v_n(t) = \alpha_n e^{-t} \sin(nt).$$

Exercise 17.4 It is sufficient to check that the given functions satisfy the equation. For the sake of completeness we explain a heuristic method for finding this solution. We first consider the case $n \geq 1$. We look for solutions of the form

$$f_n(r) = \sum_{k=-\infty}^{\infty} a_k r^k.$$

Differentiating twice the function f_n and substituting it into the equation $(r^2 f''(r) + r f'(r) - n^2 f(r) = 0)$ leads to

$$\sum_{k=-\infty}^{\infty} a_k (k^2 - n^2) r^k = 0.$$

Since the equation is satisfied for every $r > 0$, we deduce that if $n \neq |k|$, then $a_k = 0$. If $n = k$, then a non-trivial solution to the equation is given by

$$a_n r^n + a_{-n} r^{-n}.$$

When $n = 0$, the above method gives only the solution a_0. However, integrating the equation $r f'' + f' = 0$ gives

$$a_0 + b_0 \log r.$$

To summarize, we have

$$f_n(r) = \begin{cases} a_n r^n + a_{-n} r^{-n} & \text{if } n = 1, 2, 3 \cdots \\ a_0 + b_0 \log r & \text{if } n = 0. \end{cases}$$

Exercise 17.5 (i) Since f is even and 2π−periodic, we find $b_n = 0$ and

$$a_n = \frac{1}{\pi} \int_0^{2\pi} f(x) \cos(nx)\, dx = \frac{1}{\pi} \int_{-\pi}^{\pi} |x| \cos(nx)\, dx = \frac{2}{\pi} \int_0^{\pi} x \cos(nx)\, dx.$$

We therefore have $a_0 = \pi$ and, for $n \geq 1$, we get

$$a_n = \frac{2}{\pi} \left[\frac{x \sin(nx)}{n} \right]_0^\pi - \frac{2}{\pi} \int_0^\pi \frac{\sin(nx)}{n}\, dx = \frac{2}{\pi} \left[\frac{\cos(nx)}{n^2} \right]_0^\pi$$

$$= \frac{2}{\pi n^2} [(-1)^n - 1] = \begin{cases} \dfrac{-4}{\pi n^2} & \text{if } n \text{ is odd} \\ 0 & \text{if } n \text{ is even.} \end{cases}$$

We thus obtain

$$Ff(x) = \frac{\pi}{2} - \frac{4}{\pi} \sum_{k=0}^{\infty} \frac{\cos((2k+1)x)}{(2k+1)^2}.$$

(ii) We are looking for solutions of the form

$$y(x) = \frac{a_0}{2} + \sum_{n=1}^{\infty} (a_n \cos(nx) + b_n \sin(nx)).$$

We know that

$$y^{(4)}(x) = \sum_{n=1}^{\infty} n^4 (a_n \cos(nx) + b_n \sin(nx)).$$

Returning to the equation, we find

$$\frac{d^4 y(x)}{dx^4} - \alpha y(x) = -\alpha \frac{a_0}{2} + \sum_{n=1}^{\infty} (n^4 - \alpha)(a_n \cos(nx) + b_n \sin(nx))$$

$$= \frac{\pi}{2} - \frac{4}{\pi} \sum_{k=0}^{\infty} \frac{\cos((2k+1)x)}{(2k+1)^2}.$$

We identify the coefficients (which is possible only if $\alpha \neq 0$ and $\alpha \neq (2k+1)^4$, $k \in \mathbb{N}$) and we get

$$a_0 = -\frac{\pi}{\alpha}, \quad a_{2k} = 0, \quad b_{2k} = b_{2k+1} = 0$$

$$a_{2k+1} = -\frac{4}{\pi} \frac{1}{(2k+1)^2 \left((2k+1)^4 - \alpha\right)}.$$

The corresponding solution is given by

$$y(x) = -\frac{\pi}{2\alpha} - \frac{4}{\pi} \sum_{k=0}^{\infty} \frac{\cos((2k+1)x)}{(2k+1)^2 \left((2k+1)^4 - \alpha\right)}.$$

Exercise 17.6 We are looking for a solution of the form

$$x(t) = \frac{a_0}{2} + \sum_{n=1}^{\infty} (a_n \cos(nt) + b_n \sin(nt)).$$

Differentiating the function x leads to

$$x'\left(t - \frac{\pi}{2}\right) = \sum_{n=1}^{\infty} \left[-na_n \sin\left(n\left(t - \frac{\pi}{2}\right)\right) + nb_n \cos\left(n\left(t - \frac{\pi}{2}\right)\right)\right].$$

Since

$$\sin\left(n\left(t - \frac{\pi}{2}\right)\right) = \sin(nt)\cos\left(n\frac{\pi}{2}\right) - \cos(nt)\sin\left(n\frac{\pi}{2}\right)$$

$$\cos\left(n\left(t - \frac{\pi}{2}\right)\right) = \cos(nt)\cos\left(n\frac{\pi}{2}\right) + \sin(nt)\sin\left(n\frac{\pi}{2}\right),$$

we have to consider four cases: $n = 4k$, $n = 4k - 1$, $n = 4k - 2$ and $n = 4k - 3$. We have respectively

$$\sin\left(n\left(t - \tfrac{\pi}{2}\right)\right) = \sin\left(4kt\right) \qquad \cos\left(n\left(t - \tfrac{\pi}{2}\right)\right) = \cos\left(4kt\right)$$

$$\sin\left(n\left(t - \tfrac{\pi}{2}\right)\right) = \cos\left((4k - 1)t\right) \qquad \cos\left(n\left(t - \tfrac{\pi}{2}\right)\right) = -\sin\left((4k - 1)t\right)$$

$$\sin\left(n\left(t - \tfrac{\pi}{2}\right)\right) = -\sin\left((4k - 2)t\right) \qquad \cos\left(n\left(t - \tfrac{\pi}{2}\right)\right) = -\cos\left((4k - 2)t\right)$$

$$\sin\left(n\left(t - \tfrac{\pi}{2}\right)\right) = -\cos\left((4k - 3)t\right) \qquad \cos\left(n\left(t - \tfrac{\pi}{2}\right)\right) = \sin\left((4k - 3)t\right).$$

We, hence, get

$$x\left(t\right) = \frac{a_0}{2} + \sum_{k=1}^{\infty}\left\{A_{4k-3} + A_{4k-2} + A_{4k-1} + A_{4k}\right\}$$

where

$$A_{4k-3} = a_{4k-3}\cos\left((4k - 3)t\right) + b_{4k-3}\sin\left((4k - 3)t\right)$$

$$A_{4k-2} = a_{4k-2}\cos\left((4k - 2)t\right) + b_{4k-2}\sin\left((4k - 2)t\right)$$

$$A_{4k-1} = a_{4k-1}\cos\left((4k - 1)t\right) + b_{4k-1}\sin\left((4k - 1)t\right)$$

$$A_{4k} = a_{4k}\cos\left(4kt\right) + b_{4k}\sin\left(4kt\right).$$

We, moreover, find

$$x'\left(t - \frac{\pi}{2}\right) = \sum_{k=1}^{\infty}\left\{B_{4k-3} + B_{4k-2} + B_{4k-1} + B_{4k}\right\}$$

where

$$B_{4k-3} = (4k - 3)\,a_{4k-3}\cos\left((4k - 3)t\right) + (4k - 3)\,b_{4k-3}\sin\left((4k - 3)t\right)$$

$$B_{4k-2} = (4k - 2)\,a_{4k-2}\sin\left((4k - 2)t\right) - (4k - 2)\,b_{4k-2}\cos\left((4k - 2)t\right)$$

$$B_{4k-1} = -(4k - 1)\,a_{4k-1}\cos\left((4k - 1)t\right) - (4k - 1)\,b_{4k-1}\sin\left((4k - 1)t\right)$$

$$B_{4k} = -4k\,a_{4k}\sin\left(4kt\right) + 4k\,b_{4k}\cos\left(4kt\right).$$

The given equation becomes

$$x'\left(t - \frac{\pi}{2}\right) + x\left(t\right)$$

$$= \frac{a_0}{2} + \sum_{k=1}^{\infty}\left\{C_{4k-3} + C_{4k-2} + C_{4k-1} + C_{4k}\right\}$$

where

$$C_{4k-3} = (4k-2)\, a_{4k-3} \cos\left((4k-3)\,t\right) + (4k-2)\, b_{4k-3} \sin\left((4k-3)\,t\right)$$

$$
\begin{aligned}
C_{4k-2} = \;& \left(a_{4k-2} - (4k-2)\, b_{4k-2}\right) \cos\left((4k-2)\,t\right) \\
& + \left(b_{4k-2} + (4k-2)\, a_{4k-2}\right) \sin\left((4k-2)\,t\right)
\end{aligned}
$$

$$C_{4k-1} = (2-4k)\, a_{4k-1} \cos\left((4k-1)\,t\right) + (2-4k)\, b_{4k-1} \sin\left((4k-1)\,t\right)$$

$$C_{4k} = (a_{4k} + 4k\, b_{4k}) \cos\left(4kt\right) + (b_{4k} - 4k\, a_{4k}) \sin\left(4kt\right).$$

Identifying the coefficients in the equation

$$x'\left(t - \frac{\pi}{2}\right) + x(t) = 1 + 2\cos t + \sin(2t),$$

we infer that $a_0 = 2$ and

$$(4k-2)\, a_{4k-3} = \begin{cases} 2 & \text{if } k = 1 \\ 0 & \text{if } k \geq 2 \end{cases} \qquad (4k-2)\, b_{4k-3} = 0 \; \forall k \geq 1$$

$$a_{4k-2} - (4k-2)\, b_{4k-2} = 0 \quad \forall k \geq 1$$

$$b_{4k-2} + (4k-2)\, a_{4k-2} = \begin{cases} 1 & \text{if } k = 1 \\ 0 & \text{if } k \geq 2 \end{cases}$$

$$(2-4k)a_{4k-1} = (2-4k)b_{4k-1} = 0 \quad \forall k \geq 1$$

$$a_{4k} + 4k\, b_{4k} = b_{4k} - 4k\, a_{4k} = 0 \quad \forall k \geq 1.$$

We deduce

$$a_1 = 1, \quad a_{4k-3} = 0 \; \forall k \geq 2, \quad b_{4k-3} = 0 \; \forall k \geq 1$$

$$a_2 = \frac{2}{5}, \quad b_2 = \frac{1}{5}, \quad a_{4k-2} = b_{4k-2} = 0 \; \forall k \geq 2$$

$$a_{4k-1} = b_{4k-1} = 0 \; \forall k \geq 1, \quad a_{4k} = b_{4k} = 0 \; \forall k \geq 1$$

and finally

$$x(t) = 1 + \cos t + \frac{2}{5} \cos(2t) + \frac{1}{5} \sin(2t).$$

Exercise 17.7 We are looking for a solution of the form

$$x(t) = \frac{a_0}{2} + \sum_{n=1}^{\infty} (a_n \cos(nt) + b_n \sin(nt)).$$

Differentiating the function x leads to

$$x'(t) = \sum_{n=1}^{\infty} (-na_n \sin(nt) + nb_n \cos(nt))$$

$$x'(t - \pi) = \sum_{n=1}^{\infty} (-na_n \sin(nt - n\pi) + nb_n \cos(nt - n\pi))$$

$$= \sum_{n=1}^{\infty} (-na_n (-1)^n \sin(nt) + nb_n (-1)^n \cos(nt))$$

$$= \sum_{n=1}^{\infty} n(-1)^n (b_n \cos(nt) - a_n \sin(nt))$$

$$x''(t) = \sum_{n=1}^{\infty} (-n^2 a_n \cos(nt) - n^2 b_n \sin(nt)).$$

Returning to the equation, we obtain

$$x''(t) + 5x'(t - \pi) - x(t)$$

$$= -\frac{a_0}{2} + \sum_{n=1}^{\infty} \left\{ \left[-a_n + 5n(-1)^n b_n - n^2 a_n \right] \cos(nt) \right.$$

$$+ \left. \left[-b_n - 5n(-1)^n a_n - n^2 b_n \right] \sin(nt) \right\}.$$

Since

$$x''(t) + 5x'(t - \pi) - x(t) = 2 + \cos t - 3\sin(2t),$$

we find that $a_0 = -4$ and

$$a_n = b_n = 0 \quad \forall n \geq 3$$

$$\begin{cases} -a_1 - 5b_1 - a_1 = 1 \\ -b_1 + 5a_1 - b_1 = 0 \end{cases} \quad \text{and} \quad \begin{cases} -a_2 + 10b_2 - 4a_2 = 0 \\ -b_2 - 10a_2 - 4b_2 = -3. \end{cases}$$

We, hence, have

$$a_0 = -4, \quad a_1 = -\frac{2}{29}, \quad b_1 = -\frac{5}{29}, \quad a_2 = \frac{6}{25}, \quad b_2 = \frac{3}{25}$$

and thus

$$x\left(t\right) = -2 - \frac{2}{29}\cos t - \frac{5}{29}\sin t + \frac{6}{25}\cos\left(2t\right) + \frac{3}{25}\sin\left(2t\right).$$

Exercise 17.8 (i) We saw in Exercise 14.9 that (f is continuous)

$$f\left(t\right) = Ff\left(t\right) = \frac{2}{3\pi} + \frac{12}{\pi}\sum_{k=1}^{\infty}\frac{\cos\left(2kt\right)}{9 - 4k^2}.$$

(ii) We are looking for a solution of the form

$$x\left(t\right) = \frac{\alpha_0}{2} + \sum_{n=1}^{\infty}\left(\alpha_n\cos\left(nt\right) + \beta_n\sin\left(nt\right)\right)$$

(since we want x to be even, we must have $\beta_n = 0$) and thus

$$x(t - \pi) = \frac{\alpha_0}{2} + \sum_{n=1}^{\infty}\left[(-1)^n\,\alpha_n\cos\left(nt\right)\right].$$

We therefore get

$$x\left(t\right) - 2x(t - \pi) = -\frac{\alpha_0}{2} + \sum_{n=1}^{\infty}\left[\alpha_n(1 - 2\left(-1\right)^n)\cos\left(nt\right)\right]$$

$$= \frac{2}{3\pi} + \frac{12}{\pi}\sum_{n\ even}\frac{\cos\left(nt\right)}{9 - n^2}.$$

We then deduce that $\alpha_n = 0$ if n is odd. On the other hand, if n is even, we have

$$\alpha_0 = -\frac{4}{3\pi} \quad\text{and}\quad \alpha_n = -\frac{12}{\pi(9 - n^2)}.$$

The solution is therefore given by

$$x\left(t\right) = -f\left(t\right) = -\frac{2}{3\pi} - \frac{12}{\pi}\sum_{k=1}^{\infty}\frac{\cos\left(2kt\right)}{9 - 4k^2}.$$

Exercise 17.9 (i) We are again looking for solutions of the form

$$x\left(t\right) = \frac{a_0}{2} + \sum_{n=1}^{\infty}\left(a_n\cos\left(nt\right) + b_n\sin\left(nt\right)\right).$$

We therefore deduce

$$x(t - \pi) = \frac{a_0}{2} + \sum_{n=1}^{\infty} (a_n (-1)^n \cos(nt) + b_n (-1)^n \sin(nt))$$

$$x'(t) = \sum_{n=1}^{\infty} n(-a_n \sin(nt) + b_n \cos(nt)).$$

Writing the Fourier series of f, we find

$$f(t) = \frac{\alpha_0}{2} + \sum_{n=1}^{\infty} (\alpha_n \cos(nt) + \beta_n \sin(nt)).$$

Returning to the given equation, we obtain

$$x'(t) + 2x(t - \pi)$$

$$= a_0 + \sum_{n=1}^{\infty} [(nb_n + 2(-1)^n a_n) \cos(nt) + (-na_n + 2(-1)^n b_n) \sin(nt)]$$

$$= \frac{\alpha_0}{2} + \sum_{n=1}^{\infty} (\alpha_n \cos(nt) + \beta_n \sin(nt)) = f(t).$$

Identifying the coefficients leads to $a_0 = \alpha_0/2$,

$$\alpha_n = nb_n + 2(-1)^n a_n \quad \text{and} \quad \beta_n = -na_n + 2(-1)^n b_n.$$

We therefore have $a_0 = \alpha_0/2$,

$$a_n = \frac{2(-1)^n \alpha_n - n\beta_n}{4 + n^2} \quad \text{and} \quad b_n = \frac{n\alpha_n + 2(-1)^n \beta_n}{4 + n^2}.$$

(ii) If $f(t) = 1 + 4\sin(6t)$, we find

$$\alpha_i = 0 \ \forall i \neq 0, \quad \alpha_0 = 2, \quad \beta_i = 0 \ \forall i \neq 6 \quad \text{and} \quad \beta_6 = 4.$$

The solution is then

$$x(t) = \frac{1}{2} - \frac{3}{5} \cos(6t) + \frac{1}{5} \sin(6t).$$

Exercise 17.10 We write the Fourier series of f (since f is one of the data, we have that the coefficients a_n and b_n of f are known)

$$f(t) = \frac{a_0}{2} + \sum_{n=1}^{\infty} \left[a_n \cos\left(\frac{2\pi n}{T} t\right) + b_n \sin\left(\frac{2\pi n}{T} t\right) \right].$$

We then look for a solution of the form (the coefficients A_n and B_n are unknown)

$$x\left(t\right) = \frac{A_0}{2} + \sum_{n=1}^{\infty} \left[A_n \cos\left(\frac{2\pi n}{T}t\right) + B_n \sin\left(\frac{2\pi n}{T}t\right) \right]$$

and thus

$$x'\left(t\right) = \sum_{n=1}^{\infty} \left[\frac{2\pi n}{T} B_n \cos\left(\frac{2\pi n}{T}t\right) - \frac{2\pi n}{T} A_n \sin\left(\frac{2\pi n}{T}t\right) \right].$$

Returning to the equation, we have

$$x'\left(t\right) + \alpha x\left(t\right) = \frac{\alpha A_0}{2} + \sum_{n=1}^{\infty} \left[\left(\frac{2\pi n}{T} B_n + \alpha A_n \right) \cos\left(\frac{2\pi n}{T}t\right) \right.$$
$$\left. + \left(\alpha B_n - \frac{2\pi n}{T} A_n \right) \sin\left(\frac{2\pi n}{T}t\right) \right].$$

Identifying the coefficients leads to $A_0 = a_0/\alpha$ and

$$a_n = \frac{2\pi n}{T} B_n + \alpha A_n \quad \text{and} \quad b_n = \alpha B_n - \frac{2\pi n}{T} A_n$$

and thus $A_0 = a_0/\alpha$ while

$$A_n = \frac{\alpha a_n - \dfrac{2\pi n}{T} b_n}{\left(\dfrac{2\pi n}{T}\right)^2 + \alpha^2} \quad \text{and} \quad B_n = \frac{\dfrac{2\pi n}{T} a_n + \alpha b_n}{\left(\dfrac{2\pi n}{T}\right)^2 + \alpha^2}.$$

We next apply the above result to the case where $T = 2\pi$, $\alpha = 1$ and f is the given function. The Fourier coefficients of f are therefore given by

$$a_0 = \frac{1}{\pi} \int_0^\pi \left(x - \frac{\pi}{2}\right)^2 dx + \frac{1}{\pi} \int_\pi^{2\pi} \left[-\left(x - \frac{3\pi}{2}\right)^2 + \frac{\pi^2}{2} \right] dx = \frac{\pi^2}{2}$$

and for $n \geq 1$

$$a_n = \frac{1}{\pi} \int_0^\pi \left(x - \frac{\pi}{2}\right)^2 \cos\left(nx\right) dx + \frac{1}{\pi} \int_\pi^{2\pi} \left[-\left(x - \frac{3\pi}{2}\right)^2 + \frac{\pi^2}{2} \right] \cos\left(nx\right) dx$$

namely

$$a_n = \frac{1 - \cos\left(n\pi\right)}{\pi} \int_0^\pi \left(x - \frac{\pi}{2}\right)^2 \cos\left(nx\right) dx$$
$$= \frac{1 - \cos\left(n\pi\right)\left(1 + \cos\left(n\pi\right)\right)\pi}{n^2} = 0.$$

In a similar way, we have

$$b_n = \frac{1}{\pi} \int_0^\pi \left(x - \frac{\pi}{2}\right)^2 \sin(nx)\, dx$$

$$+ \frac{1}{\pi} \int_\pi^{2\pi} \left[-\left(x - \frac{3\pi}{2}\right)^2 + \frac{\pi^2}{2}\right] \sin(nx)\, dx$$

and thus

$$b_n = \frac{1 - (-1)^n}{\pi} \int_0^\pi \left(x - \frac{\pi}{2}\right)^2 \sin(nx)\, dx + \frac{\pi}{2} \int_\pi^{2\pi} \sin(nx)\, dx$$

$$= \frac{-4(1 - (-1)^n)}{\pi n^3}.$$

We therefore obtain

$$A_0 = \frac{\pi^2}{2}, \quad A_{2k} = 0, \quad A_{2k+1} = \frac{8}{\pi (2k+1)^2 \left((2k+1)^2 + 1\right)}$$

$$B_{2k} = 0 \quad \text{and} \quad B_{2k+1} = \frac{-8}{\pi (2k+1)^3 \left((2k+1)^2 + 1\right)}.$$

Exercise 17.11 Using the table of Fourier transforms, we find

$$\widehat{f}(\alpha) = \sqrt{\frac{2}{\pi}} \frac{1}{\alpha^2 + 1} \quad \text{and} \quad \widehat{g}(\alpha) = \frac{-i\alpha}{2\sqrt{2}} e^{-\alpha^2/4}.$$

If we denote the Fourier transform of the unknown function y by $\widehat{y}(\alpha)$, we get

$$\mathfrak{F}(y'')(\alpha) = -\alpha^2\, \widehat{y}(\alpha)$$

and thus

$$\mathfrak{F}\left(\int_{-\infty}^{+\infty} [y''(\tau) - y(\tau)] f(t - \tau) d\tau\right)(\alpha) = \mathfrak{F}((y'' - y) * f)(\alpha)$$

$$= \sqrt{2\pi}\, \mathfrak{F}((y'' - y))(\alpha)\, \mathfrak{F}(f)(\alpha)$$

$$= \sqrt{2\pi}\, (-\alpha^2 - 1)\, \widehat{y}(\alpha)\, \widehat{f}(\alpha)$$

$$= -2\widehat{y}(\alpha).$$

Applying Fourier transform to both sides of the equation leads to

$$3\widehat{y}(\alpha) - 2\widehat{y}(\alpha) = \widehat{y}(\alpha) = \widehat{g}(\alpha) = \frac{-i\alpha}{2\sqrt{2}} e^{-\alpha^2/4}.$$

Using again the table of Fourier transforms, we obtain

$$y(t) = g(t) = te^{-t^2}.$$

Chapter 18

Applications to partial differential equations

Exercise 18.1 *Step 1 (Separation of variables).* This exercise corresponds to Example 18.1, with $L = \pi$, $a = 1$ and $f(x) = \cos x - \cos(3x)$. We saw that the general solution to this equation is

$$u(x,t) = \sum_{n=1}^{\infty} \alpha_n \sin(nx)\, e^{-n^2 t}.$$

Step 2 (Initial condition). We next have to choose the α_n in order that

$$u(x,0) = f(x) = \cos x - \cos(3x) = \sum_{n=1}^{\infty} \alpha_n \sin(nx).$$

We therefore obtain

$$\alpha_n = \frac{2}{\pi} \int_0^\pi [\cos x \sin(nx) - \cos(3x) \sin(nx)]\, dx$$

$$= \frac{1}{\pi} \int_0^\pi [\sin((n+1)x) + \sin((n-1)x)]\, dx$$

$$- \frac{1}{\pi} \int_0^\pi [\sin((n+3)x) + \sin((n-3)x)]\, dx.$$

If $n \neq 1$ and $n \neq 3$, then

$$\alpha_n = -\frac{1}{\pi} \left[\frac{\cos((n+1)x)}{n+1} + \frac{\cos((n-1)x)}{n-1} \right]_0^\pi$$

$$+ \frac{1}{\pi} \left[\frac{\cos((n+3)x)}{n+3} + \frac{\cos((n-3)x)}{n-3} \right]_0^\pi$$

$$= \frac{1}{\pi} \left[\frac{1+(-1)^n}{n+1} + \frac{1+(-1)^n}{n-1} - \frac{1+(-1)^n}{n+3} - \frac{1+(-1)^n}{n-3} \right].$$

We therefore obtain that $\alpha_n = 0$ if n is odd ($n \neq 1$ and $n \neq 3$). If n is even, we find

$$
\begin{aligned}
\alpha_n &= \frac{2}{\pi} \left[\frac{1}{n+1} + \frac{1}{n-1} - \frac{1}{n+3} - \frac{1}{n-3} \right] \\
&= \frac{4n}{\pi} \left[\frac{1}{n^2-1} - \frac{1}{n^2-9} \right] = \frac{-32n}{\pi(n^2-1)(n^2-9)}.
\end{aligned}
$$

We furthermore have

$$
\alpha_1 = \frac{1}{\pi} \int_0^\pi \left[\sin(2x) - \sin(4x) + \sin(2x) \right] dx = 0
$$

$$
\alpha_3 = \frac{1}{\pi} \int_0^\pi \left[\sin(4x) + \sin(2x) - \sin(6x) \right] dx = 0.
$$

The solution is therefore given by

$$
u(x,t) = \frac{-32}{\pi} \sum_{n \text{ even}} \frac{n}{(n^2-1)(n^2-9)} \sin(nx) e^{-n^2 t}.
$$

Exercise 18.2 *Step 1 (Separation of variables).* We are looking for solutions to

$$
\begin{cases}
u_t = u_{xx} & \text{if } x \in (0,\pi), \ t > 0 \\
u_x(0,t) = u_x(\pi,t) = 0 & \text{if } t > 0
\end{cases}
\tag{18.1}
$$

of the form $u(x,t) = v(x)w(t)$. We therefore have

$$
\begin{cases}
v(x)w'(t) = v''(x)w(t) & \Leftrightarrow \ \dfrac{v''(x)}{v(x)} = -\lambda = \dfrac{w'(t)}{w(t)} \\
u_x(0,t) = v'(0)w(t) = u_x(\pi,t) = v'(\pi)w(t) = 0
\end{cases}
$$

this means that we have to solve

$$
\begin{cases}
v''(x) + \lambda v(x) = 0 \\
v'(0) = v'(\pi) = 0
\end{cases}
\tag{18.2}
$$

and

$$
w'(t) + \lambda w(t) = 0.
\tag{18.3}
$$

The non-trivial solutions to (18.2) are given (cf. Example 17.4) by $\lambda = n^2$ with $n = 0, 1, 2, \cdots$ and

$$
v_n(x) = \begin{cases}
a_n \cos(nx) & \text{if } n = 1, 2, \cdots \\
a_0/2 & \text{if } n = 0.
\end{cases}
$$

The corresponding w (i.e. the solutions to (18.3)) are given by $w_n(t) = e^{-n^2 t}$. Combining these two results, we find that the general solution to (18.1) is given by

$$u(x,t) = \frac{a_0}{2} + \sum_{n=1}^{\infty} a_n \cos(nx) e^{-n^2 t}.$$

Step 2 (Initial condition). We also want

$$u(x,0) = \cos(2x) = \frac{a_0}{2} + \sum_{n=1}^{\infty} a_n \cos(nx).$$

This implies that $a_n = 0 \; \forall n \neq 2$ and $a_2 = 1$. The solution is thus given by

$$u(x,t) = \cos(2x) e^{-4t}.$$

Exercise 18.3 *Step 1 (Separation of variables).* We are looking for solutions of the form

$$u(x,t) = v(x) w(t)$$

which satisfy

$$\begin{cases} (2+t)\, u_t = u_{xx} & \text{if } x \in (0,\pi), \; t > 0 \\ u(0,t) = u(\pi,t) = 0 & \text{if } t > 0. \end{cases}$$

We therefore get

$$\begin{cases} (2+t)\dfrac{w'(t)}{w(t)} = -\lambda = \dfrac{v''(x)}{v(x)} \\ v(0) = v(\pi) = 0. \end{cases}$$

We, hence, need to solve the two systems

$$\begin{cases} v''(x) + \lambda v(x) = 0 \\ v(0) = v(\pi) = 0 \end{cases} \quad \text{and} \quad \frac{w'(t)}{w(t)} = -\frac{\lambda}{(2+t)}.$$

The first system has non-trivial solutions only if $\lambda = n^2$ and which are then given by

$$v(x) = a_n \sin(nx).$$

A solution to the second equation is, when $\lambda = n^2$,

$$w(t) = (2+t)^{-n^2}.$$

The general solution is thus

$$u(x,t) = \sum_{n=1}^{\infty} a_n \sin(nx)(2+t)^{-n^2}.$$

Step 2 (Initial condition). We also want that

$$u(x,0) = \sum_{n=1}^{\infty} a_n 2^{-n^2} \sin(nx) = \sin x + 4\sin(2x).$$

This implies that $a_1 = 2$, $a_2 = 2^6$ and $a_n = 0$ if $n \neq 1,2$. The solution to our problem is therefore given by

$$u(x,t) = \frac{2\sin x}{2+t} + \frac{2^6 \sin(2x)}{(2+t)^4}.$$

Exercise 18.4 *Step 1 (Separation of variables).* We are looking for solutions to

$$\begin{cases} u_{tt} = u_{xx} & \text{if } x \in (0,1), \ t > 0 \\ u_x(0,t) = u_x(1,t) = 0 & \text{if } t > 0 \end{cases}$$

of the form $u(x,t) = v(x)w(t)$. We find

$$\begin{cases} \dfrac{w''(t)}{w(t)} = -\lambda = \dfrac{v''(x)}{v(x)} \\ v'(0)w(t) = v'(1)w(t) = 0. \end{cases}$$

The two systems that we have to solve, are thus

$$\begin{cases} v''(x) + \lambda v(x) = 0 \\ v'(0) = v'(1) = 0 \end{cases} \tag{18.4}$$

and

$$w''(t) + \lambda w(t) = 0. \tag{18.5}$$

The non-trivial solutions (cf. Example 17.4) to (18.4) are given by $\lambda = (n\pi)^2$ where $n = 0,1,2,\cdots$ and $v_n = \cos(n\pi x)$. The solutions to (18.5), when $\lambda = (n\pi)^2$, are given by

$$w_n(t) = \begin{cases} \dfrac{a_0}{2} + \dfrac{b_0}{2}t & \text{if } n = 0 \\ a_n \cos(n\pi t) + b_n \sin(n\pi t) & \text{if } n \geq 1. \end{cases}$$

The general solution therefore is

$$u(x,t) = \frac{a_0}{2} + \frac{b_0}{2}t + \sum_{n=1}^{\infty} [a_n \cos(n\pi t) + b_n \sin(n\pi t)] \cos(n\pi x).$$

Step 2 (Initial conditions). Since $u(x,0) = 0$, we find that $a_n = 0$ $\forall n \geq 0$. Since

$$u_t(x,t) = \frac{b_0}{2} + \sum_{n=1}^{\infty} \left[-n\pi a_n \sin(n\pi t) + n\pi b_n \cos(n\pi t)\right] \cos(n\pi x)$$

we get

$$u_t(x,0) = \frac{b_0}{2} + \sum_{n=1}^{\infty} n\pi b_n \cos(n\pi x) = 1 + \cos^2(\pi x)$$

$$= 1 + \left(\frac{1}{2} + \frac{1}{2}\cos(2\pi x)\right) = \frac{3}{2} + \frac{1}{2}\cos(2\pi x).$$

This implies

$$b_0 = 3, \quad b_1 = 0, \quad 2\pi b_2 = \frac{1}{2} \quad \text{and} \quad b_n = 0 \,\forall n \geq 3.$$

The solution to the problem is finally given by

$$u(x,t) = \frac{3}{2}t + \frac{1}{4\pi}\sin(2\pi t)\cos(2\pi x).$$

Exercise 18.5 *Step 1 (Separation of variables).* We are looking for solutions to the system

$$\begin{cases} u_t + u_{xxxx} = 0 & \text{if } x \in (0,\pi), \ t > 0 \\ u(0,t) = u(\pi,t) = 0 & \text{if } t > 0 \\ u_{xx}(0,t) = u_{xx}(\pi,t) = 0 & \text{if } t > 0 \end{cases}$$

which are of the form

$$u(x,t) = v(x)w(t).$$

Since

$$u(0,t) = u(\pi,t) = u_{xx}(0,t) = u_{xx}(\pi,t) = 0,$$

we obtain

$$v(0) = v(\pi) = v''(0) = v''(\pi) = 0.$$

Since

$$u_t + u_{xxxx} = 0 \implies v(x)w'(t) + w(t)v^{(4)}(x) = 0,$$

we deduce that

$$\frac{v^{(4)}}{v} = \lambda = -\frac{w'(t)}{w(t)}.$$

We therefore have to solve

$$\begin{cases} v^{(4)}(x) - \lambda v(x) = 0 \\ v(0) = v(\pi) = v''(0) = v''(\pi) = 0 \end{cases} \quad \text{and} \quad w'(t) + \lambda w(t) = 0.$$

We easily see that the first problem has a non-trivial solution only if $\lambda = n^4$. This solution is then given by $v(x) = \alpha_n \sin(nx)$. The second equation has thus $w(t) = e^{-n^4 t}$ as a solution. The general solution therefore is

$$u(x,t) = \sum_{n=1}^{\infty} \alpha_n \sin(nx) e^{-n^4 t}.$$

Step 2 (Initial condition). The general solution should also verify

$$u(x,0) = 2\sin x + \sin(3x) = \sum_{n=1}^{\infty} \alpha_n \sin(nx)$$

which implies $\alpha_1 = 2$, $\alpha_3 = 1$ and $\alpha_n = 0$ otherwise. The required solution therefore is

$$u(x,t) = 2\sin x\, e^{-t} + \sin(3x)\, e^{-81t}.$$

Exercise 18.6 *Step 1 (Separation of variables).* We first solve the system

$$\begin{cases} u_t = u_{xx} + 2u_x + 2u & \text{if } x \in (0,\pi),\ t > 0 \\ u(0,t) = u(\pi,t) = 0 & \text{if } t > 0. \end{cases}$$

We are looking for a solution of the form $u(x,t) = v(x)w(t)$. We, hence, have

$$u(0,t) = v(0)w(t) = u(\pi,t) = v(\pi)w(t) = 0 \quad \forall t.$$

We thus deduce that $v(0) = v(\pi) = 0$. The equation then becomes

$$v(x)w'(t) = (v''(x) + 2v'(x) + 2v(x))w(t).$$

Separating the variables, we obtain

$$\frac{v'' + 2v' + 2v}{v} = -\lambda = \frac{w'}{w}$$

and thus

$$\begin{cases} v''(x) + 2v'(x) + (2+\lambda)v(x) = 0 \\ v(0) = v(\pi) = 0 \end{cases} \quad \text{and} \quad w'(t) + \lambda w(t) = 0.$$

In Exercise 17.3 (with $\mu = 1 + \lambda$) we found that the first problem has a non-trivial solution only if $\lambda = n^2 - 1$, which is then given by

$$v_n(x) = \alpha_n e^{-x} \sin(nx).$$

Returning to the second equation, we find $w_n(t) = e^{-(n^2-1)t}$. The general solution then is

$$u(x,t) = e^{-x} \sum_{n=1}^{\infty} \alpha_n \sin(nx) e^{-(n^2-1)t}.$$

Step 2 (Initial condition). We have to choose the α_n in order that

$$u(x,0) = e^{-x} \sin(2x) = \sum_{n=1}^{\infty} \alpha_n e^{-x} \sin(nx).$$

We immediately see that $\alpha_2 = 1$ and $\alpha_n = 0 \; \forall n \neq 2$. The solution therefore is

$$u(x,t) = e^{-3t-x} \sin(2x).$$

Exercise 18.7 *Step 1 (Separation of variables).* We first look for solutions to

$$\begin{cases} \Delta u = 0 & \text{if } x, y \in (0, \pi) \\ u_y(x,0) = u_y(x,\pi) = 0 & \text{if } x \in (0, \pi) \end{cases}$$

which are of the form $u(x,y) = v(x)w(y)$. We have

$$\begin{cases} v''(x)w(y) + v(x)w''(y) = 0 \Leftrightarrow -\dfrac{v''(x)}{v(x)} = -\lambda = \dfrac{w''(y)}{w(y)} \\ u_y(x,0) = v(x)w'(0) = u_y(x,\pi) = v(x)w'(\pi) = 0 \end{cases}$$

that is

$$\begin{cases} w''(y) + \lambda w(y) = 0 \\ w'(0) = w'(\pi) = 0 \end{cases} \tag{18.6}$$

and

$$v''(x) - \lambda v(x) = 0. \tag{18.7}$$

The non-trivial solutions to (18.6) are given by (cf. Example 17.4) $\lambda = n^2$ with $n = 0, 1, 2, \cdots$ and

$$w_n(y) = \begin{cases} a_n \cos(ny) & \text{if } n = 1, 2, \cdots \\ a_0/2 & \text{if } n = 0. \end{cases}$$

The equation (18.7) becomes

$$v_n''(x) - n^2 v_n(x) = 0 \implies v_n(x) = \begin{cases} b_1 x + b_0 & \text{if } n = 0 \\ b_n \cosh(nx) + c_n \sinh(nx) & \text{if } n \neq 0. \end{cases}$$

Therefore, the required solution to the given equation is

$$u(x,y) = \frac{a_0}{2}(b_1 x + b_0) + \sum_{n=1}^{\infty} a_n (b_n \cosh(nx) + c_n \sinh(nx)) \cos(ny).$$

Writing

$$\frac{a_0}{2} b_1 = \alpha, \quad \frac{a_0}{2} b_0 = \frac{\beta}{2}, \quad a_n b_n = A_n \quad \text{and} \quad a_n c_n = B_n$$

we get

$$u(x,y) = \alpha x + \frac{\beta}{2} + \sum_{n=1}^{\infty} (A_n \cosh(nx) + B_n \sinh(nx)) \cos(ny).$$

Step 2 (Boundary conditions). We should also have

$$u(0,y) = \cos(2y) \quad \text{and} \quad u(\pi, y) = 0$$

that is

$$\frac{\beta}{2} + \sum_{n=1}^{\infty} A_n \cos(ny) = \cos(2y).$$

$$\alpha \pi + \frac{\beta}{2} + \sum_{n=1}^{\infty} (A_n \cosh(n\pi) + B_n \sinh(n\pi)) \cos(ny) = 0.$$

The first equation gives immediately $\beta = 0$ and $A_n = 0$ if $n \neq 2$ and $A_2 = 1$. Substituting these coefficients into the second equation, we find $\alpha = 0$ and

$$A_n \cosh(n\pi) + B_n \sinh(n\pi) = 0.$$

We therefore deduce that if $n \neq 2$, then $A_n = B_n = 0$. On the other hand, if $n = 2$, we have

$$B_2 = -\frac{\cosh(2\pi)}{\sinh(2\pi)}.$$

The solution to the problem is therefore given by

$$u(x,y) = \left[\cosh(2x) - \frac{\cosh(2\pi)}{\sinh(2\pi)} \sinh(2x) \right] \cos(2y).$$

Exercise 18.8 *Step 1 (Separation of variables).* The present exercise corresponds to Example 18.5 with

$$\Omega = \{(x,y) \in \mathbb{R}^2 : x^2 + y^2 < 4\}$$

and $\varphi(x,y) = 1 + x^2 + y^3$. We saw (cf. Example 18.5) that the general solution, in polar coordinates, is

$$v(r,\theta) = \frac{a_0}{2} + \sum_{n=1}^{\infty}(a_n \cos(n\theta) + b_n \sin(n\theta))r^n.$$

Step 2 (Boundary conditions). Since $2\cos^2\theta = 1 + \cos(2\theta)$ and

$$\begin{aligned} 4\sin^3\theta &= 4\sin\theta(1 - \cos^2\theta) = 4\sin\theta - 4\sin\theta\cos^2\theta \\ &= 4\sin\theta - 2\sin\theta(\cos(2\theta) + 1) = 2\sin\theta - (-\sin\theta + \sin(3\theta)) \\ &= 3\sin\theta - \sin(3\theta), \end{aligned}$$

we find

$$\begin{aligned} v(2,\theta) = \Psi(\theta) &= 1 + 4\cos^2\theta + 8\sin^3\theta \\ &= 3 + 2\cos(2\theta) + 6\sin\theta - 2\sin(3\theta). \end{aligned}$$

The coefficients are then given by

$$\frac{a_0}{2} = 3, \quad a_2 = \frac{1}{2}, \quad a_n = 0 \quad \text{if } n \neq 0, 2$$

$$b_1 = 3, \quad b_3 = -\frac{1}{4}, \quad b_n = 0 \quad \text{if } n \neq 1, 3.$$

We can therefore write the solution in polar coordinates, namely

$$v(r,\theta) = 3 + 3r\sin\theta + \frac{r^2}{2}\cos(2\theta) - \frac{r^3}{4}\sin(3\theta).$$

Since $\sin(3\theta) = 3\sin\theta - 4\sin^3\theta$, we find

$$v(r,\theta) = 3 + 3r\sin\theta + \frac{r^2}{2}(\cos^2\theta - \sin^2\theta) - \frac{3}{4}r^3\sin\theta + r^3\sin^3\theta.$$

The solution to the problem is finally given by

$$u(x,y) = 3 + 3y + \frac{x^2 - y^2}{2} - \frac{3}{4}y(x^2 + y^2) + y^3.$$

Exercise 18.9 *Step 1 (Polar coordinates).* We recall that

$$v_r = \cos\theta\, u_x + \sin\theta\, u_y = \frac{x}{r}\, u_x + \frac{y}{r}\, u_y.$$

Observe that, if $x^2 + y^2 = 1$, we have

$$xu_x + yu_y = v_r(1,\theta) = \cos^2\theta - \frac{1}{2} = \frac{\cos(2\theta)}{2}.$$

The problem then becomes

$$\begin{cases} v_{rr} + \dfrac{1}{r} v_r + \dfrac{1}{r^2} v_{\theta\theta} = 0 & \text{if } r \in (0,1),\ \theta \in (0,2\pi) \\ v_r(1,\theta) = \frac{\cos(2\theta)}{2} & \text{if } \theta \in (0,2\pi). \end{cases}$$

Step 2 (Separation of variables). We proceed as in Example 18.5 and we find that the general solution is given by

$$v(r,\theta) = \frac{a_0}{2} + \sum_{n=1}^{\infty} (a_n \cos(n\theta) + b_n \sin(n\theta)) r^n.$$

Step 3 (Boundary conditions). Recall that the boundary condition is

$$v_r(1,\theta) = \frac{\cos(2\theta)}{2}.$$

Differentiating the general solution v with respect to r leads to

$$v_r(r,\theta) = \sum_{n=1}^{\infty} n(a_n \cos(n\theta) + b_n \sin(n\theta)) r^{n-1}$$

$$v_r(1,\theta) = \frac{\cos(2\theta)}{2} = \sum_{n=1}^{\infty} n(a_n \cos(n\theta) + b_n \sin(n\theta)).$$

We thus infer that a_0 can be arbitrary and

$$b_n = 0\ \forall n, \quad a_n = 0\ \forall n \neq 2 \quad \text{and} \quad a_2 = \frac{1}{4}$$

which implies that

$$v(r,\theta) = \frac{a_0}{2} + \frac{r^2}{4}\cos(2\theta) = \frac{a_0}{2} + \frac{r^2 \cos^2\theta - r^2 \sin^2\theta}{4}.$$

The required solution, finally, is

$$u(x,y) = \frac{a_0}{2} + \frac{x^2 - y^2}{4}.$$

Exercise 18.10 *Step 1 (Polar coordinates).* We let $x = r\cos\theta$, $y = r\sin\theta$ and $v(r,\theta) = u(r\cos\theta, r\sin\theta)$. The problem then becomes

$$\begin{cases} v_{rr} + \dfrac{1}{r}v_r + \dfrac{1}{r^2}v_{\theta\theta} = 0 & \text{if } r \in (1,2),\ \theta \in (0, 2\pi) \\ v(1, \theta) = \cos\theta & \text{if } \theta \in (0, 2\pi) \\ v(2, \theta) = 2\cos\theta - 2\sin\theta & \text{if } \theta \in (0, 2\pi). \end{cases}$$

Step 2 (Separation of variables). As in Example 18.5 we find that the general solution is

$$v(r, \theta) = \frac{a_0}{2} + b_0 \log r + \sum_{n=1}^{\infty} (a_n \cos(n\theta) + b_n \sin(n\theta)) r^n$$

$$+ \sum_{n=1}^{\infty} (a_{-n} \cos(n\theta) + b_{-n} \sin(n\theta)) r^{-n}.$$

Step 3 (Boundary conditions). We want the above solution to satisfy

$$v(1, \theta) = \cos\theta$$

$$= \frac{a_0}{2} + b_0 \log 1 + \sum_{n=1}^{\infty} [(a_n + a_{-n}) \cos(n\theta) + (b_n + b_{-n}) \sin(n\theta)]$$

and

$$v(2, \theta) = 2\cos\theta - 2\sin\theta = \frac{a_0}{2} + b_0 \log 2$$

$$+ \sum_{n=1}^{\infty} \left[\left(2^n a_n + \frac{a_{-n}}{2^n} \right) \cos(n\theta) + \left(2^n b_n + \frac{b_{-n}}{2^n} \right) \sin(n\theta) \right].$$

We deduce

$$\begin{cases} a_0 = 0 \\ a_1 + a_{-1} = 1 \\ a_n + a_{-n} = 0\ \forall n \geq 2 \\ b_n + b_{-n} = 0\ \forall n \geq 1 \end{cases} \quad \text{and} \quad \begin{cases} \dfrac{a_0}{2} + b_0 \log 2 = 0 \\ 2a_1 + \dfrac{a_{-1}}{2} = 2 \\ 2^n a_n + \dfrac{a_{-n}}{2^n} = 0\ \forall n \geq 2 \\ 2b_1 + \dfrac{b_{-1}}{2} = -2 \\ 2^n b_n + \dfrac{b_{-n}}{2^n} = 0\ \forall n \geq 2 \end{cases}$$

and thus

$$\begin{cases} a_0 = b_0 = 0 \\ a_1 = 1 \text{ and } a_{-1} = 0 \\ b_{-1} = -b_1 = 4/3 \\ a_n = a_{-n} = b_n = b_{-n} = 0\ \forall n \geq 2. \end{cases}$$

The solution, written in polar coordinates, therefore is

$$v\left(r,\theta\right) = r\cos\theta + \left(-\frac{4}{3}r + \frac{4}{3r}\right)\sin\theta.$$

The solution is, hence, in Cartesian coordinates, given by

$$u\left(x,y\right) = x - \frac{4}{3}y + \frac{4y}{3(x^2 + y^2)}.$$

Exercise 18.11 *Step 1 (Polar coordinates).* Since

$$v_r = \cos\theta\, u_x + \sin\theta\, u_y = \frac{x}{r}\, u_x + \frac{y}{r}\, u_y\,,$$

we find that the given problem becomes

$$\begin{cases} v_{rr} + \dfrac{1}{r}v_r + \dfrac{1}{r^2}v_{\theta\theta} = 0 & \text{if } r \in (1,2),\ \theta \in (0,2\pi) \\ v\left(1,\theta\right) = 0 & \text{if } \theta \in (0,2\pi) \\ v_r(2,\theta) + v(2,\theta) = \frac{1}{2} + \log 2 + 2\cos\theta - \sin\theta & \text{if } \theta \in (0,2\pi). \end{cases}$$

Step 2 (Separation of variables). Proceeding as in Example 18.5, we find that the general solution is

$$\begin{aligned} v\left(r,\theta\right) = \ & \frac{a_0}{2} + b_0 \log r + \sum_{n=1}^{\infty}\left(a_n \cos\left(n\theta\right) + b_n \sin\left(n\theta\right)\right)r^n \\ & + \sum_{n=1}^{\infty}\left(a_{-n} \cos\left(n\theta\right) + b_{-n} \sin\left(n\theta\right)\right)r^{-n}. \end{aligned}$$

Step 3 (Boundary conditions). We then find the coefficients so that

$$v\left(1,\theta\right) = 0 = \frac{a_0}{2} + \sum_{n=1}^{\infty}\left[\left(a_n + a_{-n}\right)\cos\left(n\theta\right) + \left(b_n + b_{-n}\right)\sin\left(n\theta\right)\right]$$

which leads to

$$a_0 = 0,\quad a_n = -a_{-n} \quad \text{and} \quad b_n = -b_{-n}\,.$$

Returning to v, we have

$$v\left(r,\theta\right) = b_0 \log r + \sum_{n=1}^{\infty}\left[\left(a_n \cos\left(n\theta\right) + b_n \sin\left(n\theta\right)\right)\left(r^n - r^{-n}\right)\right].$$

and

$$v_r\left(r,\theta\right) = \frac{b_0}{r} + \sum_{n=1}^{\infty}\left[n(a_n\cos\left(n\theta\right) + b_n\sin\left(n\theta\right))\left(r^{n-1} + r^{-n-1}\right)\right].$$

We, hence, obtain

$$v_r(2,\theta) + v(2,\theta)$$

$$= \left(\frac{1}{2} + \log 2\right) + 2\cos\theta - \sin\theta = b_0\left(\frac{1}{2} + \log 2\right)$$

$$+ \sum_{n=1}^{\infty}\left(n2^{n-1} + n2^{-n-1} + 2^n - 2^{-n}\right)\left(a_n\cos\left(n\theta\right) + b_n\sin\left(n\theta\right)\right).$$

We thus find

$$b_0 = 1, \quad a_1 = \frac{8}{11}, \quad b_1 = -\frac{4}{11}, \quad a_n = b_n = 0 \;\forall n \neq 0,1$$

and the solution in polar coordinates is

$$v\left(r,\theta\right) = \log r + \frac{8}{11}r\cos\theta - \frac{4}{11}r\sin\theta - \frac{8}{11}\frac{\cos\theta}{r} + \frac{4}{11}\frac{\sin\theta}{r}.$$

We finally find that the solution, in Cartesian coordinates, has the following form

$$u\left(x,y\right) = \log\sqrt{x^2 + y^2} + \frac{8}{11}x - \frac{4}{11}y - \frac{8}{11}\frac{x}{x^2 + y^2} + \frac{4}{11}\frac{y}{x^2 + y^2}.$$

Exercise 18.12 *Step 1 (Polar coordinates).* We let $x = r\cos\theta$, $y = r\sin\theta$ and $v\left(r,\theta\right) = u\left(r\cos\theta, r\sin\theta\right)$, so that the problem becomes

$$\begin{cases} v_{rr} + \dfrac{1}{r}v_r + \dfrac{1}{r^2}v_{\theta\theta} = 0 & \text{if } r \in (0,1),\ \theta \in (0,\pi) \\ v(r,0) = v(r,\pi) = 0 & \text{if } r \in (0,1) \\ v\left(1,\theta\right) = \sin^2\theta = \dfrac{1}{2} - \dfrac{1}{2}\cos\left(2\theta\right) & \text{if } \theta \in (0,\pi). \end{cases}$$

Step 2 (Separation of variables). We first consider the system

$$\begin{cases} v_{rr} + \dfrac{1}{r}v_r + \dfrac{1}{r^2}v_{\theta\theta} = 0 & \text{if } r \in (0,1),\ \theta \in (0,\pi) \\ v(r,0) = v(r,\pi) = 0 & \text{if } r \in (0,1). \end{cases}$$

We are looking for a solution of the form $v\left(r,\theta\right) = f\left(r\right)g\left(\theta\right)$. The equation then becomes

$$f''g + \frac{1}{r}f'g + \frac{1}{r^2}fg'' = 0 \;\Rightarrow\; \frac{r^2 f'' + rf'}{f} = \lambda = -\frac{g''}{g}.$$

We, hence, get

$$\begin{cases} g''(\theta) + \lambda g(\theta) = 0 \\ g(0) = g(\pi) = 0 \end{cases} \quad \text{and} \quad r^2 f''(r) + r f'(r) - \lambda f(r) = 0.$$

The non-trivial solutions to the first system are given by $\lambda = n^2$ and

$$g_n(\theta) = \alpha_n \sin(n\theta).$$

The solutions to the second equations are, for $n > 0$,

$$f_n(r) = \beta_n r^n.$$

(Note that there also exist solutions of the form r^{-n} but, as $(0,0) \in \overline{\Omega}$, these solutions are singular and we therefore ignore them.) The general solution is (writing $a_n = \alpha_n \beta_n$)

$$v(r,\theta) = \sum_{n=1}^{\infty} a_n r^n \sin(n\theta).$$

Step 3 (Boundary conditions). We still have to determine the Fourier coefficients of the function $\sin^2 \theta$. When $n = 2$ we find

$$a_2 = \frac{2}{\pi} \int_0^{\pi} \sin^2 \theta \sin(2\theta) \, d\theta = 0.$$

If $n \neq 2$, we find

$$a_n = \frac{2}{\pi} \int_0^{\pi} \sin^2 \theta \sin(n\theta) \, d\theta$$

$$= \frac{2}{\pi} \left[-\frac{\cos(n\theta)}{2n} + \frac{\cos((n+2)\theta)}{4(n+2)} + \frac{\cos((n-2)\theta)}{4(n-2)} \right]_0^{\pi}$$

and thus

$$a_n = \frac{1}{2\pi} \left[2 \frac{1 - \cos(n\pi)}{n} + \frac{\cos(n\pi) - 1}{n-2} + \frac{\cos(n\pi) - 1}{n+2} \right]$$

$$= \frac{4}{\pi} \frac{\cos(n\pi) - 1}{n(n^2 - 4)}.$$

Therefore, if n is even, we have $a_n = 0$, while if n is odd, we get

$$a_n = -\frac{8}{\pi n(n^2 - 4)}.$$

We can finally write the solution in polar coordinates, namely

$$v\left(r,\theta\right) = -\frac{8}{\pi}\sum_{\substack{n=1 \\ n\ odd}}^{\infty}\frac{r^{n}\sin\left(n\theta\right)}{n\left(n^{2}-4\right)}.$$

Exercise 18.13 *Step 1 (Fourier transform).* We denote by $v\left(\alpha,y\right)$ the Fourier transform (in x) of $u\left(x,y\right)$, i.e.

$$v\left(\alpha,y\right) = \mathfrak{F}(u)\left(\alpha,y\right) = \frac{1}{\sqrt{2\pi}}\int_{-\infty}^{+\infty}u\left(x,y\right)e^{-i\alpha x}\,dx.$$

(By abuse of notations, we denote this transformation in x by $\mathfrak{F}(u)$.) We find (cf. Table of Fourier transforms), for $f\left(x\right) = e^{-x^{2}}$,

$$v(\alpha,0) = \mathfrak{F}\left(f\right)\left(\alpha\right) = \widehat{f}\left(\alpha\right) = \frac{1}{\sqrt{2}}e^{-\frac{\alpha^{2}}{4}}.$$

Observe that

$$\begin{cases} \mathfrak{F}(u_{xx})\left(\alpha,y\right) = (i\alpha)^{2}\mathfrak{F}(u)\left(\alpha,y\right) = -\alpha^{2}v\left(\alpha,y\right) \\ \mathfrak{F}\left(u_{yy}\right)\left(\alpha,y\right) = v_{yy}\left(\alpha,y\right). \end{cases}$$

Returning to the problem and applying Fourier transform (in x) to both sides of the equations, we get

$$\begin{cases} v_{yy}\left(\alpha,y\right) - \alpha^{2}v\left(\alpha,y\right) = 0 & \text{if } \alpha \in \mathbb{R},\ y > 0 \\ v(\alpha,0) = \widehat{f}\left(\alpha\right) & \text{if } \alpha \in \mathbb{R} \\ v\left(\alpha,y\right) \to 0 & \text{if } \alpha \in \mathbb{R},\ y \to +\infty. \end{cases}$$

Considering α as a parameter and using the suggestion, we obtain

$$v\left(\alpha,y\right) = \widehat{f}\left(\alpha\right)e^{-|\alpha|y}$$

and thus

$$v\left(\alpha,y\right) = \frac{1}{\sqrt{2}}e^{-\frac{\alpha^{2}}{4}-|\alpha|y}.$$

Step 2 (Final solution). We find the solution applying the inverse Fourier transform, i.e.

$$\begin{aligned} u\left(x,y\right) &= \frac{1}{\sqrt{2\pi}}\int_{-\infty}^{+\infty}\frac{1}{\sqrt{2}}e^{-\frac{\alpha^{2}}{4}-|\alpha|y}e^{i\alpha x}\,d\alpha \\ &= \frac{1}{2\sqrt{\pi}}\int_{-\infty}^{+\infty}e^{-\frac{\alpha^{2}}{4}-|\alpha|y}\cos\left(\alpha x\right)\,d\alpha \\ &\quad + i\frac{1}{2\sqrt{\pi}}\int_{-\infty}^{+\infty}e^{-\frac{\alpha^{2}}{4}-|\alpha|y}\sin\left(\alpha x\right)\,d\alpha. \end{aligned}$$

Since $\alpha \to e^{-\frac{\alpha^2}{4} - |\alpha| y} \sin(\alpha x)$ is odd and

$$\alpha \to e^{-\frac{\alpha^2}{4} - |\alpha| y} \cos(\alpha x)$$

is even, we deduce that

$$u(x, y) = \frac{1}{\sqrt{\pi}} \int_0^{+\infty} e^{-\frac{\alpha^2}{4} - \alpha y} \cos(\alpha x) \, d\alpha.$$

Exercise 18.14 *Step 1 (Conformal mapping).* We are looking for a conformal mapping

$$f : \{z \in \mathbb{C} : \operatorname{Re} z > 0\} \to \{\zeta \in \mathbb{C} : |\zeta| < 1\}.$$

We easily find that such a mapping and its inverse are given by

$$f(z) = \frac{z - 1}{z + 1} \quad \text{and} \quad f^{-1}(\zeta) = \frac{\zeta + 1}{1 - \zeta}.$$

If $z = x + iy$ and $\zeta = \alpha + i\beta$, we, hence, have

$$f(x + iy) = \frac{x + iy - 1}{x + 1 + iy} = \frac{x^2 + y^2 - 1}{(x + 1)^2 + y^2} + i \frac{2y}{(x + 1)^2 + y^2}$$

$$f^{-1}(\alpha + i\beta) = \frac{\alpha + 1 + i\beta}{1 - \alpha - i\beta} = \frac{1 - \alpha^2 - \beta^2}{(\alpha - 1)^2 + \beta^2} + i \frac{2\beta}{(\alpha - 1)^2 + \beta^2}.$$

The boundary condition becomes (recall that $\alpha^2 + \beta^2 = 1$)

$$\psi(\alpha, \beta) = \varphi\left(f^{-1}(\alpha, \beta)\right) = \varphi\left(\frac{1 - \alpha^2 - \beta^2}{(\alpha - 1)^2 + \beta^2}, \frac{2\beta}{(\alpha - 1)^2 + \beta^2}\right)$$

$$= \varphi\left(0, \frac{\beta}{1 - \alpha}\right) = \left(1 + \frac{\beta^2}{(1 - \alpha)^2}\right)^{-2}$$

and thus

$$\psi(\alpha, \beta) = \frac{(1 - \alpha)^4}{\left((1 - \alpha)^2 + \beta^2\right)^2} = \frac{(1 - \alpha)^4}{\left(1 + \alpha^2 + \beta^2 - 2\alpha\right)^2} = \frac{1}{4}(1 - \alpha)^2.$$

We therefore have to find a solution to

$$\begin{cases} \Delta v = 0 & \text{if } \alpha^2 + \beta^2 < 1 \\ v(\alpha, \beta) = \frac{1}{4}(1 - \alpha)^2 & \text{if } \alpha^2 + \beta^2 = 1. \end{cases}$$

Step 2 (Polar coordinates). We let $\alpha = r \cos \theta$, $\beta = r \sin \theta$ and $w(r, \theta) = v(r \cos \theta, r \sin \theta)$. The boundary condition becomes

$$w(1, \theta) = \frac{1}{4}(1 - \cos \theta)^2 = \frac{1}{4} + \frac{1}{4} \cos^2 \theta - \frac{1}{2} \cos \theta$$

$$= \frac{3}{8} + \frac{1}{8} \cos(2\theta) - \frac{1}{2} \cos \theta$$

and the problem we have to solve is

$$\begin{cases} w_{rr} + \dfrac{1}{r} w_r + \dfrac{1}{r^2} w_{\theta\theta} = 0 & \text{if } r \in (0, 1),\ \theta \in (0, 2\pi) \\ w(1, \theta) = \dfrac{3}{8} - \dfrac{1}{2} \cos \theta + \dfrac{1}{8} \cos(2\theta) & \text{if } \theta \in (0, 2\pi). \end{cases}$$

We next find (cf. Example 18.5) that the general solution is

$$w(r, \theta) = \frac{a_0}{2} + \sum_{n=1}^{\infty} (a_n \cos(n\theta) + b_n \sin(n\theta)) r^n.$$

The boundary condition

$$w(1, \theta) = \frac{3}{8} - \frac{1}{2} \cos \theta + \frac{1}{8} \cos(2\theta)$$

allows us to determine the coefficients

$$\frac{a_0}{2} = \frac{3}{8}, \quad a_1 = -\frac{1}{2}, \quad a_2 = \frac{1}{8}, \quad a_n = 0 \ \forall n \geq 3 \quad \text{and} \quad b_n = 0 \ \forall n.$$

The solution in polar coordinates is therefore given by

$$w(r, \theta) = \frac{3}{8} - \frac{r}{2} \cos \theta + \frac{r^2 \cos(2\theta)}{8} = \frac{3}{8} - \frac{r}{2} \cos \theta + \frac{r^2 \cos^2 \theta - r^2 \sin^2 \theta}{8},$$

while in Cartesian coordinates we have

$$v(\alpha, \beta) = \frac{3}{8} - \frac{\alpha}{2} + \frac{1}{8}(\alpha^2 - \beta^2).$$

Step 3 (Final solution). Returning to the given problem, we find that its solution is given by $u = v \circ f$, that is

$$u(x, y) = v\left(\frac{x^2 + y^2 - 1}{(x+1)^2 + y^2}, \frac{2y}{(x+1)^2 + y^2} \right)$$

$$= \frac{3}{8} - \frac{1}{2} \frac{x^2 + y^2 - 1}{(x+1)^2 + y^2} + \frac{1}{8} \frac{\left(x^2 + y^2 - 1\right)^2 - 4y^2}{\left((x+1)^2 + y^2\right)^2}.$$

Exercise 18.15 *Step 1 (Conformal mapping).* Let f be such that it maps Ω onto the unit disk D. Since the boundary of Ω is a straight line and the boundary of D is a circle, we can take f as a Möbius transformation. It is therefore sufficient to choose three points on $\partial\Omega$ and their images on ∂D to characterize such a transformation (up to inversion). For example, we can take

$$f(0) = -1, \quad f(1) = -i \quad \text{and} \quad f(-1) = i.$$

In this case we find that the map is given by

$$f(z) = \zeta = \alpha + i\beta = \frac{z - i}{z + i}$$

and

$$f^{-1}(\zeta) = z = x + iy = i\frac{1 + \zeta}{1 - \zeta}.$$

This leads to

$$\alpha = \frac{x^2 + y^2 - 1}{x^2 + (y + 1)^2} \qquad \beta = \frac{-2x}{x^2 + (y + 1)^2}$$

$$x = \frac{-2\beta}{(\alpha - 1)^2 + \beta^2} \qquad y = \frac{1 - \alpha^2 - \beta^2}{(\alpha - 1)^2 + \beta^2}.$$

The boundary condition then becomes (recall that on the boundary of D we have $\alpha^2 + \beta^2 = 1$)

$$\psi(\alpha, \beta) = \varphi\left(f^{-1}(\alpha, \beta)\right) = \varphi\left(\frac{-2\beta}{(\alpha - 1)^2 + \beta^2}, \frac{1 - \alpha^2 - \beta^2}{(\alpha - 1)^2 + \beta^2}\right)$$

$$= \varphi\left(\frac{\beta}{\alpha - 1}, 0\right) = \frac{8\dfrac{\beta^2}{(\alpha - 1)^2}}{\left[1 + \dfrac{\beta^2}{(\alpha - 1)^2}\right]^2}$$

and thus

$$\psi(\alpha, \beta) = \frac{8\beta^2 (\alpha - 1)^2}{\left[(\alpha - 1)^2 + \beta^2\right]^2} = 2\beta^2.$$

We, hence, have to solve

$$\begin{cases} \Delta v = 0 & \text{if } \alpha^2 + \beta^2 < 1 \\ v(\alpha, \beta) = 2\beta^2 & \text{if } \alpha^2 + \beta^2 = 1. \end{cases}$$

Step 2 (Polar coordinates). We let $\alpha = r\cos\theta$, $\beta = r\sin\theta$ and $w(r,\theta) = v(r\cos\theta, r\sin\theta)$. We then get

$$\begin{cases} w_{rr} + \dfrac{1}{r}w_r + \dfrac{1}{r^2}w_{\theta\theta} = 0 & \text{if } r \in (0,1),\ \theta \in (0,2\pi) \\ w(1,\theta) = 2\sin^2\theta = 1 - \cos(2\theta) & \text{if } \theta \in (0,2\pi). \end{cases}$$

Separating variables we find (cf. Example 18.5) that the general solution is

$$w(r,\theta) = \frac{a_0}{2} + \sum_{n=1}^{\infty}(a_n\cos(n\theta) + b_n\sin(n\theta))\,r^n.$$

Since $w(1,\theta) = 1 - \cos(2\theta)$, we deduce that all the coefficients are zero except $a_0 = 2$ and $a_2 = -1$. The solution in polar coordinates is therefore given by

$$w(r,\theta) = 1 - r^2\cos(2\theta) = 1 - r^2\cos^2\theta + r^2\sin^2\theta$$

while the one in Cartesian coordinates is

$$v(\alpha,\beta) = 1 - \alpha^2 + \beta^2.$$

Step 3 (Final solution). Since the solution of our problem is given by

$$u(x,y) = v(f(x,y)),$$

we deduce

$$\begin{aligned} u(x,y) &= 1 - \left(\frac{x^2 + y^2 - 1}{x^2 + (y+1)^2}\right)^2 + \left(\frac{-2x}{x^2 + (y+1)^2}\right)^2 \\ &= 4\,\frac{y(1+y)^2 + x^2(2+y)}{((1+y)^2 + x^2)^2}. \end{aligned}$$

Exercise 18.16 *Step 1 (Conformal mapping).* We easily find that the map

$$f(z) = \frac{2}{z - (2+2i)}$$

is such that

$$f : \Omega \to D = \{\zeta \in \mathbb{C} : |\zeta| < 1\}.$$

Indeed, since $\Omega = \{z \in \mathbb{C} : |z - (2+2i)| > 2\}$, we immediately have that

$$|\zeta| = |f(z)| = \frac{2}{|z - (2+2i)|} < 1.$$

We, hence, obtain

$$f(x,y) = \zeta = \alpha + i\beta = \frac{2}{x + iy - 2 - 2i} = \frac{2}{(x-2) + i(y-2)}$$

$$= \frac{2(x-2)}{(x-2)^2 + (y-2)^2} + i\frac{-2(y-2)}{(x-2)^2 + (y-2)^2}$$

and

$$f^{-1}(\alpha, \beta) = z = x + iy = 2 + 2i + \frac{2}{\zeta}$$

$$= 2\frac{\alpha + (\alpha^2 + \beta^2)}{\alpha^2 + \beta^2} + 2i\frac{(\alpha^2 + \beta^2) - \beta}{\alpha^2 + \beta^2}.$$

Since $\alpha^2 + \beta^2 = 1$ on ∂D, the boundary condition can be written as

$$\psi(\alpha, \beta) = \varphi\left(f^{-1}(\alpha, \beta)\right) = \varphi\left(2\frac{\alpha}{\alpha^2 + \beta^2} + 2, \ 2 - \frac{2\beta}{\alpha^2 + \beta^2}\right)$$

$$= \varphi(2\alpha + 2, 2 - 2\beta) = (2 + 2\alpha)^2 + 2(2 - 2\beta)^2$$

$$= 4\left[1 + 2\alpha + \alpha^2 + 2 - 4\beta + 2\beta^2\right] = 4\left[4 + 2\alpha - 4\beta + \beta^2\right].$$

We therefore have to solve

$$\begin{cases} \Delta v = 0 & \text{if } \alpha^2 + \beta^2 < 1 \\ v(\alpha, \beta) = 4\left[4 + 2\alpha - 4\beta + \beta^2\right] & \text{if } \alpha^2 + \beta^2 = 1. \end{cases}$$

Step 2 (Polar coordinates). We let $\alpha = r\cos\theta$, $\beta = r\sin\theta$ and $w(r, \theta) = v(r\cos\theta, r\sin\theta)$. The boundary condition can be written

$$w(1, \theta) = 4\left(4 + 2\cos\theta - 4\sin\theta + \sin^2\theta\right)$$

$$= 4\left(4 + 2\cos\theta - 4\sin\theta + \frac{1}{2} - \frac{1}{2}\cos(2\theta)\right)$$

$$= 18 + 8\cos\theta - 2\cos(2\theta) - 16\sin\theta$$

and the system becomes

$$\begin{cases} w_{rr} + \frac{1}{r}w_r + \frac{1}{r^2}w_{\theta\theta} = 0 & \text{if } r \in (0,1), \ \theta \in (0, 2\pi) \\ w(1, \theta) = 18 + 8\cos\theta - 2\cos(2\theta) - 16\sin\theta & \text{if } \theta \in (0, 2\pi). \end{cases}$$

Separating the variables we find (cf. Example 18.5) that the general solution is given by

$$w(r, \theta) = \frac{a_0}{2} + \sum_{n=1}^{\infty} (a_n \cos(n\theta) + b_n \sin(n\theta)) r^n.$$

Using the boundary condition $w(1, \theta)$ we deduce that all coefficients are zero except

$$\frac{a_0}{2} = 18, \quad a_1 = 8, \quad a_2 = -2 \quad \text{and} \quad b_1 = -16.$$

The solution in polar coordinates is therefore given by

$$w(r, \theta) = 18 + 8r \cos \theta - 16r \sin \theta - 2r^2 \cos(2\theta)$$
$$= 18 + 8r \cos \theta - 16r \sin \theta - 2r^2 \cos^2 \theta + 2r^2 \sin^2 \theta$$

and the one in Cartesian coordinates is

$$v(\alpha, \beta) = 18 + 8\alpha - 16\beta + 2\beta^2 - 2\alpha^2.$$

Step 3 (Final solution). Since the solution is given by

$$u(x, y) = v(f(x, y)),$$

we have

$$u(x, y) = 18 + \frac{16(x - 2)}{(x - 2)^2 + (y - 2)^2} + \frac{32(y - 2)}{(x - 2)^2 + (y - 2)^2}$$
$$+ \frac{8(y - 2)^2}{((x - 2)^2 + (y - 2)^2)^2} - \frac{8(x - 2)^2}{((x - 2)^2 + (y - 2)^2)^2}.$$

Exercise 18.17 Differentiating twice with respect to t gives

$$u_t = \frac{c}{2}[f'(x + ct) - f'(x - ct)] + \frac{1}{2}[g(x + ct) + g(x - ct)]$$
$$u_{tt} = \frac{c^2}{2}[f''(x + ct) + f''(x - ct)] + \frac{c}{2}[g'(x + ct) - g'(x - ct)].$$

A similar calculation leads to

$$u_{xx} = \frac{1}{2}[f''(x + ct) + f''(x - ct)] + \frac{1}{2c}[g'(x + ct) - g'(x - ct)]$$

i.e.

$$u_{tt} = c^2 u_{xx}.$$

We, moreover, have

$$u(x, 0) = \frac{1}{2}[f(x) + f(x)] = f(x)$$
$$u_t(x, 0) = \frac{1}{2}[g(x) + g(x)] = g(x).$$

Bibliography

[1] L. V. Ahlfors, *Complex analysis*, third edition, McGraw-Hill, Inc., 1979.

[2] K. Arbenz - A. Wohlhauser, *Compléments d'analyse*, PPUR, 1981.

[3] K. Arbenz - A. Wohlhauser, *Variables complexes*, PPUR, 1981.

[4] S. D. Chatterji, *Cours d'Analyse 1, analyse vectorielle*, PPUR, 1997.

[5] S. D. Chatterji, *Cours d'Analyse 2, analyse complexe*, PPUR, 1997.

[6] S. D. Chatterji, *Cours d'Analyse 3, équations différentielles ordinaires et aux dérivées partielles*, PPUR, 1998.

[7] W. Fleming, *Functions of several variables*, second edition, Springer-Verlag, 1977.

[8] N. Fusco - P. Marcellini - C. Sbordone, *Analisi matematica 2*, Liguori Editore, 1996.

[9] E. Giusti, *Analisi matematica 2*, Bollati Boringhieri, 1992.

[10] E. Kreyszig, *Advanced engineering mathematics*, sixth edition, Wiley, 1988.

[11] M. H. Protter - C. B. Morrey, *A first course in real analysis*, first edition, Springer-Verlag, 1977.

[12] C. S. Rees - S. M. Shah - Č. V. Stanojević, *Theory and applications of Fourier analysis*, Dekker, 1981.

[13] C. D. Sogge, *Fourier integrals in classical analysis*, Cambridge University Press, 1993.

[14] E. M. Stein - R. Shakarchi, *Fourier analysis*, Princeton University Press, 2003.

[15] E. M. Stein - R. Shakarchi, *Complex analysis*, Princeton University Press, 2003.

[16] D. V. Widder, *The Laplace transform*, Princeton University Press, 1946.

Table of Fourier Transform

	$f(y)$	$\mathfrak{F}(f)(\alpha) = \widehat{f}(\alpha)$
1	$f(y) = \begin{cases} 1 & \text{if } |y| < |b| \\ 0 & \text{otherwise} \end{cases}$	$\sqrt{\dfrac{2}{\pi}} \dfrac{\sin(|b|\,\alpha)}{\alpha}$
2	$f(y) = \begin{cases} 1 & \text{if } b < y < c \\ 0 & \text{otherwise} \end{cases}$	$\dfrac{e^{-i\alpha b} - e^{-i\alpha c}}{i\alpha\sqrt{2\pi}}$
3	$f(y) = \begin{cases} e^{-\omega y} & \text{if } y > 0 \\ 0 & \text{otherwise} \end{cases} \quad (\omega > 0)$	$\dfrac{1}{\sqrt{2\pi}} \dfrac{1}{(\omega + i\alpha)}$
4	$f(y) = \begin{cases} e^{-\omega y} & \text{if } b < y < c \\ 0 & \text{otherwise} \end{cases}$	$\dfrac{1}{\sqrt{2\pi}} \dfrac{e^{-(\omega+i\alpha)b} - e^{-(\omega+i\alpha)c}}{(\omega + i\alpha)}$
5	$f(y) = \begin{cases} e^{-i\omega y} & \text{if } b < y < c \\ 0 & \text{otherwise} \end{cases}$	$\dfrac{1}{i\sqrt{2\pi}} \dfrac{e^{-i(\omega+\alpha)b} - e^{-i(\omega+\alpha)c}}{\omega + \alpha}$
6	$\dfrac{1}{y^2 + \omega^2} \quad (\omega \neq 0)$	$\sqrt{\dfrac{\pi}{2}} \dfrac{e^{-|\omega\alpha|}}{|\omega|}$
7	$\dfrac{e^{-|\omega y|}}{|\omega|} \quad (\omega \neq 0)$	$\sqrt{\dfrac{2}{\pi}} \dfrac{1}{\omega^2 + \alpha^2}$
8	$e^{-\omega^2 y^2} \quad (\omega \neq 0)$	$\dfrac{1}{\sqrt{2}\,|\omega|} e^{-\alpha^2/4\omega^2}$
9	$ye^{-\omega^2 y^2} \quad (\omega \neq 0)$	$\dfrac{-i\alpha}{2\sqrt{2}|\omega|^3} e^{-\frac{\alpha^2}{4\omega^2}}$
10	$\dfrac{4y^2}{(\omega^2 + y^2)^2} \quad (\omega \neq 0)$	$\sqrt{2\pi}(\tfrac{1}{|\omega|} - |\alpha|)e^{-|\omega\alpha|}$

Table of Laplace Transform

	$f(t)$	$\mathfrak{L}(f)(z) = F(z)$	
1	$f_\alpha(t) = \begin{cases} 1/\alpha & \text{if } t \in [0, \alpha] \\ 0 & \text{otherwise} \end{cases}$ (Dirac mass)	$\dfrac{1 - e^{-\alpha z}}{\alpha z} \xrightarrow[\alpha \to 0]{} 1$	$\forall z$
2	$\varepsilon(t) = \begin{cases} 1 & \text{if } t \geq 0 \\ 0 & \text{if } t < 0 \end{cases}$	$\dfrac{1}{z}$	$\operatorname{Re} z > 0$
3	$e^{-\alpha t}$	$\dfrac{1}{z + \alpha}$	$\operatorname{Re} z > -\alpha$
4	$\dfrac{t^n}{n!}$	$\dfrac{1}{z^{n+1}}$	$\operatorname{Re} z > 0$
5	$t\,e^{-\alpha t}$	$\dfrac{1}{(z + \alpha)^2}$	$\operatorname{Re} z > -\alpha$
6	$\sin(\omega t)$	$\dfrac{\omega}{z^2 + \omega^2}$	$\operatorname{Re} z > 0$
7	$\cos(\omega t)$	$\dfrac{z}{z^2 + \omega^2}$	$\operatorname{Re} z > 0$
8	$e^{\alpha t}\sin(\omega t)$	$\dfrac{\omega}{(z - \alpha)^2 + \omega^2}$	$\operatorname{Re} z > \alpha$
9	$e^{\alpha t}\cos(\omega t)$	$\dfrac{z - \alpha}{(z - \alpha)^2 + \omega^2}$	$\operatorname{Re} z > \alpha$
10	$\sinh(\omega t)$	$\dfrac{\omega}{z^2 - \omega^2}$	$\operatorname{Re} z > |\omega|$
11	$\cosh(\omega t)$	$\dfrac{z}{z^2 - \omega^2}$	$\operatorname{Re} z > |\omega|$
12	$e^{\alpha t}\sinh(\omega t)$	$\dfrac{\omega}{(z - \alpha)^2 - \omega^2}$	$\operatorname{Re} z > \alpha + |\omega|$
13	$e^{\alpha t}\cosh(\omega t)$	$\dfrac{z - \alpha}{(z - \alpha)^2 - \omega^2}$	$\operatorname{Re} z > \alpha + |\omega|$
14	$t\cos(\omega t)$	$\dfrac{z^2 - \omega^2}{(z^2 + \omega^2)^2}$	$\operatorname{Re} z > 0$

Index